Advances in Polymer Science

Volume 289

Aims and Scope

The series Advances in Polymer Science presents critical reviews of the present and future trends in polymer and biopolymer science. It covers all areas of research in polymer and biopolymer science including chemistry, physical chemistry, physics, and material science.

The thematic volumes are addressed to scientists, whether at universities or in industry, who wish to keep abreast of the important advances in the covered topics.

Advances in Polymer Science enjoys a longstanding tradition and good reputation in its community. Each volume is dedicated to a current topic, and each review critically surveys one aspect of that topic, to place it within the context of the volume. The volumes typically summarize the significant developments of the last 5 to 10 years and discuss them critically, presenting selected examples, explaining and illustrating the important principles, and bringing together many important references of primary literature. On that basis, future research directions in the area can be discussed. Advances in Polymer Science volumes thus are important references for every polymer scientist, as well as for other scientists interested in polymer science - as an introduction to a neighboring field, or as a compilation of detailed information for the specialist.

Review articles for the individual volumes are invited by the volume editors. Single contributions can be specially commissioned.

Readership: Polymer scientists, or scientists in related fields interested in polymer and biopolymer science, at universities or in industry, graduate students.

Gert Heinrich • Reinhold Kipscholl • Radek Stoček
Editors

Degradation of Elastomers in Practice, Experiments and Modeling

With contributions by

R. Behnke · A. K. Bhowmick · J. J. C. Busfield · J. Chanda · V. Dietze · P. Ghosh · R. Gieré · U. Giese · G. Heinrich · J. Jungk · M. Kaliske · S. Kautz · R. Kipscholl · M. Klüppel · O. Kratina · T.-T. Mai · Y. Morishita · R. Mukhopadhyay · S. Nakano · K. Oßwald · O. Peter · M. Pöschl · J. Ramier · K. Reincke · T. Saha · C. Schwarzendahl · B. Shaw · R. Stoček · T. Tada · S. Thust · K. Tsunoda · K. Urayama · P. Zádrapa · T. Zushi

Springer

Editors
Gert Heinrich
Technische Universität Dresden
Dresden, Germany

Reinhold Kipscholl
Coesfeld GmbH and Co. KG
Dortmund, Germany

Radek Stoček
PRL Polymer Research Lab s.r.o.
Zlín, Czech Republic

ISSN 0065-3195 ISSN 1436-5030 (electronic)
Advances in Polymer Science
ISBN 978-3-031-15163-7 ISBN 978-3-031-15164-4 (eBook)
https://doi.org/10.1007/978-3-031-15164-4

This Springer imprint is published by the registered company Springer Nature Switzerland AG
The registered company address is: Gewerbestrasse 11, 6330 Cham, Switzerland

Preface

Elastomers are subject to degradation of rubber properties over time due to mechanical stress, friction processes, and a number of other external factors such as exposure to light, oxygen (ozone), and heat. The molecular changes caused by these environmental conditions can significantly impair the mechanical properties and thus the service life of rubber products such as tires, seals, O-rings, bushing elements, conveyor belts, shoes as well as body implants. The changes of rubber material properties are also well known to every car driver, bicyclist, or even pedestrians as well as users of kitchen aid appliances such as capsule coffee machines. Abrasion is obviously an environmental problem that has only been studied more systematically in recent years, but is becoming considerably more important in the emerging era of climate change and the current energy and raw materials crisis.

This book provides up-to-date research results on identifying and quantifying the abundance of tire wear particles in the different environmental compartments, and on determining their toxicity and reducing their environmental emission and immission (Gieré et al.). Oxidative and thermal aging of rubber is accelerated by stress and reactive gases such as ozone. The addition of antioxidants, UV stabilizers, and anti-ozonates can slow or prevent these problems. However, the basic processes of rubber friction and tribology, internal material fatigue, and structural thermo-oxidative controlled changes in the material rubber, which cause and accompany the deterioration of the material, are extremely complex and partly not understood.

One aim of this book is to provide an up-to-date overview of certain systematic studies on abrasion and wear phenomena and thermo-oxidative aging phenomena. This includes physical, fracture-mechanical, tribological, chemical, and physico-chemical aspects, as well as the question of which new laboratory test methods provide good correlations to practice, and how modern cross-scale simulation methods promote understanding and enable predictions.

Giese et al. report about selected methods like nanoindentation, FT-IR spectroscopy, chemiluminescence, and physical testing in combination with artificial aging to describe quantitatively the mechanistic role of the polymer's microstructure and of crosslinking during thermal-oxidative aging. Shaw et al. report about DSC, FTIR, and TGA results on HNBR confirming that during aging there is a steady evolution of the some of the ingredients in the compound such as the processing oils and the plasticiser which appears to have a significant effect on the fatigue properties of the materials. The relation between thermo-oxidative aging and mechanical fatigue under repeated loading of elastomer compounds is investigated by Klüppel et al. by referring to a wide variety of polymer types for special applications. Beside technical elastomers, typical tire tread compounds and special elastomers for high temperature resistant applications are studied. Special effects of antioxidants on the aging behaviour of natural rubber (NR) and synthetic styrene-butadiene rubber (S-SBR) are presented in the chapter of Oßwald et al. The chapter of Saha et al. summarizes the high-temperature degradation of various elastomers with the help of the reactive molecular dynamics simulation technique. Different fragmented products were identified with the help of the simulation method and experimentally verified. Mechanochemical behavior at the tire tread surface is discussed in Nakano et al. In their study, the wear mechanism at molecular scale is investigated by quantum mechanics-based atomistic simulation with focusing on the smearing wear, which is originated from mechanochemical effects. A general overview of the abrasion mechanisms occurring on automotive tires provides the chapter of Heinrich et al. Particular attention is paid to fatigue abrasion, which is especially the cause of mass loss in passenger car tires in public road traffic. A brief introduction to the fracture mechanics of rubber leading to different tire wear phenomena is presented in the chapter of Kipscholl et al. A broad overview is given of novel, high-performance technical measuring equipment with which very reliable experimental results could be obtained. The study of Ghosh et al. emphasizes on capturing the influence of material composition (rubber, rubber blend, and filler) and lab operational conditions (temperature, waveform, and R ratio) on fatigue crack growth resistance of tire rubber compounds. Velocity and crack-tip features of fast-moving cracks are discussed in Mai et al. in relation to the nonlinear elasticity and viscoelasticity of bulk elastomers. The authors describe the characterization of quasi-stationary cracks in elastomers subjected to various types of biaxial loading, providing a basis for the fracture mechanics of elastomers under multiaxial deformation. Very large-scale wear phenomena (cut and chip), when the elastomer is exposed to harsh conditions, are the topic of the chapter by Pöschl et al., in which the effect of cross-link density on the elastomer's resistance to the wear mechanism is studied for the first time. A unique contribution to the experimental and numerical description of self-heating of rubber under cyclic loading is described in the chapter by Peter et al. This contribution describes a new numerical method – within an ANSYS Workbench finite element method environment – for simulating the kinetics of the Heat Build-up process in rubber under cyclic multiaxial loading by involving novel loading principles. While different subtopics of degradation of rubber under strain, temperature, wear, tire-road abrasion, or fatigue are discussed in the present book from an experimental and simple physical and analytical point of view, the chapter of Behnke et al. mainly aims at the numerical aspects of the model setup from the material to the structural scale. The objective of this chapter is to establish the link between experimental testing on the material scale and predictive simulation on the structural scale.

This volume thus provides a thorough state-of-the-art explanation of the essential chemical, physical, and mechanical principles as well as practices of material characterization for wear prediction. It conveys novel strategies and procedures of planning effective wear test programs. It describes new insights into main aspects of rubber degradation by material's fatigue, wear and aging evolution, as well as their impact on mechanical rubber properties. The initiating factors of abrasion, the development of surface abrasion on sharp and blunt tracks, and the influence of smear and lubricants are also summarized. Thus, the present volume provides a comprehensive overview of the recent developments in this area and will be of interest to both academic researchers and industrial professionals in this field.

Dresden, Germany — Gert Heinrich
Dortmund, Germany — Reinhold Kipscholl
Zlín, Czech Republic — Radek Stoček

Contents

Adv Polym Sci (2023) 289: 1–14
https://doi.org/10.1007/12_2022_113

Published online: 25 March 2022

Basic Mechanisms and Predictive Testing of Tire-Road Abrasion

Gert Heinrich and Manfred Klüppel

Contents

Abstract The chapter provides a brief overview of the abrasion mechanisms occurring on automotive tires. Particular attention is paid to fatigue abrasion, which is especially the cause of mass loss in passenger car tires in public road traffic. The framework of the paper is a very simple physical model (Resnikowskij, Kautschuk Gummi Kunststoffe 9:33–37, 1960) of the relationship between abrasion and friction work during Hertzian contact and sliding friction of rubber over the periodical roughness of a road. Essential physical quantities of the rubbers determined in the laboratory, such as modulus, tensile elongation, coefficient of friction, and a Woehler-like fatigue parameter, give an expression for laboratory fatigue wear that correlates very well with tire wear under an outdoor test program commonly used by tire companies on public highways. This result also makes it clear why tire wear under moderate severity conditions cannot be described by abrasive wear tests in the laboratory, e.g. with DIN-Abrader.

G. Heinrich (✉)
Technische Universität Dresden, Institut für Textilmaschinen und Textile Hochleistungswerkstofftechnik, Dresden, Germany
e-mail: gheinrich@ipfdd.de

M. Klüppel
Deutsches Institut für Kautschuktechnologie e.V., Hannover, Germany
e-mail: manfred.klueppel@dikautschuk.de

Keywords Abrasion mechanisms · Fatigue wear · Predictive laboratory testing of tire performance · Tire-road abrasion

1 Introduction

Tire-road abrasion is an important economic and ecological factor that is in permanent discussion in tire industry. It is strongly influenced by tire operation conditions, because different abrasion mechanisms take place in dependence of severity, road texture, and weather conditions [1–4].

Very recently, the problem of micro-particles in the environment has reached the attention of publicity, and numerous corresponding studies were published. (see [5] and the references therein). It is well known that road traffic emissions caused by both exhaust and non-exhaust sources contribute significantly to the particulate matter (PM) concentration in an urban atmosphere [6]. Most of the PM mass is created by the abrasion of car tires, together with a road surface, and the corresponding particles have diameters ranging approximately between 1 and 10 μm [6]. The worn rubber debris consists of up to 30% of strongly degraded polymer with molar mass in the range of 100,000 g/mol [7, 8]. The tire industry and raw material suppliers have long sought to develop predictive laboratory methods to determine the core characteristics of a car tire represented by the so-called magic triangle of tire technology, which states that improvement in rolling resistance must come at the expense of wet grip and wear/durability. Both authors were team members in such projects, which were pushed some time ago, publicly funded by the companies Continental AG and Bayer AG (Rubber Application Business Unit). The core objectives of these projects were to be able to largely dispense with the construction and practical testing of tires by means of physically motivated models and new laboratory compound tests and thus to save development costs and shorten development times. Furthermore, the aim of such projects was to gain a better understanding of the mechanisms underlying elastomer friction and abrasion in order to be able to develop new, high-performance rubbers and tread materials in a more targeted manner. Comprehensive physical and chemical investigations were carried out in pursuit of this objective. The tire manufacturer carried out the production and testing of the tires to verify the predictions.

The sub-project "abrasion" aimed at predicting the road abrasion of passenger car tires with higher precision than before. Therefore, three approaches were taken:

Firstly, direct prediction on the basis of abrasion tests on laboratory test specimens, in which new test methods were used in particular that were adapted to the load conditions in operation and shed more light on aspects of dynamic crack propagation and fatigue. Secondly, predictions based on laboratory abrasion tests with modification of standard procedures, whereby the normal force was varied and the friction force was recorded. Further, abrasion tests on small model tires (solid rubber wheels) varying normal force, speed, and slip were performed. In this case, experience could be gained with the LAT100, among others. The well-known VMI LAT100 (Laboratory Abrasion & Skid Tester or "Grosch machine") rubber compound tester simulates near road conditions of the abrasion process (https://www.vmi-group.com/specifications/vmi-lat100-tire-tread-compound-tester/. Accessed 14 Oct 2021). The test machine can be employed for routine measurements, measuring both the abrasion resistance and the friction properties under a wide range of selectable conditions. Abrasion parameters like speed, load, and slip angle can be set independently to offer maximum flexibility. Notably, traction, dry or wet, is measured directly on a locked wheel, or through measuring side force while varying other parameters. Reasonably useful correlations to tire tests could be made in all cases in the mentioned project. It is noted, however, that the validity of these correlations relates solely to the specific conditions of the underlying tire tests (i.e., road surface, test procedure, temperature, etc.). A transfer to changed tire test conditions is not possible. The LAT100 laboratory conditions would then have to be readjusted again. From a scientific perspective, it should also be noted that these empirical test methods say little about the fundamental physical processes or contribute little to understanding them.

For this reason, thirdly, predictions were made in the abrasion project on the basis of an abrasion model that draws on simple physical ideas of elastic contact, elastomer friction, and the elastomer fatigue during fatigue abrasion processes. The characteristic values required for this model were determined with the help of laboratory tests. This is reported on in more detail here in this book chapter.

Before we recall previous discussions of specific wear mechanisms and their laboratory tests with correlations to tire abrasion, it must be pointed out that wear and damage phenomena on tires generally go far beyond the narrower concept of volume abrasion on which we are focusing. Tires can become damaged for a variety of reasons (see, e.g., (https://www.continental-tires.com/car/tire-knowledge/tire-damage-age-repair/tire-damages)). The most common types of damage are punctures, cuts, impacts, cracks, bulges, and irregular wear; and one distinguishes between several types of irregular wear like heel and toe wear, one-sided wear, and center wear. Heel and toe wear is a pattern caused by normal usage and suspension settings. It is the outwardly visible manifestation of various distortional forces at work on the tire tread. Cuts are the result of external influences like bad road conditions, protruding bodywork parts, or sharp, foreign objects such as stones or glass.

2 Tire-Road Abrasion Mechanisms

Efforts have been made since many years to understand and to model the tire-road abrasion mechanisms. The so-called slipping wheel models (Grosch and Schallamach [9–12], Kragelsky and Nepomnyashchi [13]) have been introduced to calculate tire slide-slip under lateral acceleration and the related frictional work [9–12, 14–18]. Muhr and Roberts [19] argued that it is not yet sufficiently possible to determine the actual frictional work or sliding path in the case of slipping tires or even in the case of certain lab experiments. Consequently, the decomposition of total slip into deformation- and slide-slip is not self-evident. This makes it difficult to establish simple correlations in terms of wear between lab and tire tests, as frictional work applies as an elementary severity variable in all wear models. Furthermore, lab experiments (in the case of pure slide-slip) yield a characteristic empirical distinction between abrasive wear (on harsh surfaces), fatigue wear (on dull surfaces), and blade wear on the basis of the wear exponent that describes the abrasion loss ΔV as a power law function of the frictional work ΔW: $\Delta V \sim \Delta W^m$. This energy dissipated per unit travelling distance is equal to the dissipative side force component $F_s \cdot \sin\alpha$ tending to slow the wheel down and responsible for abrasion (α being the slip angle). Note, the second component of the side force is $F_s \cdot \cos\alpha$ which supports the centrifugal cornering force. The side force, and hence the dissipated energy, depends on load. Hence, when abrasion is expressed as function of ΔW this includes both load and slip and can be considered as measure of severity of an abrasion test on a particular surface for a given speed.

Depending on the vulcanized rubber material, the exponent m has values of order 1 in the abrasive case and between 1 and 3 in the fatigue case, with the exponent decreasing with increasing harshness, or decreasing grain size in the case of grindstones [14–19]. In addition, blade wear, obtained from razor-blades located perpendicular to the sliding direction, correlates with tear propagation wherever exponents m between 2 and 5 are observed [19–24]. Determination of the corresponding wear exponent in tire tests could thus allow for conclusions as to the relevant wear mechanism [19, 20, 25]. One can imagine, however, that this would lead to an almost hopeless puzzle for tire developers.

Previous findings on tire wear have led to the conclusion that under moderate severity conditions, primarily fatigue wear occurs on smooth roads with low microstructure, while abrasive or cutting wear is significant with increasing road harshness and higher lateral forces [1–4, 9, 10, 14–20]. A generally intelligible definition of fatigue wear is given, for example, in (https://www.tribonet.org/wiki/fatigue-wear/). Fatigue wear is considered as a type of wear where a number of load cycles of the rubber due to the rough road surface are needed to generate debris. Often, two mechanisms of fatigue wear are distinguished: high- and low-cycle fatigue. In high-cycle fatigue, the number of cycles before failure is very large, so the component life is relatively long. Cracks in this case are generated due to pre-existing microdefects in the material, close to which the local stress may exceed the yield value, even though nominally the macroscopic contact is in the elastic regime. In the low-cycle fatigue, the number of cycles before failure is low; consequently, the corresponding rubber component fails fast.

It is known that the fatigue wear of rubber typically increases with increasing polymer glass transition temperature, T_g, increasing ambient temperature and decreasing filler concentration [1–4, 26]. *cis*-1,4-Polybutadiene is known as synthetic rubber with a very low $T_g \sim -102°C$ and, therefore, widely employed in tire treads for trucks and automobiles. The polymer is noted for its high resistance to abrasion.

According to Veith [1–4], the wear rate always increases with a rise in ambient temperature. Even seasonally related changes in the surface microstructure of the roads (harsh in winter and smooth in summer) were observed and investigated. The increase in wear rate with increasing temperature is more pronounced the lower the glass transition temperature of the rubber polymer and the lower the degree of filler reinforcement. The influence of T_g of rubber polymers can be well understood by the so-called dynamic glass temperature in the case of relevant high-frequency severity of slipping contacts between tires and road. This (high-frequency) dynamic glass transition temperature can be measured, for example, by NMR [27]. We would like to note also that a marked difference in wear is observed for indoor and outdoor tests that simulate city driving – with frequent braking, accelerating, and cornering operations – and highway driving – with little load alteration [28]. This underscores the significance of slide-slip as regards tire wear.

A recent comprehensive review about mechanisms of rubber abrasion contains the Ph.D. thesis of Wu [29]. Again, it is demonstrated in detail that different abrasion mechanisms operate under different conditions. When rubber slides against surfaces containing sharp asperities, abrasive wear is the dominant abrasion mechanism caused by tensile failure. When rubber slides against smooth rigid abrasives, the stress concentrations are much lower, generating fatigue failure than tensile failure to remove wear particles. This failure mode is called fatigue wear or adhesion wear. This kind of abrasion process can even be divided in two phases, an initiation phase and a steady state, respectively [29].

As a supplement we note that lately, there has been intensive work on the laboratory prediction of the so-called cut and chip (C&C) wear that results from the exposure of the tire to harsh severity conditions [30, 31]. Sometimes, these wear phenomena are also termed as C&C&C which stands for cut, chip, and chunk wear field conditions. In the case of tires, C&C wear refers to the detachment or breakage of rubber material pieces from their treads when riding on a rough road surface (e.g., gravel roads, roots, stalks). This is most commonly associated with off-the-road tires (OTR), agricultural, industrial, light truck (LTT), SUV as well as moto-cross/enduro (MX) tires. Moreover, this phenomenon is also observed with highway commercial truck as well as all-season passenger car and touring moto-bike tires. To avoid intensive field tests of tires, Stoček et al. introduced an advanced testing method and fully instrumented equipment labelled Instrumented Chip&Cut Analyzer (ICCA, Coesfeld GmbH & Co. KG, Germany) [32, 33]. The ICCA method is meanwhile successfully integrated as a standard lab method in the rubber respective tire industries.

3 Experimental

Five different individual tread compounds, denoted PTX (X = 1,3,4,5,6), all of different Shore A hardness (ShA), have been prepared as slick-tires without tread pattern profile. We note that corresponding more tires were also built with summer and winter tread patterns, which were used for the prediction tools for rolling resistance, wet skid resistance, and winter properties. The rubber compounds corresponded to summer and winter compounds, which differed in formulations, filler content, and filler type (carbon black, silica). The physical (mechanical, viscoelastic) properties of all compounds were extensively characterized in the laboratory. The complete presentation of these data goes beyond the scope of this article. We note that, in conjunction with physical and empirical models, these physical laboratory data were especially used for developing prediction concepts for rolling resistance, wet skid performance, and winter performance (snow, ice) of the tires. Here, we only show Shore A hardness (ShA) and rebound resilience at room temperature (R-RT) and at 70°C (R-70):

	ShA-RT	ShA-70	R-RT [%]	R-70 [%]
PT1	64.1	56.5	23.0	33.8
PT3	67.7	60.9	23.4	44.0
PT4	65.8	61.5	23.9	42.7
PT5	60.7	59.2	46.7	59.2
PT6	63.1	60.1	33.8	42.9

The entire testing program also included a compound PT2 (Sh A-RT = 52). For the abrasion road tests, the relative soft PT2 compound was not used in the program for cost reasons alone, as it has too poor abrasion resistance.

The fatigue tests have been performed with scribed strip samples by using a Tear Fatigue Analyzer (TFA, Coesfeld GmbH & Co. KG, Germany). The use of TFA allowed very fast determination of Woehler coefficients compared to the classical method of fatigue testing of un-scribed specimens.

The modulus was measured in shear with a Hydropulser (MTS) at 60°C, 100 Hz. The tensile strength was obtained with S2 samples by using a tensile testing machine (Zwick). The friction coefficient was measured on dry concrete with a customized linear friction tester (LFT) at 5°C and 3 mm/s. We used an LFT which was primarily used in friction tests with tire materials at TU Wien (Austria), Institut für Mechanik der Werkstoffe und Strukturen (https://www.imws.tuwien.ac.at/no_cache/mitarbeiter/uebersicht/profil/staffmembers/staff/detail/pichler-2/. Accessed 14 Oct 2021).

We found that the exact numerical values of the friction coefficients are in the order of ~1 for all rubber compounds used. These values have negligible influence on the goodness of correlation between laboratory abrasion prediction and tire testing in the context of the concept presented here. The tire abrasion test was carried out by Continental AG on public highways using VW Passat vehicles (front-wheel drive) with 195/65 R15 tires. The test series was completed after a mileage of 24,400 km, corresponding to a maximum wear rate of ~80%. An intermediate result at 11,200 km was used for preliminary correlations that we will not discuss here. All front axle tires showed inner shoulder bevel wear, while those on the rear axle showed saw-tooth wear. In addition, all tires exhibited belt separations of <2 mm and the tires with tread compounds PT3 and PT6 even showed occasional separations of 3–5 mm.

4 Laboratory Prediction of Fatigue Wear of Tires

In the case of fatigue wear one assumes that the abraded volume ΔV is proportional to the frictional work ΔW [13, 34]. The abrasion resistance β (Eq. 1) is defined by the ratio of the friction work to the abraded volume, whereby the friction work can be represented as the product of the friction coefficient μ, the normal force F_N, and the friction path Δl

$$\beta = \frac{\Delta W}{\Delta V} = \frac{\mu\, F_N \Delta l}{\Delta V} \tag{1}$$

It can be measured by performing wear experiments on rough substrates as visualized in Fig. 1.

Another assumption of fatigue wear is that abrasion occurs in characteristic layers of thickness δ, which separates from the rubber surface as a result of fatigue abrasion after a characteristic distance l_0 has been covered. The characteristic distance l_0 (Eq. 2) is calculated using the average distance between the asperities, z, as well as their number n in order to remove a characteristic layer, i.e. a cycle number till failure of the rubber.

$$l_0 = z\, n \tag{2}$$

Using the product of the characteristic layer thickness δ and the abrasion area A_0, the characteristic abraded volume is then calculated as:

$$\Delta V = \delta\, A_0 \tag{3}$$

Hence, we obtain for the abrasion resistance

$$\beta = \frac{\mu\, F_N\, z}{A_0}\,\frac{n}{\delta} \tag{4}$$

Kragelsky and Nepomnyashchi [13] showed that there was an analogy between the normal fatigue strength at fatigue tests (Woehler tests; the principle is visualized in Fig. 2) and the friction tests with an indenter (Fig. 3) on a rubber surface. The

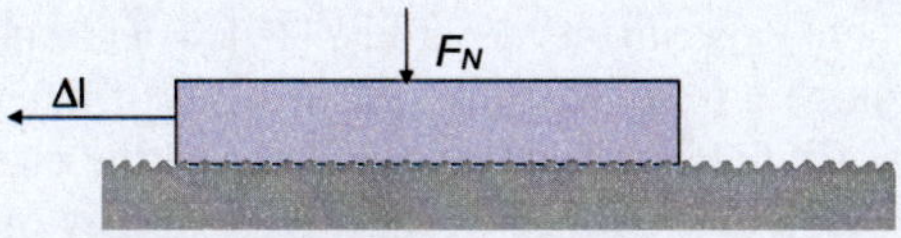

Fig. 1 Visualization of a sliding rubber against abrasives

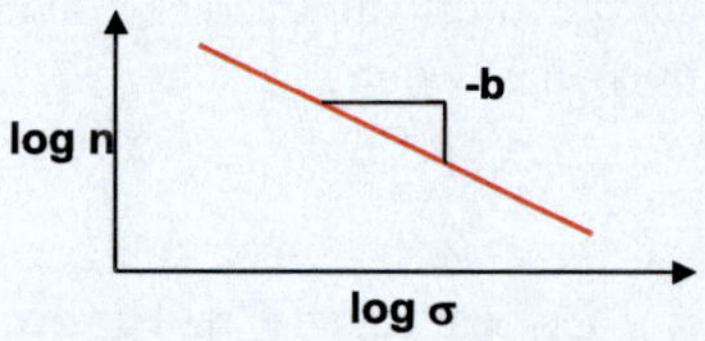

Fig. 2 Visualization of a Woehler test with characteristic exponent b

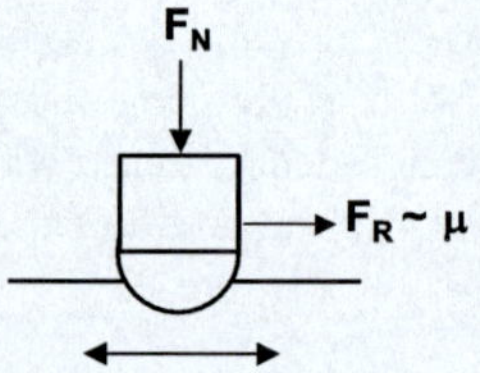

Fig. 3 Visualization of an indenter test, where the indenter is pressed with normal load F_N against a rubber surface; F_R is the friction force, and μ the friction coefficient

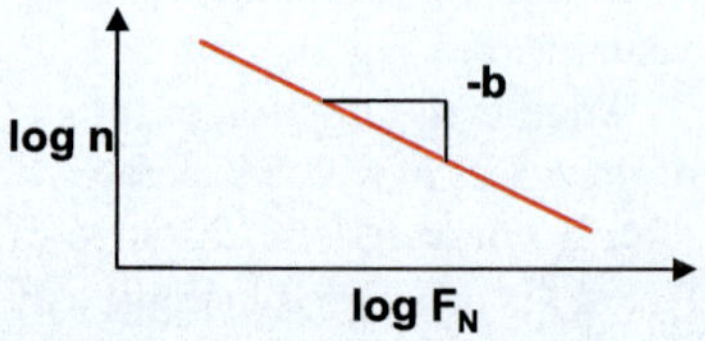

Fig. 4 Friction path, expressed via number n, that is necessary for abrasion depending on the normal force F_N

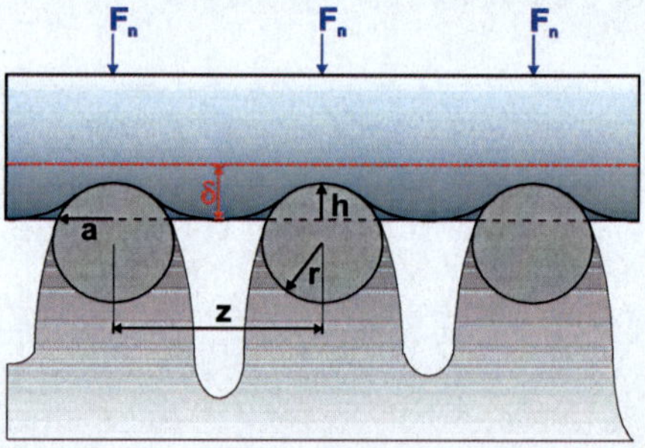

Fig. 5 Hertz theory of elastic contact with spherical asperities used for the fatigue abrasion model

fatigue strength tests show how often, expressed in cycles, n, a test specimen can be cyclically stressed depending on the maximum stress, σ, until it is destroyed.

Figure 4 demonstrates then in analogy to the Woehler test how often the indenter, depending on the normal force, F_N, has to rub over the rubber surface in order to remove a characteristic layer thickness.

It is interesting to note that the same fatigue resistance coefficient b was found in both experiments described [13]. This result would allow the approximate use of the slope b from Woehler tests instead of indenter tests.

In the simplest way we describe the elastic contact of a tire or rubber block on a rough road surface with the Hertz theory of elastic contact [35] as shown in Fig. 5. It assumes uniform asperity radii r with equal distance z, where each asperity is loaded with the equal normal force F_n (= load per asperity $F_N \cdot z^2/A_0$). The penetration depth h according to Eq. 5 for incompressible rubber (Poisson ratio $\nu = 0.5$) is

$$h = (3/4)^{4/3} E^{-2/3} F_n^{2/3} r^{-1/3} \tag{5}$$

where E is the Young's modulus. Accordingly, a penetration depth h results in a penetration radius a of

$$a^2 = h\,(2\,r - h) \approx 2\,h\,r \tag{6}$$

The average stress σ_m in the deformed rubber material, as a result of the penetration of the asperities, is calculated according to Eq. 7:

$$\sigma_m = F_n/\left(\pi\, a^2\right) = k'' \left(\frac{N}{A_0}\right)^{1/3} E^{2/3} \left(\frac{z}{r}\right)^{2/3} \tag{7}$$

Here, the relationship for calculating the normal force acting on a single asperity (Eq. 8) is used.

$$F_n = F_N \frac{z^2}{A_0} \tag{8}$$

The numerical constant in Eq. 7 is $1/(2\pi \cdot 0.68) \approx 0.23$.

Another assumption of the model is that the characteristic abrasion layer thickness $\delta \simeq h$ corresponds to the order of magnitude of the penetration depth h. From this simplification follows Eq. 9. In this formula the weighting factor $(\pi\, r^2)/z^2$ is found, which relates the penetration depth via the distances of the asperities and their radiuses:

$$\delta = k' \frac{\pi\, r^2}{z^2}\, h \tag{9}$$

where $k' \approx 1$. Using these formulas and the Hertz theory, the characteristic abrasion layer thickness can be calculated as follows:

$$\delta = 0.68\, \pi\, k' \left(\frac{r}{z}\right)^2 E^{-2/3} \left(\frac{F_N\, z^2}{A_0}\right)^{2/3} r^{-1/3} \tag{10}$$

The above introduced dimensionless fatigue resistance exponent b can be obtained from a Woehler curve in which the service life n is plotted as a function of stress σ. The negative slope in the double logarithmic plot corresponds to the fatigue resistance exponent b and the (extrapolated) stress value σ_0 is read at a life of one cycle (see Fig. 2), i.e.:

$$n(\sigma) = \left(\frac{\sigma}{\sigma_0}\right)^{-b} \tag{11}$$

The corresponding "abrasion lifetime" can now be calculated with the average stress (Eq. (7)) obtained from the Hertz theory as follows:

$$n = \left(\frac{\sigma_0}{k''}\right)^b E^{-2b/3} \left(\frac{F_N}{A_0}\right)^{-b/3} \left(\frac{z}{r}\right)^{-2b/3} \tag{12}$$

where n is the cycle number till failure. After inserting the abrasion resistance according to Eq. 4 and the previously obtained formulas, the overall Eq. 13 is obtained as

$$\beta = \frac{\mu}{0.68\,\pi\,k'} \left(\frac{\sigma_0}{k''}\right)^b E^{2(1-b)/3} \left(\frac{F_N}{A_0}\right)^{(1-b)/3} \left(\frac{z}{r}\right)^{(5-2b)/3} \tag{13}$$

Resnikowskij [34] has tested this formula to show the normal force dependence of the abrasion resistance and the coefficient of friction by combining almost all values of the rubber properties and the friction surface into one constant:

$$\frac{\beta}{\mu} = \text{const.} F_N^{(1-b)/3} \tag{14}$$

He showed results of a corresponding experiment for an unfilled styrene-butadiene (30:70) rubber, where a clear linear dropping dependence of abrasion resistance log (β/μ) on the normal force log F_N was observed. The fatigue resistance coefficient b could be determined via Eq. 14 as $b = 2.2$.

Further simplification of Eq. 13 is based on the suggestion of Resnikowskij [34] to substitute extrapolated stress value and normal load with the tensile strength f_z, which is easily estimated as standard physical ultimate value of the rubber. Obviously, this is in line with the typical high-cycle fatigue regime for rubbers, where the number of cycles before failure is high. The cracks in this case are generated due to pre-existing micro defects in the material, close to which the local stress may exceed the yield value, even though nominally the macroscopic contact is in the elastic regime. Considering the incompressible character of the rubber, we may additionally substitute the elastic modulus by the shear modulus $G \approx E/3$. Then we obtain:

$$\beta = \mu\, f_z^{\,b}\, G^{2(1-b)/3}\, K_\alpha^{\,b} \tag{15}$$

This simplified abrasion resistance formula includes the Woehler exponent b, the shear modulus G, the tensile strength f_z, and the friction coefficient μ. In particular, there is only one parameter (K_α) that can be freely adjusted.

5 Results and Discussion

Based on the fatigue wear theory of Kragelsky and Nepomnyashchi [13, 34] it is possible to perform lab-predictions of tire abrasion under moderate severity conditions and compare this to real outdoor tire test. Therefore, tire abrasion tests were carried out for five different tire tread compounds by Continental AG on public highways with 195/65 R15 tires. The development of the wear rate of the front tires is shown in Fig. 6. It demonstrates that an almost linear increase of the wear rate is obtained and clear differences are realized for the model compounds used in this study.

Table 1 shows a summary of the abrasion results after 22,400 km. In this, it is seen that the PT3 compound suffers the highest abrasion and would accordingly provide the shortest mileage. Mixture PT5 showed the lowest wear over the entire test period and would thus provide the highest mileage. The second best result was

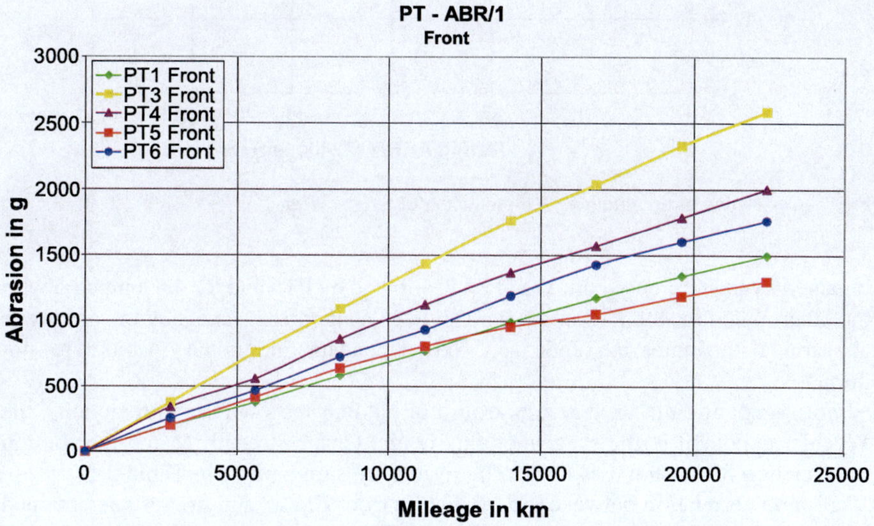

Fig. 6 Wear rate vs. mileage outdoor tests for the front tires

Table 1 Results of tire abrasion tests

	Front weightloss [g]	Front rating	Rear weightloss [g]	Rear rating	F+R weightloss [g]	F+R rating
PT1	1,495	100	730	100	2,225	100
PT3	2,586	57.8	1,247	58.5	3,833	58.0
PT4	1,999	74.8	839	87.0	2,838	78.4
PT5	1,296	115.4	612	119.3	1,908	116.6
PT6	1,761	84.9	816	89.5	2,577	86.3

Rating > 100 means higher abrasion resistance, i.e. lower weight loss

Table 2 Tensile strength f_z, Woehler exponent b, shear modulus G, and friction coefficient μ of the composites

	Tensile strength f_z [MPa]	Woehler exponent b [-]	Shear modulus G [MPa]	Friction coefficient μ [-]
PT1	19.96	8.88	3.5	1.04
PT3	16.28	8.31	3.5	0.90
PT4	18.32	6.66	2.7	1.05
PT5	18.24	6.79	2.2	1.06
PT6	17.52	7.47	2.9	1.29

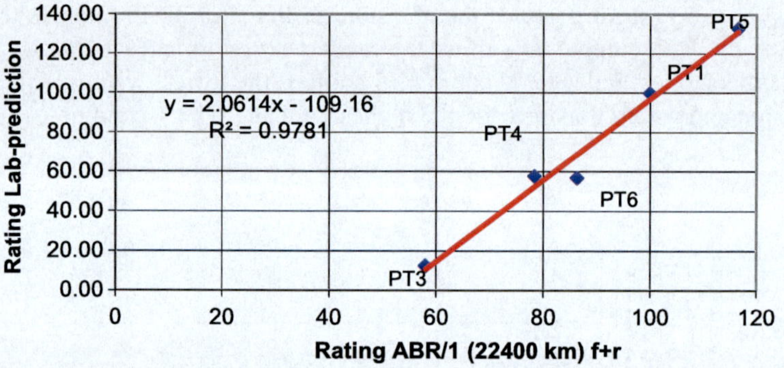

Fig. 7 Correlation of tire rating vs. lab-prediction of fatigue wear

obtained by the reference compound PT1 followed by PT6 and PT 4. Compared with the other materials, the PT5 compound showed a significant tendency toward lower abrasion with increasing mileage. Axle-dependent individual results are not discussed here.

For comparing the wear results from tire outdoor tests with lab-predictions, the Woehler exponent b, the shear modulus G, the tensile strength f_z, and the friction coefficient μ have been measured. The results are summarized in Table 2.

A good correlation between lab-prediction from Eq. 15 and tire tests is obtained in Fig. 7 if the physical parameters from Table 2 are used for each rubber. A high correlation of $R^2 = 0.978$ is found with $K_\alpha = 0.179$ when using the specific friction coefficients estimated with a Linear Friction Tester (LFT) when the rubbers slide on concrete at 5°C and a velocity of 3 mm/s.

By applying the simplified formula to calculate abrasion resistance, even without using a specific coefficient of friction (i.e., we assume $\mu = 1$ for all rubbers), gives a good correlation between Rating β and Rating tire testing ABR (22,400 km; front + rear) with regression coefficient $R^2 = 0.92$ and $K_\alpha \approx 0.213$. However, we point out that no correlation is found if the tire tests are compared to DIN-abrasion results [36]. This demonstrates that simple abrasive wear processes cannot describe tire wear under moderate severity conditions.

6 Conclusions

It is shown that tire wear obtained by an outdoor test program under moderate severity conditions correlates well with laboratory predictions based on a fatigue abrasion model for tires [34]. The model assumes that abrasion loss takes place in characteristic layers of degraded polymer, which result from the repeated deformation of the tire tread by road asperities. These layers detach from the surface by propagating cracks if the number of repeated deformations (cycles) exceeds the lifetime of the material. The layer thickness is determined by referring to Hertzian contact of spherical asperities with a flat rubber surface. The lifetime obtained from Woehler tests enters into the model via a characteristic fatigue-exponent; and extrapolated stress values and normal load were substituted via the tensile strength of the used tread rubbers. A physically comprehensible very good correlation between laboratory prediction and tire results is achieved. Furthermore, it is argued that abrasive laboratory tests (e.g., DIN-abrasion test) do not provide any correlation to the corresponding tire test.

Acknowledgement We thank our former project partners Continental AG and Bayer AG for the support of this work within the past joint project "Predictive Laboratory Testing of Tire Performance." We thank Continental Reifen Deutschland GmbH for their agreement to publish this article.

References

1. Veith AG (1992) Rubber Chem Technol 65:601
2. Veith AG (1987) Polymer Testing 7:177
3. Veith AG (1995) Tire Sci Technol 23:212
4. Veith AG (1973) Rubber Chem Technol 46(801):821
5. Gieré R et al. Tire abrasion as a major source of microplastics in the environment, present book volume
6. Penkala M, Ogrodnik P, Rogula-Kozlowska W (2018) Particulate matter from the road surface abrasion as a problem of non-exhaust emission control. Environments 5:9. https://doi.org/10.3390/environments5010009
7. Cadle SH, Williams RL (1978) Rubber Chem Technol 51:7
8. Cadle SH, Williams RL (1979) Rubber Chem Technol 52:147
9. Schallamach A (1952) J Polym Sci 9:385
10. Schallamach A (1968) Rubber Chem Technol 41:209
11. Grosch KA, Schallamach A (1961) Wear 4:356
12. Schallamach A, Turner DM (1960) Wear 3:1
13. Kragelsky IV, Nepomnyashchi EF (1965) Wear 8:303
14. Grosch KA (1992) Rubber Chem Technol 65:1
15. Grosch KA, Schallamach A (1969) Kautschuk Gummi Kunstst 22:288
16. Grosch KA, Schallamach A (1970) Rubber Chem Technol 43:701
17. Grosch KA, Schallamach A (1965) Trans Inst Rubber Ind 41:80
18. Grosch KA, Schallamach A (1966) Rubber Chem Technol 39:287
19. Muhr AH, Roberts AD (1992) Wear 158:213

20. Gent AN, Nah C (1996) Rubber Chem Technol 69:819
21. Southern E, Thomas AG (1978) Plast Rubber Matter Appl 3:133
22. Thomas AG (1974) J Polym Sci Polym Symp 48:145
23. Pulford CTR (1983) J Appl Polym Sci 28:709
24. Gent AN, Pulford CTR (1983) J Appl Polym Sci 28:943
25. Sakai H (1996) Tire Sci Technol 24:252
26. Kienle RN, Dizon ES, Brett TJ, Eckert CF (1971) Rubber Chem Technol 44:996
27. Heinrich G, Rennar N, Dumler H (1996) Kautschuk Gummi Kunststoffe 49:32
28. Stalnaker D, Turner J, Parekh D, Whittle B, Norton R (1996) Tire Sci Technol 24:94
29. Wu G (2016) The mechanisms of rubber abrasion, Ph.D. thesis, Queen Mary Univ. of London
30. Wunde M, Klüppel M, Vatterott C, Tschimmel J, Lacayo-Pineda J, Schulze A, Heinrich G (2019) Verbesserung der Laborvorhersagen zum Risswachstum und Verschleiß von LKW-Reifenlaufflächen. Kautschuk Gummi Kunststoffe 72:72–78
31. Stoček R, Heinrich G, Schulze A, Wunde M, Klüppel M, Vatterott C, Tschimmel J, Lacayo-Pineda J, Kipscholl R (2020) Chip & cut wear of truck tire treads: comparison between laboratory and real tire testing. Kautschuk Gummi Kunststoffe 73(6):51–55
32. Stoček R, Mars WV, Robertson CG, Kipscholl R (2018) Characterizing rubber's resistance against chip and cut behaviour. Rubber World 257:38–40
33. Stoček R, Mars WV, Kipscholl R, Robertson CG (2019) Characterisation of cut and chip behaviour for NR, SBR and BR compounds with an instrumented laboratory device. Plast Rubber Compos 48:14–23
34. Resnikowskij M (1960) Kautschuk Gummi Kunststoffe 9:33–37
35. Landau LD, Lifschitz EM (1970) Lehrbuch der theoretischen Physik. Band 7. Elastizitätstheorie, Akademie-Verlag, Berlin
36. Abraham F, Alshuth T (2015) Marvalová, Petriková (eds) A new fatigue wear simulation method for road tyre wear, proceedings of constitutive models for rubber IX Taylor & Francis Group, London, pp 123–126, 978-1-138-02873-9

Adv Polym Sci (2023) 289: 15–48
https://doi.org/10.1007/12_2022_114

Published online: 27 April 2022

Thermo-Oxidative Aging and Mechanical Fatigue of Elastomer Compounds Used in Various Fields of Rubber Industry

M. Klüppel and J. Jungk

Contents

M. Klüppel (✉)
Deutsches Institut für Kautschuktechnologie e.V., Hannover, Germany
e-mail: manfred.klueppel@dikautschuk.de

J. Jungk
Freudenberg Technology Innovation SE & Co. KG, Weinheim, Germany

Abstract The relation between thermo-oxidative aging and mechanical fatigue under repeated loading of cured elastomer compounds is investigated by referring to a wide variety of polymer types for special applications. Beside technical elastomers, typical tire tread compounds and special elastomers for high temperature resistant applications are studied. On one side, the lifetime for various loads is evaluated, whereby mechanical relaxation effects and the evolution of surface temperature are investigated. Furthermore, the internal temperature profile is calculated in relation to the measured surface temperature by referring to heat conduction theory in order to get information about the full heat history of the samples during thermo-mechanical aging. On the other side, the effect of thermal aging at various temperatures on mechanical and ultimate properties is investigated and compared to the lifetime characteristics of the samples without thermal aging. Significant differences between the various polymer types and the applied curing systems and filler types have been found. Thermal aging effects during mechanical aging are unlikely to affect the lifetime for most of the samples due to their insensitivity against thermal aging or due to relatively short aging times combined with small temperature increases. By means of the measured stress decrease during mechanical fatigue tests, several aspects influencing the lifetime of rubber compounds have been identified and analyzed.

Keywords Crack propagation · Lifetime prediction · Mechanical fatigue · Polymer degradation · Thermo-oxidative aging

1 Introduction

Aging and fatigue of elastomers are important economic factors for the rubber industry. Exposed to heat, dynamic load, oxygen or potentially further media elastomeric parts change their mechanical properties tremendously over time and hence cannot fulfill their original tasks. These phenomena explain the occurrence of wear or cut&chip in tire treads as well as in example leakage for sealing applications.

To avoid warranty issues and to efficiently develop long-lasting rubber goods the understanding of aging and fatigue mechanism and the development of prediction methods is crucial in order to guarantee performance of rubber parts over the whole lifetime [1]. An important topic of research is therefore the physical and technical understanding of aging phenomena in order to estimate the importance of these aging effects on mechanical properties [2]. However, the chemical and physical mechanism leading to aging and fatigue of rubber compounds involve a complex interplay of changes in crosslinking density, breaking of chains, changes in the polymer-filler coupling as well as the damage of the filler network [2]. This complex situation makes it difficult even today to predict reliably the lifetime of rubber compound with acceptable effort.

Significant research was already performed over decades reaching out to a better understanding of the mechanism and reasons for the material property changes over time. Most researchers separate hereby the temperature influenced changes of material properties usually termed as thermal aging and the mechanically induced property changes termed as fatigue.

Thermal aging mechanisms typically have been evaluated by oven aging. The characterization of physical properties before and after thermal aging allows for an evaluation of the influence of recognized and investigated aging processes on the economic important properties and the estimation of their significance. Depending on the type of rubber, the defined temperature conditions and the aging time either no effect [3], small changes [4, 5], or tremendous changes [6] of the mechanical properties have been found. The aging at elevated temperatures is also used to predict the elastomer performance under realistic application conditions by using an Arrhenius law [7]. However, this approach has also disadvantages since small variations in the lifetime at elevated temperatures lead to exponentially higher variations for the application temperatures and therefore bare a large risk for deviations.

Most investigations cover oxidative aging as one of the very important driving forces for aging of rubber parts. Bolland and Gee [8] explained the oxygen induced aging by a complex cycle of radical and autocatalytical chemical reactions. The reaction can lead to a build-up or degradation of chemical bonds leading to stiffening or softening, respectively. Several authors also evaluated aging processes under nitrogen and [4, 9] and typically found significant less influences of temperature on the materials properties at comparable conditions [10–12].

Significant progress was made in recent years by predicting aging phenomena based on the influence of oxygen. The importance of oxygen and the measurement of the oxygen consumption [13–15] allowed the quantification of chemical-oxidative reactions. In this context also the diffusion and dissolving of oxygen within the rubber matrix has become increasingly important. The Diffusion Limited Oxidation (DLO) has been proven to significantly influence the spatial distribution of aging within rubber parts. Here, in essence oxygen diffuses in the material and starts to oxidize the rubber matrix. Due to the shorter distance oxidation starts at the sample surface. As a side effect the oxidized material at the sample surface increases in density and hence the diffusion speed of oxygen into the core of the rubber parts is reduced. As a result, less oxygen can reach the sample core. This approach explains very well that especially large rubber parts can have a very strongly oxidized surface whereas the properties inside of the part are nearly unaffected. Modeling approaches combined the newly developed analysis methods with chemically and physically originated network models to model the forming and destruction of polymer bonds taking the DLO effect into consideration [15–17]. This modeling allows a significant improvement in the understanding and prediction of chemical-oxidative reaction mechanisms. Nevertheless, the mechanisms in real applications are in most applications not only connected to oxidative reactions, especially under dynamic loads.

Therefore, further research focuses on the interplay of oxidative aging mechanisms and the crack growth rate of rubber compounds. Typically, crack growth rates under nitrogen or vacuo are significantly smaller [9, 18, 24] or even not measurable [19]. Mars and Fatemi [20] explained the influence of oxygen on the fatigue behavior by the diffusion and dissolving von oxygen into the rubber matrix and hence the locally changing properties leading to higher crack growth rates.

In addition, it can be shown that the effect of oxygen is depending on the crack growth rate itself. Typically, at high enough growth rate the effect of oxygen is negligible. It is assumed that the effect of oxygen in thermal aging significantly differs from the effect of oxygen under mechanical load. Presumably stress-activated oxidative chain scission plays here an important role and relates the fatigue process also to the initiation of cracks. Several groups could show that this effect is closely related to the presence and size of flaws [21, 22] and hence the homogeneity of the material.

Since the mechanical fatigue is based on a variety of physical and chemical mechanisms the combination of different analyzing methods is an essential part of research for the prediction of lifetime for rubber goods. Several reliable methods can be used for the analysis of mechanical fatigue phenomena. The existence of flaws (non-dispersed filler agglomerates) can be analyzed by dispersion measurements using light microscopic techniques [23]. The heat build-up and therefore the influence of the internal sample temperature during the mechanical aging process can be estimated, e.g., by a novel experimental device [33]. Alternatively, the full temperature profile can be estimated by a numerical method, which has to be adapted to measurements of the surface temperature of the samples, e.g., by an infrared sensor.

It is the aim of this paper to analyze the aging and fatigue processes during cyclic mechanical and/or thermal treatment for a large variety of rubber compounds. We will argue that the aging causes several effects such as degradation of the polymer network, polymer-filler coupling, and filler network. Accordingly, the influence of polymer and filler type as well as filler content on the different aging processes will be investigated. The damage of both types of networks and the resulting mechanical properties are analyzed and described by referring to physical models. Ideally, a separation of degradation phenomena by thermal and mechanical aging is achieved. With these investigations, a contribution for a better understanding of the specific aging and wear processes of filler reinforced elastomers shall be achieved.

1.1 Materials

1.2 Sample Pool

To provide wide-ranging investigations, the sample pool covers the field of tire tread rubbers, technical and special elastomers. As reinforcing fillers carbon black and silica are used. The following combinations have been selected:

- Tire tread compounds: S-SBR and NR with carbon black; S-SBR and BR/S-SBR blends with silica

- Technical elastomers: EPDM, ACM, and CR with carbon black
- High temperature resistant elastomers: HNBR and FKM with carbon black

The abbreviations for the polymers used here are specified in the appendix. The sample pool with all ingredients is summarized in Table 1. All ingredients and abbreviation used in Table 1 are also explained in detail in the appendix.

1.3 Sample Preparation

All investigated materials have been mixed using a 1.5 l internal mixer (Werner & Pfleiderer, Germany). Polymer, filler, oil and anti-aging ingredients have been mixed during 5–7 min at around 70°C in the internal mixer. The vulcanization chemicals have been added at the two-roll mill (Berstorff, Germany) at lower temperatures. The samples have been vulcanized by melt curing with respect to the t_{90}-time determined with a vulcameter (MDR 2000E) at the special vulcanization temperature for each system. The ACM compounds have been cured due to the instructions of the supplier and tempered after curing for 4 h at 180°C. Test samples were produced in the following shapes: 2 mm plates, 6 mm plates, 10 mm plates, and dumbbells.

2 Experimental Techniques

2.1 Physical Properties

All investigated systems have been characterized by the hardness measurement (Shore A, DIN 53505), density measurement (DIN 53279, ISO 1183), tensile strength (DIN 53504), rebound (DIN 53512, ISO 4662), and the evaluation of filler dispersion [5]. Additionally, tear strength measurements have been performed with trousers specimens after DIN 53515 (ISO 34-1).

2.2 Mechanical Fatigue

Mechanical aging or fatigue tests were carried out using a servo-hydraulic system (MTS Biaxial Test-line). Here, a special sample holder (see Fig. 1) was used, which allows for the parallel investigations of four dumbbells at one time in order to provide optimal statistics. The samples were loaded in constant strain mode at a frequency of 2 Hz. The used strains were 300%, 200%, 150%, 100%, and 60%. At least three different strains have been loaded for each material. The *R*-factor, defined as maximum to minimum strain $R = \varepsilon_{min}/\varepsilon_{max}$, was kept constant at the value

Table 1 Composition of the materials used in this study; brackets "()" indicate the amount in phr (per hundred rubber), the commercially used terms are explained in the appendix

Sample	Polymer	Filler (phr)	Silane (phr)	Oil (phr)	Antiaging (phr)	Vulcanization system (phr)		
ACM 1	HyTemp 4051	N550 (50)		WB222 (2)	Naugard 445(1), Diuron (2)	St.acid (1)	NS70(6)	
ACM 2	HyTemp AR-71	N550 (50)		WB222 (2)	Naugard 445(1)	St.acid (1)	NS70(6)	S(0.3)
CR 1	Neoprene GRT	N660 (50)		DOP (10)	IPPD (3)	ZnO (4) St.acid (0.5)	ETU (1) MgO (5)	TMTD (0.5)
CR 2	Neoprene GRT	N234 (50)		DOP (10)	IPPD (3)	ZnO (4) St.acid (0.5)	ETU (1) MgO (5)	TMTD (0.5)
EPDM 1	Nordel 4570	N234 (50)			TMQ (1,5)	ZnO (4) St.acid (1)	Peroxide (5)	
EPDM 2	Keltan 5580	N234 (50)			TMQ (1,5)	ZnO (4) St.acid (1)	Peroxide (5)	
FKM 1A	Tecnoflon P459	N660 (20)				Ca(OH)2 (2,5)	TAIC (4)	Dicup(50%) (2.5)
FKM 2A	Tecnoflon P459	N660 (10) N234 (10)				Ca(OH)2 (2,5)	TAIC (4)	Dicup(50%) (2.5)
HNBR 1	Zetpol 2020	N234 (50)			Naugard 445 (1,5) Vanox ZMTI (1)	ZnO (5) St.acid	(0.5) TOTM (5)	VulCup40 KE (8) HVA-2 (2)
NR 1	RSS1	N234 (90)		DAE (30)	6PPD (2) TMQ (2)	ZnO (3) St.acid (1)	TBBS (2)	DPG (2) Sulfur (2)
NR 2	RSS1	N234 (50)			6PPD (2) TMQ (2)	ZnO (3) St.acid (1)	TBBS (2)	DPG (2) Sulfur (2)
BR/SBR 1	CB24/VSL5025–0; 30: 70	GR7000 (90)	Si69 (7.2)	DAE (30)	6PPD (2) TMQ (2)	ZnO (3) St.acid (1)	TBBS (2)	DPG (2) Sulfur (2)

BR/SBR 2	CB24/VSL5025–0; 30: 70	GR7000 (50)	Si69 (4)	DAE (30)	6PPD (2) TMQ (2)	ZnO (3) St.acid (1)	TBBS (2)	DPG (2) Sulfur (2)
SBR 1	VSL 5025-0	N234 (90)		DAE (30)	6PPD (2) TMQ (2)	ZnO (3) St.acid (1)	TBBS (2)	DPG (2) Sulfur (2)
SBR 2	VSL 5025-0	N234 (50)			6PPD (2) TMQ (2)	ZnO (3) St.acid (1)	TBBS (2)	DPG (2) Sulfur (2)
SBR 3	VSL 5025-0	GR7000 (90)	Si69 (7.2)	DAE (30)	6PPD (2) TMQ (2)	ZnO (3) St.acid (1)	TBBS (2)	DPG (2) Sulfur (2)

Fig. 1 Sample holder of MTS Biaxial Test-line for four dumbbell samples

$R = 0.1$. The samples have been mechanically loaded until all four dumbbells were broken.

For a part of the sample pool, the samples were additionally loaded in constant force mode at a frequency of 2 Hz. Other than in the strain-controlled mode, in this setup only one dumbbell could be measured at one time. Three different forces were taken for one material with statistics of three samples per force. The forces were chosen to obtain strains between 60% and 200%. The ratio of the maximum to the minimum force F_{min}/F_{max} was kept constant at the same R-value $R = 0.1$ used for the strain-controlled mode. The mechanical aging at different strains in constant strain mode until breakdown results in well-known Wöhler diagrams where the logarithm of the maximum strain is plotted against the logarithm of the (averaged) number of cycles at break. However, these common Wöhler curves are not well suited for a comparison of different compounds that differ significantly in hardness. Therefore, a new analyzing method with the energy density W_0, given by the product of maximum stress and strain, as parameter is applied instead of stress or strain. This appears to be a natural choice, because fracture mechanics is generally based on energy criteria, like the Griffith Criterion. Note that also the tearing energy used for the evaluation of crack propagation rates scales with the energy density stored in the samples [1].

2.3 Temperature Measurement During Mechanical Aging

During mechanical aging, the temperature of the samples is usually increasing due to energy dissipation mechanisms. Therefore, the temperature has been measured while a sample was mechanically aged. The mechanical aging was performed with 100% strain at a frequency of 2 Hz. During mechanical aging, the surface temperature was measured by the infrared sensor VarioTHERM (InfraTec GmbH, 50 Hz frequency, temperature resolution (at 30°C) < 0.1 (0.01 mA); temperature range: -25 ... +1,200°C). The camera was operated with an emission factor of $\varepsilon = 0.9$, which is close to the value of black body radiation and refers to the black color of the samples. The distinction between the different samples is neglected.

2.4 Thermal Aging

The thermal aging in the oven has been performed for periods of 1, 3, 6, and 10 days. The aging temperature has been chosen specifically for each material. The following temperatures have been applied: 100°C for SBR, BR/SBR, NR and CR, 130°C for EPDM, 150°C for HNBR and ACM, 180°C for the FKM compounds. Physical properties have been measured before aging and after each aging period.

3 Results and Discussion

In the following sections the test results will be summarized and discussed. The results of purely thermal aging and mechanical fatigue are presented in Sects. 3.1–3.3 for the different compound classes, respectively. In Sects. 3.4–3.6 we will analyse and discuss the results by providing further in-depth analysis and cross correlations between the different analysis methods with respect to the different material classes.

3.1 Tire Tread Compounds (SBR, NR, BR/SBR)

3.1.1 High Filler Loading (90 phr Filler +30 phr Oil)

The results of thermal aging of highly filled tire tread compounds are summarized in Table 2. We observe that hardness values are increasing significantly for all SBR and SBR/BR compounds, whereas the hardness of NR is only mildly affected. The same effect but in reverse direction can be observed for the measurement of rebound. The

Table 2 Thermal aging behavior after 3, 6, and 10 days of the tire tread systems (high loading) SBR, NR, and BR/SBR at 100°C

Specimen	Temp. °C	Density g/cm^3	Shore A	Disp. %	Rebound	Elongation at break %	Tensile strength MPa
BR/SBR 1		1.197	61	92	33.9 ± 2.1	350	16.4
BR/SBR 1 3d	100	–	70	–	45.3 ± 0.6	216	13.0
BR/SBR 1 6d	100	–	71	–	43.1 ± 0.4	181	12.4
BR/SBR 1 10d	100	–	75	–	42.6 ± 1.2	135	10.5
NR 1		1.164	80	97	27.9 ± 0.3	368	19.3
NR 1 3d	100	1.174	81	–	28.7 ± 0.4	309	17.1
NR 1 6d	100	1.173	82	–	27.9 ± 0.3	185	12.0
NR 1 10d	100	1.188	83	–	26.9 ± 0.4	115	9.3
SBR 1		1.179	80	97	12.6 ± 0.1	415	21.3
SBR 1 3d	100	1.188	84	–	14.1 ± 0.2	262	18.2
SBR 1 6d	100	1.186	86	–	14.4 ± 0.3	196	16.6
SBR 1 10d	100	1.191	88	–	14.5 ± 0.2	155	16.5
SBR 3		1.185	65	98	15.1 ± 0.8	293	16.9
SBR 3 3d	100	–	72	–	22.2 ± 0.2	219	16.2
SBR 3 6d	100	–	75	–	19.2 ± 0.2	193	15.3
SBR 3 10d	100	–	74	–	19.3 ± 0.1	165	15.2

rebound value is increasing for SBR compounds and blends but stays constant for Natural Rubber.

The results of mechanical fatigue of highly filled tire tread compounds are summarized by the "Wöhler" curves in Fig. 2. The following results can be stated: NR provides the best resistance against fatigue. For SBR, there are only slight differences whether carbon black (SBR 1) or silica (SBR 3) is used. All samples have comparable slope, except BR/SBR 1 with a significantly larger slope. Compared to SBR 3 (both filled with silica), BR/SBR 1 shows shorter lifetimes for small energies but larger lifetimes for larger energies. This shows that the addition of BR (30%) can improve the fatigue performance of SBR compounds.

3.1.2 Low Filler Loading (50 phr Filler)

For the low loading tire tread systems (Table 3) we can observe a stepwise tendency in hardness as for the highly filled composites. First it grows and stays then nearly constant. Rebound decreases in the first aging step and stays in all further steps nearly constant. For the systems without oil and SBR 2 with oil, the elongation at

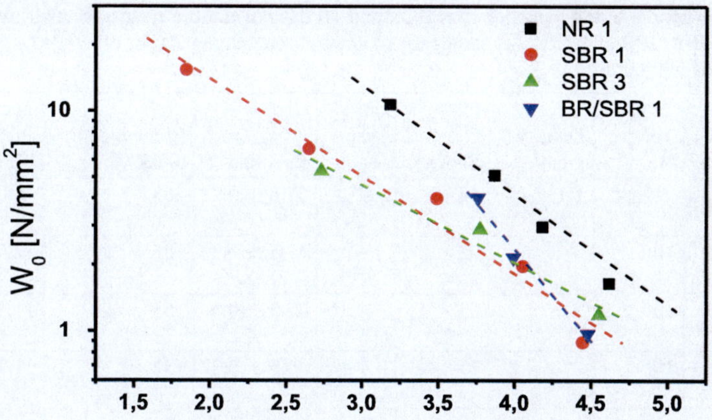

Fig. 2 "Wöhler" curves of the highly filled systems in the energy density representation

break and the tensile strength decrease significantly in the first and decrease further in the following aging steps. For NR 2 with oil both values decrease as well but not with that immense drop at the first step.

From the Wöhler curves in Fig. 3, the following results are obtained: NR 1 and NR 2 have the highest resistance against aging of all investigated materials in the sample pool. Increase in filler concentration results in higher stability against fatigue at large energies. Only for NR, the influence of the filler amount is small. BR/SBR 2 (without oil) showed a significantly lower lifetime compared to BR/SBR 2 with oil, although the dispersion is worse for BR/SBR 2 with oil.

3.1.3 Influence of Filler Type and Amount for Tire Tread Compounds

The influence of the amount of filler can clearly be seen for the samples NR 1 and NR 2, where NR 1 contains 90 phr of N234 and 30 phr oil and NR 2 contains 50 phr of N234. The lower filled NR 2 has higher rebound, elongation at break and tensile strength, but a lower hardness (Tables 3 and 4). For SBR 1 and SBR 2 (containing the same ratio of filler as the NR pair) these relations have been turned around for elongation at break and tensile strength, they are for the low filled system smaller. The high filled SBR 1 has a higher value of hardness and a lower one for rebound than the SBR 2.

In the silica systems SBR 3 and BR/SBR 1 containing 90 phr of GR7000, 7.2 phr Si69 and 30 phr oil and BR/SBR 2 containing 50 phr of GR7000 and 4 phr Si69, the lower filled blend has the higher rebound and hardness but lower elongation at break and tensile strength.

Table 3 Thermal aging behavior after 3, 6, and 10 days of the tire tread systems (low loading) SBR, NR, BR/SBR at 100°C. (+) stands for compounds containing 30 phr oil while (–) stands for compounds without oil

Specimen	Temp. °C	Density g/cm^3	ShoreA	Disp. %	Rebound	Elongation at break %	Tensile strength MPa
BR/SBR 2 (–)		1.137	64	91	51.6 ± 0.2	258	15.0
BR/SBR 2 3d	100	–	72	–	47.5 ± 0.3	185	12.9
BR/SBR 2 6d	100	–	73	–	47.5 ± 0.6	153	11.3
BR/SBR 2 10d	100	–	75	–	46.5 ± 0.2	132	11.2
BR/SBR 2 (+)		1.116	48	75	52 ± 1.1	474	14.6
NR 2 (–)		1.112	76	99	45.4 ± 0.4	487	31.4
NR 2 3d	100	–	81	–	48.9 ± 0.3	318	21.4
NR 2 6d	100	–	82	–	46.7 ± 0.5	222	15.6
NR 2 10d	100	–	83	–	46.5 ± 0.3	148	12.2
NR 2 (+)		1.092	61	70	45.1 ± 0.3	545	17.7
NR 2 3d	100	1.096	64	–	44.0 ± 0.3	490	19.1
NR 2 6d	100	1.103	–	–	–	294	10.6
NR 2 10d	100	1.111	65	–	40.4 ± 0.3	191	7.1
SBR 2 (–)		1.133	76	99	14.2 ± 0.2	304	18.8
SBR 2 3d	100	–	83	–	15.6 ± 0.6	168	19.9
SBR 2 6d	100	–	85	–	15.2 ± 0.6	180	18
SBR 2 10d	100	–	89	–	15.0 ± 0.5	166	19.6
SBR 2 (+)		1.109	59	61	13.8 ± 0.0	602	26.6
SBR 2 3d	100	1.113	64	–	13.2 ± 0.3	380	14.4
SBR 2 6d	100	1.115	–	–	–	352	15.5
SBR 2 10d	100	1.120	68	–	12.6 ± 0.2	291	14.2

3.2 *Technical Rubbers (HNBR, CR, EPDM)*

The thermal aging behavior of all technical rubbers is summarized in Table 4. All technical rubbers show under thermal aging a continuous increase in hardness, where always the first aging stage (3 days) gives the greatest rise. From the same magnitude but with reverse tendency, the rebound decreases, elongation at break and tensile strength decrease with each aging step.

The energy based "Wöhler" curves are summarized in Fig. 4. It is found that the different materials differ in their load dependency. In example, the chloroprene

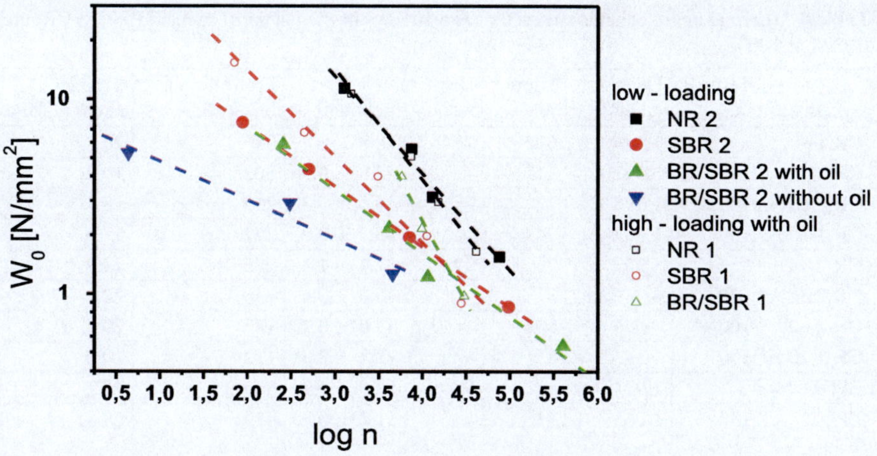

Fig. 3 "Wöhler" curves of the tire tread systems in the energy density representation

rubbers improve their relative performance at higher loads (CR1) or lower loads (CR2) compared to the two EPDM types and the HNBR which behave somewhat similar. Especially, the surprisingly larger lifetime at high loading of the sample CR 1 with the inactive filler (50 phr N660) is astonishing since the dispersion of CR 1 is worse than CR 2. The sample EPDM 2 with the semi-crystalline polymer results in a better resistance against mechanical aging than EPDM 1. HNBR has the worst lifetime of all samples considered in this section.

However, in all measurements each of the four dumbbells aged at the same strain breaks at rather different cycle numbers. Averaging of such varying values can falsify the results and the "Wöhler" curve in Fig. 4 should be considered very carefully. The large scattered values are perhaps a hint that in some dumbbells a huge particle initiates early a crack, whereas other dumbbells have no such defects. It is known that already one single large particle in a dumbbell is sufficient for reducing the lifetime [21, 22]. Dispersion measurements, which have been performed in order to ensure comparable homogeneity, can only show a limited image of possible crack initiators due to the statistics of dispersion measurements.

3.3 *High Temperature Resistant Rubbers (ACM, FKM)*

ACM 1 and ACM 2 show both a constant value of tensile strength independent of aging time but differ in all other parameters. ACM 1 (polymer: Hy Temp 4051) shows an increase in hardness and rebound and a decrease in the maximal reachable elongation. In contrast, ACM 2 (polymer: Hy Temp AR71) shows the inverse behavior.

Table 4 Thermal aging behavior after 3, 6, and 10 days of the technical rubbers HNBR, CR, and EPDM at 100°C

Specimen	Temp. °C	Density g/cm³	Shore A	Disp. %	Rebound	Elongation at break %	Tensile strength MPa
CR 1		1.374	69	77	48.8 ± 0.4	387	20.2
CR 1 3d	100	1.392	71	–	47.9 ± 0.5	302	19.2
CR 1 6d	100	1.398	72	–	46.6 ± 0.3	256	18.4
CR 1 10d	100	1.407	74	–	45.7 ± 0.2	222	17.6
CR 2		1.357	76	92	37.4 ± 1.1	285	23.8
CR 2 3d	100	–	81	–	45.5 ± 0.4	230	22.0
CR 2 6d	100	–	83	–	43.6 ± 0.2	189	20.5
CR 2 10d	100	–	85	–	43.8 ± 0.3	172	20.0
EPDM 1		1.074	69	98	53.0 ± 1.2	345	26.2
EPDM 1 3d	130	1.065	71	–	51.1 ± 0.7	439	28.5
EPDM 1 3d	130	–	72	–	52.8 ± 2.6	332	20.1
EPDM 1 6d	130	–	72	–	52.5 ± 2.7	239	15.9
EPDM 1 10d	130	–	80	–	56.4 ± 0.2	339	23.0
EPDM 2		1.068	80	97	53.6 ± 0.3	353	29.3
EPDM 2 3d	130	1.072	82	–	51.6 ± 0.2	355	31.3
EPDM 2 6d	130	1.062	82	–	51.0 ± 0.2	347	31.2
EPDM 2 10d	130	1.071	82	–	51.9 ± 0.3	327	27.5
HNBR 1		1.150	76	98	29.8 ± 0.3	225	26.9
HNBR 1 3d	150	–	82	–	32.2 ± 0.2	191	25.0
HNBR 1 6d	150	–	84	–	35.7 ± 0.3	159	23.3
HNBR 1 10d	150	–	85	–	32.8 ± 0.2	115	20.8

FKM 1A and FKM 2A stay unimpressed from heat treatment for rebound, hardness, and elongation at break. Tensile strength remains constant for FKM 1A (20 phr N660) but decreases for FKM 2A (Table 5).

Considering the "Wöhler" curves in Fig. 5, the following results are obtained: For FKM 1A and FKM 2A similar lifetimes are found, indicating that the influence of filler type is almost negligible. ACM 1 has a larger slope than ACM 2 and a smaller lifetime in the measured range.

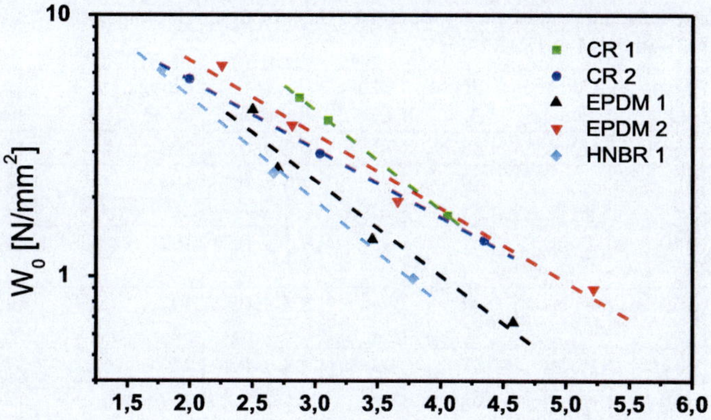

Fig. 4 "Wöhler" curves of the technical rubbers in the energy density representation

3.4 *Analysis of Force Decreases During Mechanical Aging*

The lifetime of materials under loading is dependent on several factors. First of all, the existence of flaws and the crack resistance of the material itself play an important role. Additionally, also the heat build-up in the sample due to energy dissipation mechanisms influences the lifetime. Due to all these aspects, the force decreases during mechanical aging at constant strain as seen exemplary in Fig. 6 for NR 2.

In the first few cycles, the breakage of clusters takes place. Depending on the applied strain, a certain part of clusters will break. This cluster breakage results in a force decrease in the first cycles. After 5–20 cycles a dynamical balance will be obtained in which breakage and reaggregation of clusters during each cycle are in equilibrium. However, as depicted in Fig. 6, the force is decreasing as well after a higher number of cycles until a plateau after about 7,000 cycles is reached. This is probably a temperature effect due to smaller modulus at larger temperatures.

3.4.1 Relation of Sample Temperature to Dissipated Energy

The sample temperature increases due to energy dissipation mechanisms. The direct consequence is the decrease of the elastic modulus with aging cycles. Indeed, the temperature influences probably also other material properties that are important for mechanical aging. Depending on the sensitivity of the material, effects due to thermal aging are additionally possible.

The stress-strain cycles during mechanical aging have been evaluated to get information about the variation of stiffness and hysteresis of the samples. Representative stress-strain cycles for NR 1 and CR 1 are shown in Fig. 7. The drop of the stiffness with aging time, reflecting the characteristic force decrease considered above, becomes apparent. For NR 1, which contains a high amount of active carbon

Table 5 Thermal aging results after 3, 6, and 10 days of the high temperature resistant rubbers ACM at 150°C and FKM at 180°C

Specimen	Temp. °C	Density g/cm^3	Shore A	Disp. %	Rebound	Elongation at break %	Tensile strength MPa
ACM 1		1.250	58	95	6.2 ± 0.0	203	11.1
ACM 1 3d	150	–	59	–	7.5 ± 0.1	198	11.5
ACM 1 6d	150	–	61	–	9.3 ± 0.2	197	11.8
ACM 1 10d	150	–	64	–	8.5 ± 0.1	191	11.5
ACM 2		1.271	61	99	6.9 ± 0.1	246	11.9
ACM 2 3d	150	–	62	–	6.5 ± 0.2	267	12.5
ACM 2 6d	150	–	57	–	6.3 ± 0.2	283	12.6
ACM 2 10d	150	–	56	–	6.3 ± 0.2	289	12.2
FKM unfilled		1.843	56	–	8.2 ± 0.1	–	–
FKM 1A		1.872	80	93	7.2 ± 0.0	157	22.9
FKM 1A 3d	180	–	82	–	7.1 ± 0.2	169	24.4
FKM 1A 6d	180	–	82	–	7.2 ± 0.2	165	23.4
FKM 1A 10d	180	–	82	–	7.3 ± 0.2	158	22.5
FKM 2A		1.868	80	93	8.6 ± 0.0	137	22.4
FKM 2A 3d	180	–	85	–	8.6 ± 0.2	124	19.5
FKM 2A 6d	180	–	86	–	8.8 ± 0.2	97	14.9
FKM 2A 10d	180	–	85	–	9.0 ± 0.3	92	15

black, the energy dissipation (area of stress-strain cycle) is larger compared to CR 1, which contains less carbon black with lower activity.

A fair correlation between the dissipated energy, obtained from numerical integration of the area within stress-strain cycles, and the measured surface temperature for all materials is shown in Fig. 8. Most materials can be described by a single sigmoidal curve. This seems to be trivial since the dissipated energy is causing heat build-up in the sample. However, the measured surface temperature is usually significantly smaller than the internal sample temperature. The surface temperature depends on heat conducting phenomena and the heat capacity, which are different for the investigated samples. The deviations from the fit line for the EPDM-systems and the highly filled tread compounds (with oil), NR 1 and SBR 1, are supposed to be related to these effects.

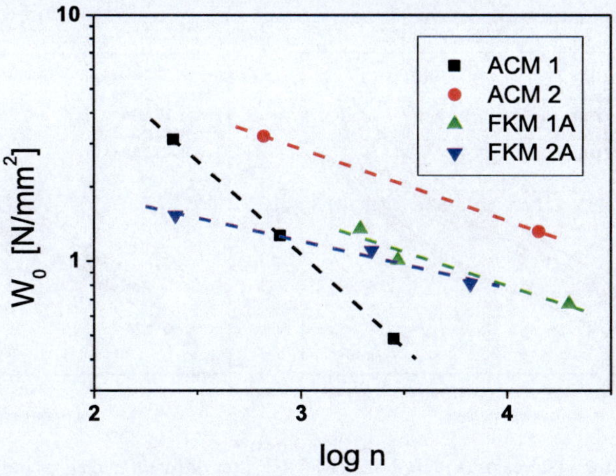

Fig. 5 "Wöhler" curves of the high temperature rubbers in the energy density representation

Fig. 6 Analysis of force development during mechanical loading. The different aging effects are indicated. The example shows NR 2 at 100% strain

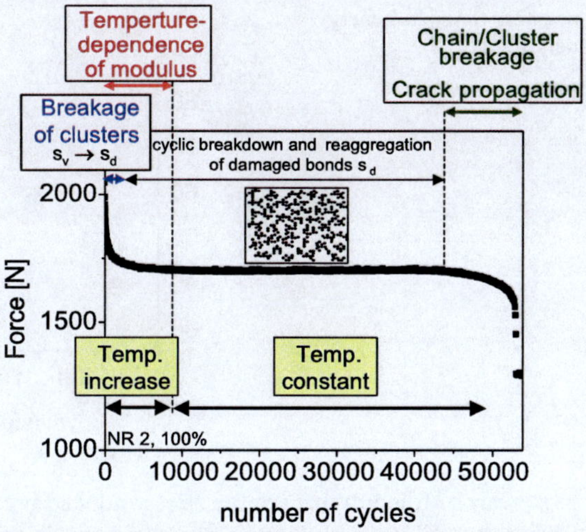

Since the energy dissipation is closely connected to the breakdown and reaggregation of filler clusters, it can be expected that the filler amount in the sample is an important parameter for the temperature increase during mechanical aging. In Fig. 9, the surface temperature is plotted versus the effective filler volume fraction, which considers the non-spherical structure of the filler particles. Accordingly, the effective filler volume fraction is larger than the real volume fraction, since part of the polymer, the so-called occluded rubber hidden in voids, is not deformed and acts like additional filler. Here, the effective filler volume fraction is obtained by fitting quasistatic stress-strain cycles of unaged samples to the Dynamic Flocculation Model (DFM) [25–28]. The DFM is a micro-mechanical model that relates the

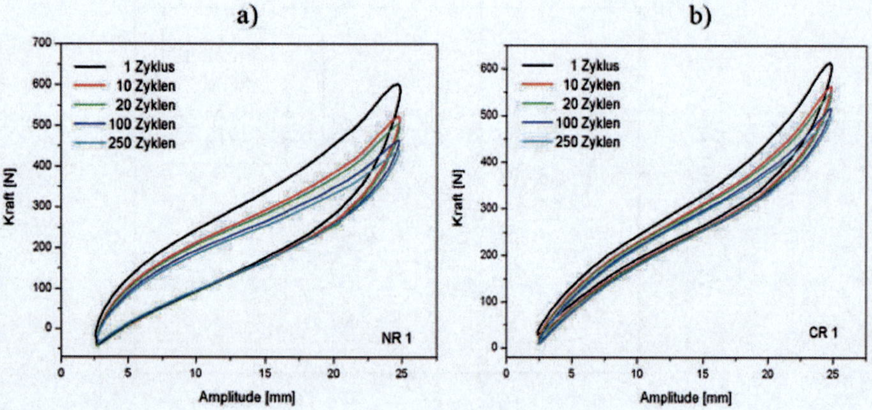

Fig. 7 Stress-strain curves of (**a**) NR 1 and (**b**) CR 1 after different cycles, as indicated

Fig. 8 Correlation between the surface temperature and averaged dissipated energy per cycle

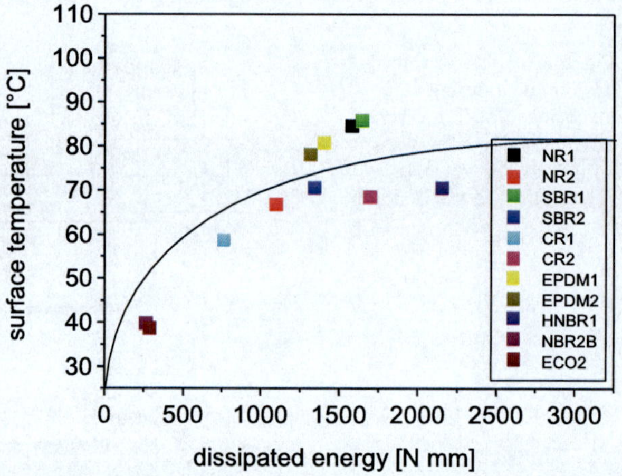

hysteresis of filled rubbers to the stress-induced breakdown and reaggregation of the filler network. Recently, it was demonstrated that the DFM delivers valuable information on the thermo-oxidative aging of EPDM compounds [28]. We will not go into details here but refer to Ref. [28] where a detailed description of the DFM is given.

Although the data points in Fig. 9 cannot be described by a single line, a clear tendency is seen that increasing effective filler volume fraction leads to larger temperatures. This correlation shows that the fillers are indeed the main source for energy dissipation. Besides the filler amount, many factors play a role such as the polymer–filler interaction. Therefore, it is understandable that the values in Fig. 9 show a large spreading for the different polymers. Comparing only samples with the same polymer and filler type (NR 1 and NR 2; SBR 1 and SBR 2, BR/SBR 1 and BR/SBR 2), the higher filled sample has always a larger surface temperature.

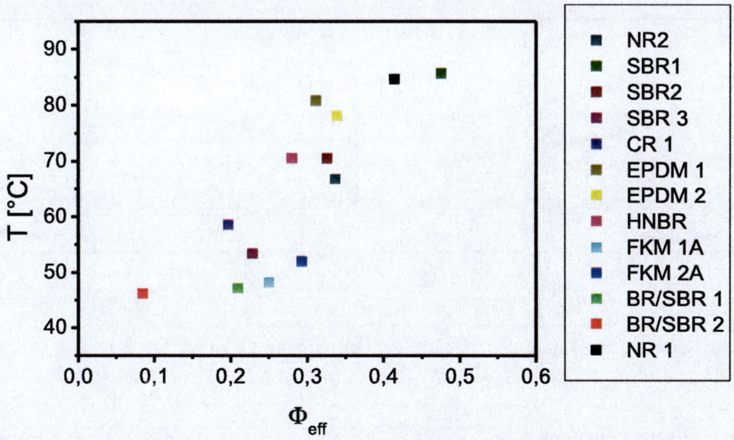

Fig. 9 Correlation between surface temperature and effective filler volume fraction

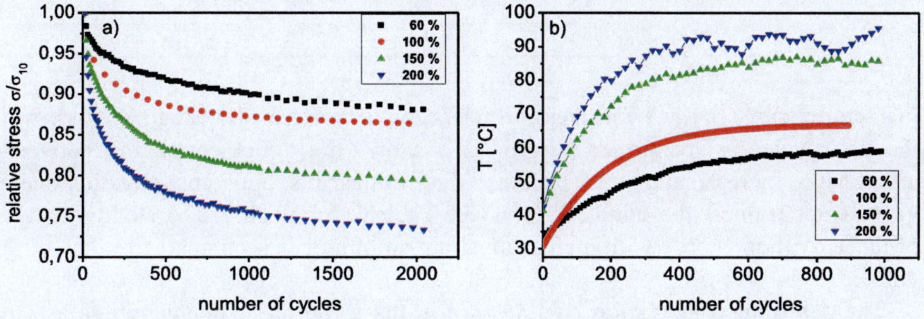

Fig. 10 **(a)** Stress decrease and **(b)** surface temperature increase during mechanical loading for NR 2 at different strain amplitudes

3.4.2 Relation Between Force Decrease and Sample Temperature

In order to investigate the temperature influence on the mechanical forces, surface temperature and stress have been measured simultaneously while the dumbbells were mechanically aged. The temperature increase and the stress decrease are seen for the NR 2 sample in Fig. 10 for different strain amplitudes. Both seem to be connected since the temperature increase is stronger and the stress decrease is more pronounced with higher strain amplitude.

To test this assumption, the temperature increase of the sample, here the surface temperature was chosen, has to be connected to the stress decrease. This can be done since the temperature as well as the stress in dependence on the number of cycles is known. With this information, the stress at each measured temperature can be obtained as shown in Fig. 10a. Using an Arrhenius plot with the logarithm of stress against $1/T$ (see Fig. 11b), a nearly linear behavior is found beside the stress drop at

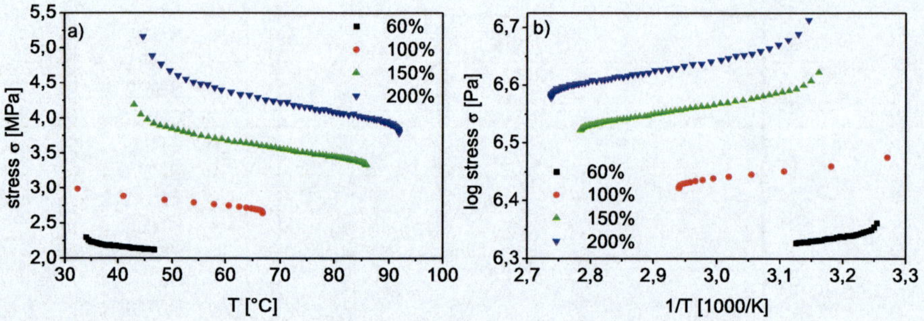

Fig. 11 (**a**) Stress-temperature dependence and (**b**) Arrhenius plot for NR 2 aged at various strain amplitudes

Table 6 Activation energies for NR 2 aged at different elongations

Strain amplitude (%)	Activation energy E_A (kJ/mol)
60	3.19
100	2.92
150	3.46
200	3.59

low temperatures (large $1/T$ values), which is connected to the breakage of clusters in the first cycles as mentioned before. This shows the direct correlation between temperature increase and stress decrease. From linear fits, apparent activation energies are determined of around 3 kJ/mol (see Table 6) for all strains. A slight increase of the activation energy with increasing strain amplitude is seen apart from the 60% value.

The determined activation energies are in the same order of magnitude as the value obtained by dynamic mechanical analysis (around 8 kJ/mol for NR 2) though in dynamic mechanical measurements only small deformations (0.5%) are applied. Unfortunately, for most materials the stress decrease due to heat build-up is superimposed by degradation effects in particular at large strains. For these materials, no Arrhenius dependence is found and the activation energy cannot be determined.

3.4.3 Comparison of Tire Tread Compounds (NR, SBR, BR/SBR)

Figure 12 shows the increase in surface temperature and the associated stress decrease during mechanical aging for the high filled tire tread compounds. For the carbon black filled systems (NR 1 and SBR 1), the temperature reaches much larger values than for silica filled samples (SBR 3 and BR/SBR 1). This makes sense since the temperature increase is caused by energy dissipation (see Sect. 3.4.1), which is smaller for the silica filled samples. The more the temperature increases, the more the stress decreases. Furthermore, both silica filled samples with small temperature increases have also less pronounced stress decreases than the carbon black filled

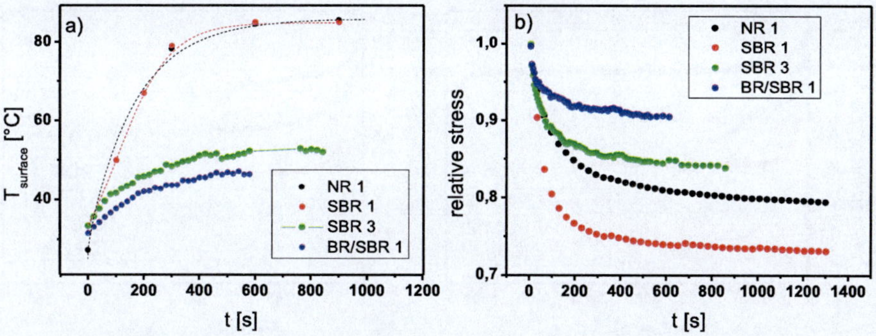

Fig. 12 (**a**) Surface temperature increase and (**b**) relative stress decrease during mechanical aging at 100% strain for the highly filled tire tread systems. The stress is related to the first measured data point after 5 s

Fig. 13 Temperature increase during mechanical aging for the low loading tread systems

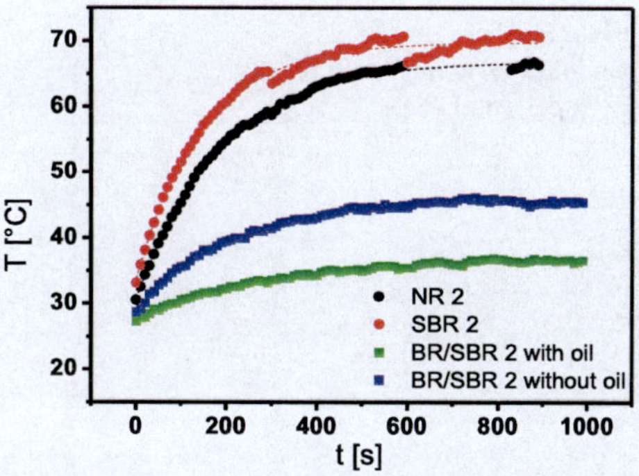

ones. Besides the breakdown of the filler network in the first cycles, the stress decrease is mainly a temperature effect due to decreasing stiffness at larger temperatures. Although NR 1 and SBR 1 have almost the same temperature values, the stress decrease is more pronounced for SBR 1 than for NR 1. This can be explained by the larger change of the stiffness for SBR 1 with increasing temperature that can also be seen in dynamic mechanical measurements. The apparent activation energy describing the temperature dependence of the shear modulus is larger for SBR 1 (16.27 kJ/mol) than for NR 1 (11.56 kJ/mol).

For the lowly filled tire compounds the temperature increase during mechanical aging is shown in Fig. 13. The behavior is quite similar to the highly filled systems only the temperatures are all in all lower.

3.4.4 Technical Rubbers (HNBR, CR, EPDM)

EPDM 1 and EPDM 2 dumbbells show almost the same surface temperature during mechanical aging (~80°C), but the force decrease is much more pronounced for

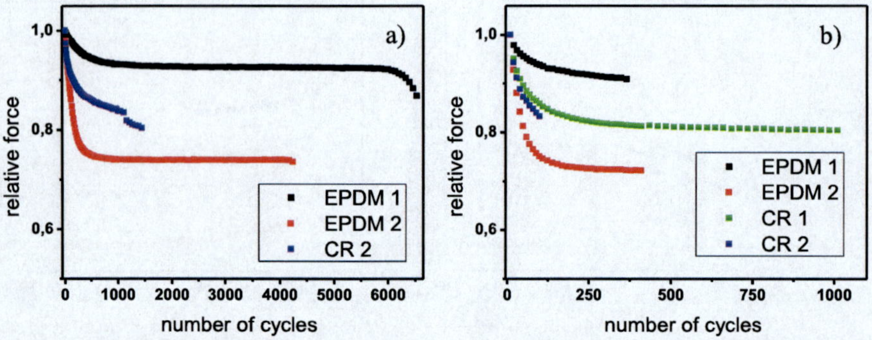

Fig. 14 Force decrease during Wöhler experiments at (**a**) 60% and (**b**) 150% for the technical rubbers. The force is related to the first measured data point after 10 cycles. Data are shown till the first dumbbell is broken, only

Fig. 15 Shear modulus vs. temperature of EPDM 1 and EPDM 2

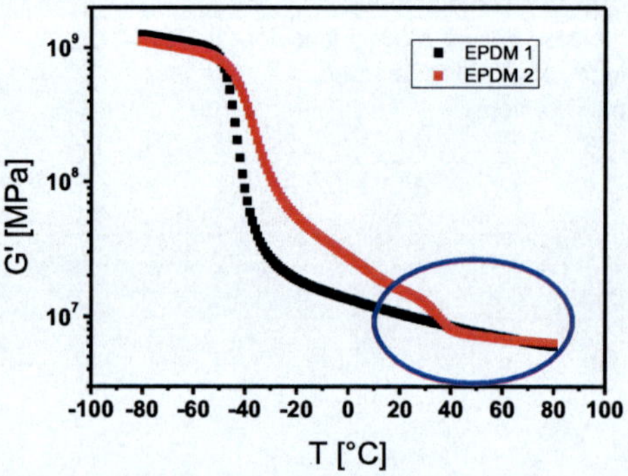

EPDM 2 (see Fig. 14). This finding can be explained by a strong temperature dependence of the modulus of EPDM 2 due to the semi-crystallinity resulting from the high ethylene content of the polymer (Keltan 5580, Lanxess, Germany) [34]. This strong temperature effect in the stiffness is consistent with dynamic mechanical measurements (see Fig. 15). Probably, the crystallinity is also the reason for the different behavior of both EPDM samples under loading.

The surface temperature of CR 2 dumbbells during mechanical aging is obtained at 68°C and is larger than for CR 1 with 59°C. This is an implication of the more active filler type, causing a higher energy dissipation. The larger temperature increase during loading causes a more pronounced force decrease as seen in Fig. 14b due to the decreased elastic modulus with increased temperature.

3.4.5 High Temperature Resistant Compounds (ACM, FKM)

The temperature increase and the related force decrease of the FKM samples are shown in Fig. 16. FKM 2A has slightly higher temperature values and a more

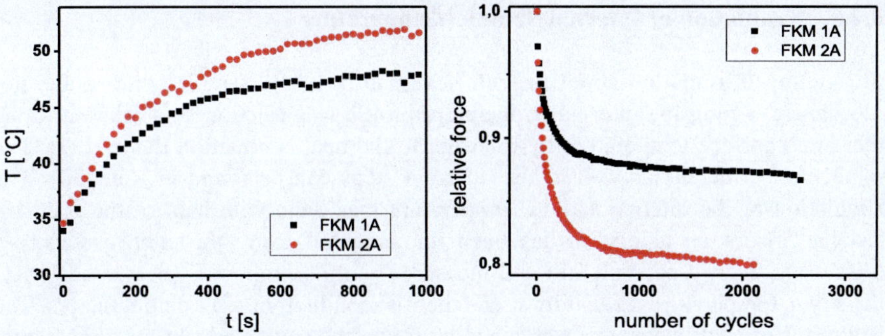

Fig. 16 (**a**) Temperature increase and (**b**) force decrease of FKM at 40% strain

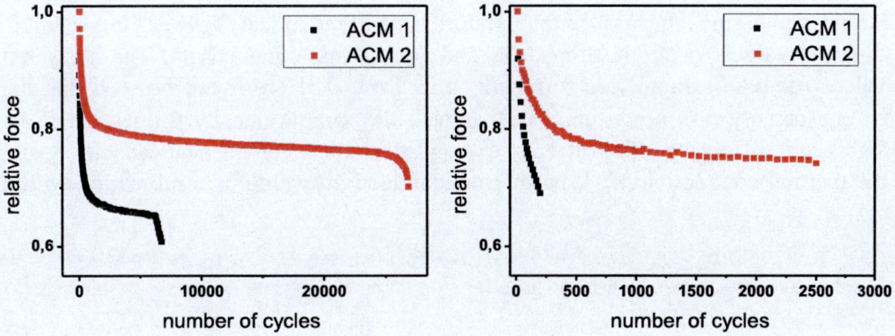

Fig. 17 Force decrease of ACM at (**a**) 60% and (**b**) 100% strain

pronounced force decrease due to the more active filler causing higher energy dissipation (for explanations to the correlation between temperature/energy dissipation and temperature/force decrease, see Sects. 3.4.1 and 3.4.2). Obviously, these slight differences don't influence the mechanical aging much. Due to the high temperature resistance of FKM, also visible by the results in Sect. 3.3, thermal aging effects due to the increased temperature can be excluded for the FKM samples.

The force decrease of ACM samples is shown in Fig. 17. A more pronounced drop of the force and earlier break-off are found for ACM 1. The characteristics of the force drop before break-off indicate for both samples stable crack propagation at 60% and catastrophic crack propagation at 100% strain. Both systems are near the percolation threshold and changes in filler network properties due to loading are large.

3.4.6 Evaluation of Internal Sample Temperature

For getting information about the full heat history of the samples during thermo-mechanical aging, the internal temperature profile was calculated in relation to the measured surface temperature by referring to the heat conduction theory [29]. The mathematical details as well as the different steps can be found in Ref. [30]. The simulation of the internal sample temperature was done with help of the program "Maple." Here an algorithm has been implemented using the energy density $\dot{f}$ (dissipated energy per volume) and measured sample properties (heat capacity c_p, density ρ, temperature diffusivity κ, and thermal conductivity λ) as input factors. The heat-transfer coefficient α was obtained by fitting the simulated surface temperature to measured values. The simulated curves for the samples SBR 1 and NR 1 are shown in Fig. 18.

Conspicuous is the significant difference in the internal sample temperature for both samples since the surface temperature is nearly identical. The reason lies in the different values for the heat capacity and the thermal conductivity. The measured values for both samples are summarized in Table 7. Here it can be seen that the energy density $\dot{f}$ is nearly identical, as well the sample density ρ does not show significant differences. The difference is significant between the heat capacity c_p and the thermal conductivity λ. Low thermal conductivity value in combination with a

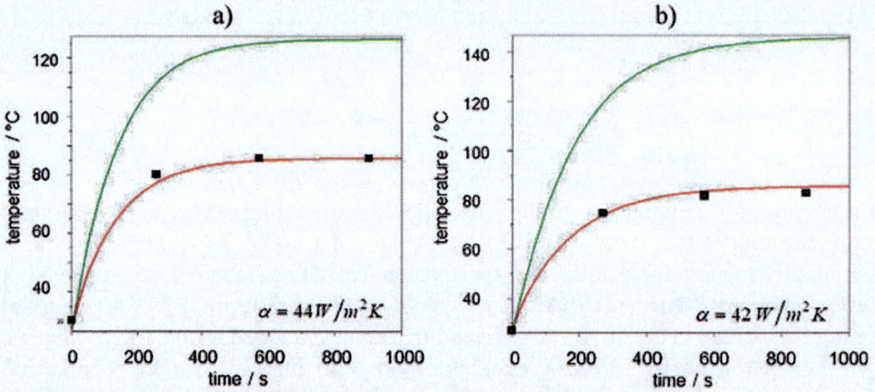

Fig. 18 Simulated temperature curves for the samples (**a**) SBR 1 and (**b**) NR 1. Red lines are surface temperatures, which are fitted to experimental data (squares) to obtain the heat-transfer coefficient $\boldsymbol{\alpha}$, as indicated. Green lines are internal temperatures in the center of the samples

Table 7 Input parameters for temperature simulations of the samples NR 1 and SBR 1

Sample	$\dot{f}$ (energy density) 10^5 W/m^3	c_p (heat capacity) J/g	ρ (sample density) g/cm^3	λ (therm. conductivity) W/(m K)
NR 1	7.2	1.179	1.164	0.169
SBR 1	7.5	1.088	1.179	0.250

Table 8 Comparison of the simulated and the measured internal temperature

Sample	Surface temperature [°C]	Simulated internal temperature [°C]	Measured internal temperature [°C]
NR 1	85	145	130
SBR 1	86	127	110

high value for the heat capacity results in higher internal temperature in the middle of the samples, which can be seen for the sample NR 1 in Table 8.

The simulated internal temperatures have been verified using a specular temperature sensor made of stainless steel connected to a multimeter. The measurement is not very accurate since the temperature decreases very fast due to the temperature sensor. But taking this into account the simulated temperatures have been nearly verified (Table 8). The temperature difference in both cases lays around 15°C.

In conclusion, this method allows the simulation of the internal temperature of the sample during mechanical testing. The examples show that the internal temperatures are much larger than the surface temperatures and the internal temperatures can be different for samples having the same surface temperatures.

3.5 Crack Propagation Analysis

Considering again Fig. 6, it is obvious that the stress values are decreasing further after a large number of cycles for which the temperature is almost constant. Here, damage effects like crack propagation in the sample itself are the reason. At a certain moment, a defect, which can be a large particle, initiates a crack. Therefore, the dispersion and the size of particles are important aspects for mechanical aging. After initiation, the crack propagates in the sample with a velocity depending strongly on material properties as well as on measuring conditions like the applied strain. Consequently, a further important aspect influencing mechanical aging is the crack propagation resistance.

The crack propagation can be classified into two regimes: the stable and the catastrophic one. The different behaviors are reflected in the characteristics of the force decrease shortly before the break-off as seen in Fig. 19 for SBR 3. The force shows a slow, continuous decrease for 100%, while it decreases abruptly for 200%. The resulting fracture surfaces also differ strongly. At small strain a rough "hilly" fracture pattern is found whereas at large strain relatively flat surfaces are obtained. Consequently, the characteristics in force decrease together with the fracture surfaces indicate that the crack propagation is stable at small strains and catastrophic at large strains for SBR 3.

For certain samples, pictures from the fracture surface have been recorded with a digital microscope as seen in Fig. 20 for SBR 1. Here, the fracture patterns can clearly be distinguished. Also, for this material, the fracture patterns indicate stable crack propagation at small strain and catastrophic propagation at large strain. In

Fig. 19 Force decrease during mechanical aging (4 dumbbells in the sample holder) for SBR 3 at (**a**) 100% and (**b**) 200%

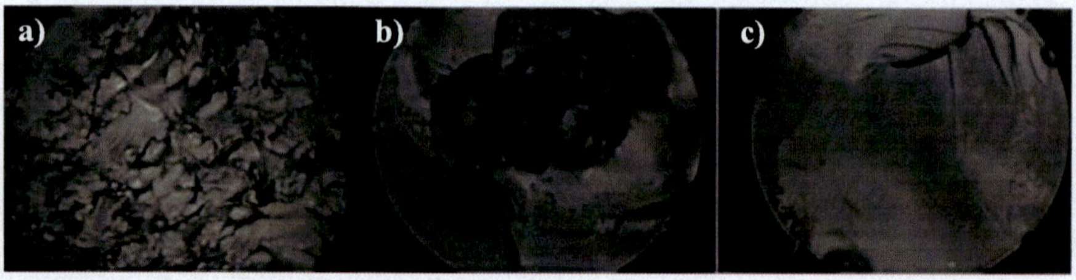

Fig. 20 Fracture surface for SBR 1 recorded with digital microscopy for different strain values: (**a**) 100% strain, (**b**) 200% strain and (**c**) 300% strain

between at 200%, it seems that the crack was initiated somewhere inside the dumbbell, the propagation was first stable and switched later over to a catastrophic behavior. This is similar to results, which are often found in tear analyzer measurements. Here, transitions from stable propagation for small crack lengths to catastrophic one at large crack lengths are often observable.

In Fig. 21, the tolerated number of cycles at constant energy density 2 N/mm^2, obtained from the Wöhler evaluation, is plotted versus the crack propagation resistance measured with trousers specimens. A rough tendency can be found that with increasing resistance against crack propagation the lifetime of the material gets larger. This shows the importance of this parameter concerning mechanical aging phenomena. Note that the measuring conditions are rather different for the tear strength measurements and the Wöhler experiments.

Fig. 21 Crack propagation resistance obtained by tear strength measurements with trousers specimens vs. lifetime obtained from Wöhler diagrams at constant energy density $W_0 = 2$ N/mm

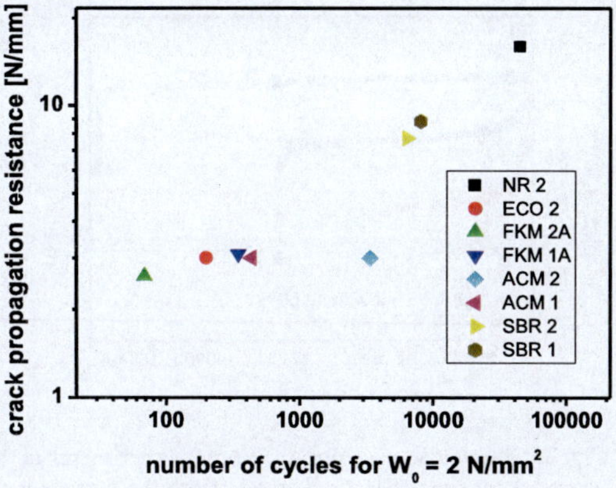

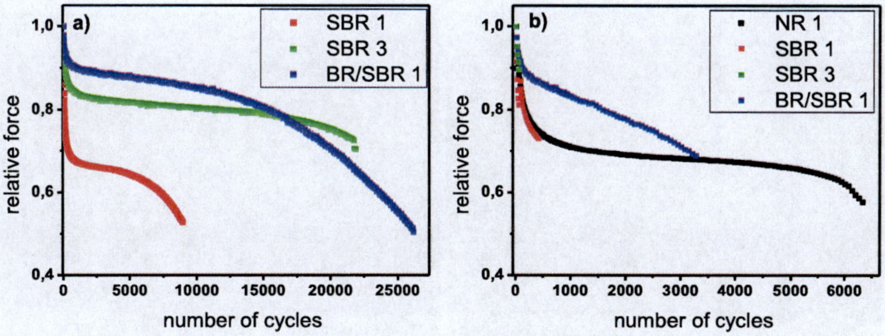

Fig. 22 Force decrease during Wöhler experiments at (**a**) 100% and (**b**) 200% for the highly filled tire tread systems. The force is related to the first measured data point after 10 cycles

3.5.1 Crack Growth Behavior for Selected Material Classes

The entire development of the force until the first dumbbell is broken is shown for the high filled tire tread compounds in Fig. 22. SBR 1 and SBR 3 show similar behavior. The force drop shortly before the break-off indicates stable crack propagation for 100% strain whereas the abrupt force hints to a catastrophic one for 200% strain. Note that also the Wöhler curves are almost the same. For the highly filled SBR it can therefore be concluded that the lifetime is mainly affected by the polymer. Furthermore, the larger temperature increase for SBR 1 has obviously no influence on the aging behavior and temperature effects like the thermal aging during loading can be excluded.

NR 1 shows even at large strain of 200% stable crack propagation. This might be related to the high crack resistance found for NR in tear strength measurements. The large resistance is probably one reason for the long lifetime of NR 1 seen in the Wöhler curve. The crack propagation and other aging factors are certainly influenced

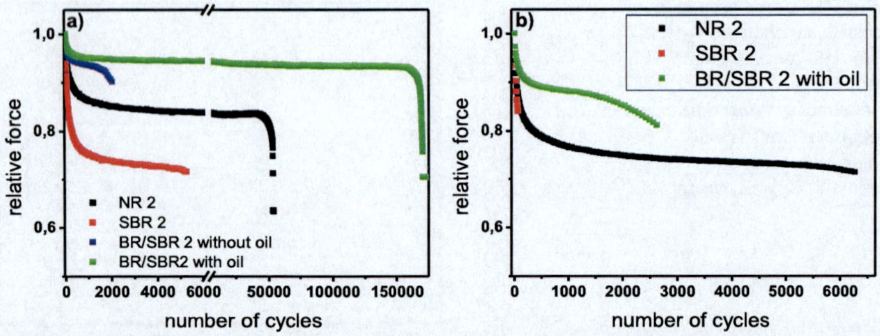

Fig. 23 Force decrease during Wöhler experiments at (**a**) 100% and (**b**) 200% for the low loading tire tread systems. The force is related to the first measured data after 10 cycles

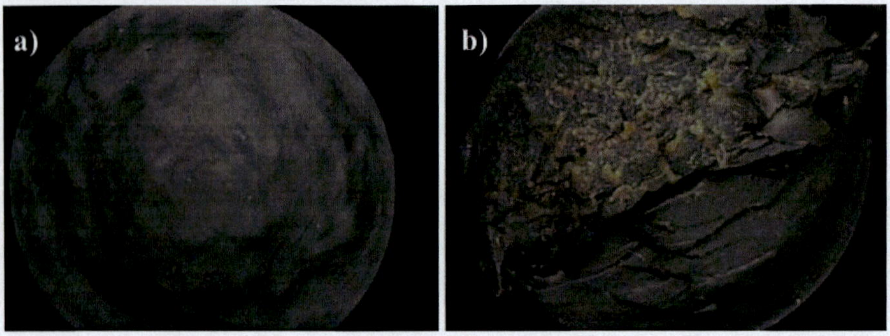

Fig. 24 Fracture surface of BR/SBR 2 at 150% strain, (**a**) without oil and (**b**) with oil

by the large stability of the filler network described above and strain-induced crystallization expected at large strains. It is striking that the force decrease of BR/SBR 1 behaves different from the other samples. Comparing SBR 3 and BR/SBR 1 (both 90 phr silica), it seems that the accelerated decline starts roughly at the same cycle number (~10^4 at 100% and 10^3 at 200%), but BR/SBR 1 survives much more cycles. The force decline before break-off is extremely long and relatively flat. It remains unclear whether this is related to extreme degradation effects of the polymer network or results from a slowed down crack propagation in the blend system.

The force decrease of the low loading tire tread systems is compared in Fig. 23 for 100% and 200% strain, respectively. The force drop before the break-off shows similar behavior for NR 2 and BR/SBR 2 with oil. In addition, for both systems the lifetime is relatively large, but this is mainly related to the low stiffness of BR/SBR 2. Comparing both BR/SBR samples, the characteristics of the force curves show no noticeable differences except the early breakdown of the sample without oil. Also, the fracture patterns look similar as seen in Fig. 24. Interestingly, the oil can clearly be identified in the picture that suggests a micro-phase separation. It appears that the large oil amount in the crack has a positive effect on aging properties, which increases the lifetime despite the worse dispersion.

3.6 Summarizing Mechanical and Thermal Influences

Table 9 summarizes the maximum surface temperatures obtained at 100% strain. In order to investigate the influence of thermal aging caused by the larger temperatures during mechanical loading, all samples were thermally aged (see Sects. 3.1 to 3.3). In all cases the aging temperatures are larger than the surface temperatures, though the internal temperatures can be significantly larger (see Sect. 3.4). Considering the changes of material properties during thermal aging, the materials can be classified in

Table 9 Table indicating the influence of thermal aging on mechanical fatigue

	Surface temperature [°C]	Aging temperature [°C]	Sensitive to thermal aging	Medium life time at 100% [h]
NBR 2B	40	100		
HNBR 1	70	130		
EPDM 1	81	130		
EPDM 2	78	130		
SBR 1	86	100	x	1.25
CR 1	59	100	x	1.26
CR 2	68	100	x	0.15
SBR 2	70	100	x	0.98
NR 1	85	100	x	5.29
NR 2	67	100	x	10.54
SBR 3	53	100	x	4.91
BRSBR 1	47	100	x	4.16
BRSBR 2 with oil	37	100	x	55.50
BRSBR 2 without oil	46	100	x	3.98
FKM 1A (at 40 %)	48	180		
FKM 2A (at 40 %)	52	180		

For gray marked materials, effects of thermal aging can be excluded due to their stability against thermal aging. Orange marked materials have only small temperature increases and in addition with the short aging times thermal aging effects are unlikely. Red marked materials are sensitive to thermal aging effects, have strong temperature increases, and thermal aging effects are possible sensitive and stable materials against thermal aging as marked in Table 9. Effects of thermal aging can be excluded for the stable materials. For the sensitive materials, the medium lifetimes at 100% loading are listed in the table. Considering the small surface temperatures in combination with the short aging times, for most of the samples, marked orange in the table, thermal aging effects are unlikely. Only for SBR 1, SBR 2, NR 1, and NR 2 thermal aging effects are likely due to the large temperature during mechanical loading.

We point out that the combination of heat and stress leads to an increased reaction or chain scission rate. This can be described by referring to the idea of Kramer's escape rate (see, e.g., [31]). Accordingly, the energy barrier for chain scission is reduced by the mechanical energy due to stretching of the samples. The energy barrier for C-C chain scission is of the order 300 kJ/mol. The mechanically stored energy density at large amplitudes is of the order 10 N/mm^2 = 10^{-2} J/mm^3 (compare Wöhler curves). This implies a rough estimate of 1 kJ/mol for the maximum molar energy, if the bond volume is taken as 1 nm^3, i.e. the molar volume of C-C bonds is evaluated as 10^5 mm^3/mol. This shows that the mechanical work gives only small contributions to the scission of polymer chains. More important is the accumulation of stress in front of the crack tips [32]. In the so-called process zone in front of the tip the stress is significantly higher implying also higher temperatures (flash temperature effect). Under such conditions chain scission can take place, though a lot of viscoelastic energy dissipation around the crack tip lowers the energy available for crack propagation [32].

4 Conclusions

The relation between thermo-oxidative aging and mechanical fatigue has been analyzed for a broad sample pool, which includes numerous different materials as tire tread, technical and high temperature rubbers. A basic characterization of unaged and thermally aged samples has been performed for the entire sample pool by measuring ShoreA hardness, rebound, density, dynamic mechanical properties, and stress-strain diagrams. The dispersion has been measured in order to obtain a picture about the flaws.

Mechanical fatigue tests have been performed by cyclic loading under constant strain. The analysis of Wöhler curves has been optimized by calculation of the energy density, which is related to the modulus of the material. Therefore, in the new representation of Wöhler curves the stiffness of the material was taken into account. Significant differences between the various types of polymer as well as the applied curing systems and filler types have been found.

The heat build-up in the sample during mechanical aging has been quantified by measuring the surface temperature. It has been shown that the heat build-up correlates with the dissipated energy calculated from the hysteresis of the measured stress-strain cycles. The input energy and the resulting temperature increase depended strongly on filler type and filler fraction. Thermo-oxidative aging effects during mechanical aging are unlikely to affect the lifetime for most of the samples due to their insensitivity against thermal aging or due to relatively short aging times combined with small temperature increase.

By means of the stress decrease during mechanical aging, several aspects influencing the fatigue behavior have been identified and analyzed. The decrease during the first cycles can be related to the breakage of clusters. Then, temperature effects play a role due to the increasing temperature and therewith decreasing modulus. In this range, a correlation between temperature and stress has clearly been shown for a few selected samples. Finally, crack initiation and propagation affect the stress decrease. The characteristic behavior before the break-off seems to be related to the fracture patterns. Slow stable crack propagation results in rough fracture patterns whereas samples with catastrophic propagation have relatively flat surfaces.

Acknowledgements We thank our former project partners, Continental AG, ContiTech AG, Felix Böttcher GmbH & Co. KG, Freudenberg Technology Innovation SE & Co. KG, Goodyear SA, Optibelt GmbH and Zeon Chemicals L.P. for the support of this work. The continuous help and valuable discussions with Dr. D. Steinhauser, Dr. N. Stübler, and Dr. E. Geberth are appreciated.

Appendix: Summary of Used Materials

Polymers

HyTemp 4051: ACM Elastomer, Zeon Chemicals L.P.
HyTemp AR-71: ACM Elastomer, Zeon Chemicals L.P.
Neoprene GRT: Chloroprene Copolymer, DuPont.
Nordel 4570: Amorphous EPDM Copolymer, Dow Inc.
Keltan 5580: Partly crystalline EPDM Copolymer, Lanxess
Tecnoflon P459: High fluorine Fluoroelastomer, Solvay
Zetpol 2020: HNBR, Zeon Chemicals L.P.
RSS1: Natural Rubber RSS1 grade
CB24: High-cis BR, Lanxess
VSL 5025-0: High vinyl S-SBR, Lanxess

Fillers

N550: Carbon Black type, ASTM
N660: Carbon Black type, ASTM
N234: Carbon Black type, ASTM
GR7000: High Surface silica, Evonik
Si69, Bifunctional silane, Evonik

Plasticizer and Processing Aids

DAE: Distillate Aromatic Extract
DOP: Dioctyl Phthalate
TOTM: Trioctyl trimellitate
WB222: Release agent by Struktol

Anti-Aging Chemicals

6PPD: N-1,3-dimethylbutyl-N′-phenyl-p-phenylendiamin
TMQ: Poly(1,2-dihydro-2,2,4-trimethyl-quinoline
Naugard 445: aromatic amine type antioxidant
Diuron: 3-(3,4-Dichlorphenyl)-1,1-dimethylharnstoff
Vanox ZMTI: antioxidative additive

Vulcanization Chemicals

St.acid: Stearic Acid
ZnO: Zinc Oxide
TBBS: N-tert.-Butyl-2-benzothiazyl-sulfenamid
TMTD: Tetramethylthiuram disulfide
DPG: Diphenylguanidine
S: Sulfur
MgO: Magnesium oxide
Ca(OH)2: Calcium hydroxide
NS70: 70% active dispersion of sodium stearate in an acrylic elastomer binder
ETU: Ethylene thiourea
MgO: Magnesium oxide
Peroxide: DiCumyl peroxide
Dicup(50%): DiCumyl peroxide (50%)
VulCup40 KE: organic peroxide-based curing agent
HVA-2: coagent for peroxide systems
TAIC: Triallyl isocyanurate

References

1. Gent AN (2012) Engineering with rubber: how to design rubber components. Carl Hanser Verlag, Munich

2. Mars W, Fatimi A (2002) A literature survey on fatigue analysis approaches for rubber. Int J Fatigue 24:949–961
3. Jones WN (1925) A low-temperature combustion method for oxidation of rubber. Ind Eng Chem 17:871
4. Kohman GT (1929) The absorption of oxygen by rubber. J Phys Chem 33:226–243. https://doi.org/10.1021/j150296a006
5. Reed MC (1929) Ind Eng Chem 21:316
6. Shelton JR, Winn H (1944) Aging of GR-S vulcanizates in air, oxygen, and nitrogen. Ind Eng Chem 36:728–730. https://doi.org/10.1021/ie50416a013
7. ISO (2011) Rubber, vulcanized or thermoplastic. Accelerated ageing and heat resistance tests (No. BS ISO 188:2011)
8. Bolland JL, Gee G (1946) Kinetic studies in the chemistry of rubber and related materials. II. The kinetics of oxidation of unconjugated olefins. Trans Faraday Soc 42:236–243. https://doi.org/10.1039/TF9464200236
9. Kim SG, Lee S-H (1994) Effect of crosslink structures on the fatigue crack growth behavior of NR vulcanizates with various aging conditions. Rubber Chem Technol 67:649–661. https://doi.org/10.5254/1.3538700
10. Aoyagi Y (2018) Aging processes of sealing materials, PhD thesis, Gottfried Wilhelm Leibniz Universität, Hannover, https://doi.org/10.15488/4458
11. Aoyagi Y, Giese U, Jungk J, Beck K (2018) The analysis of aging processes of EPDM elastomers using low field NMR with inverse Laplace transform and stress relaxation measurements. Kautschuk Gummi Kunststoffe 71(7–8):26–35
12. Aoyagi Y, Giese U, Kreiselmaier R (2017) Characterization of stress relaxation of cured EPDM using chemiluminescence. Kautschuk Gummi Kunststoffe 70(9):19–25
13. Gillen K, Celina M, Bernstein R (2004) Review of the ultrasensitive oxygen consumption method for making more reliable extrapolated predictions of polymer lifetimes. Annu Tech Conf Proc 2(2004):2289–2293
14. Giese U (2014) Aging Behav Elastomers. https://doi.org/10.1007/978-3-642-36199-9_284-1
15. (2016) Chemisch-mechanisch gekoppelte Modellierung und Simulation oxidativer Alterungsvorgänge in Gummibauteilen. https://nbn-resolving.org/urn:nbn:de:bsz:ch1-qucosa-222075
16. Schlomka C, Ihlemann J, Naumann C (2017) Simulation of oxidative aging processes in elastomer components using a dynamic network model. In: Constitutive models for rubber X, pp 77–82
17. Gil-Negrete N, Nieto J, Apezetxea IS, Murua J (2019) Quasi-static and dynamic properties of aged natural rubber compounds for transport applications. In: Constitutive models for rubber XI, pp 560–565. https://doi.org/10.1201/9780429324710-99
18. Soma P, Tada N, Uchida M, Nakahara K, Taga Y (2010) A fracture mechanics approach for evaluating the effects of heat aging on fatigue crack growth of vulcanized natural rubber. J Solid Mech Mater Eng 4:727–737. https://doi.org/10.1299/jmmp.4.727
19. Lake GJ, Thomas AG, In: Gent, A.N. (2012) Engineering with rubber, chapter 5: strength. Carl Hanser Verlag, Munich
20. Mars WV, Fatemi A (2004) Factors that affect the fatigue life of rubber: a literature survey. Rubber Chem Technol 77:391–412. https://doi.org/10.5254/1.3547831
21. Ludwig M (2017) Entwicklung eines Lebensdauervorhersagekonzepts für Elastomerwerkstoffe unter Berücksichtigung der Fehlstellenstatistik, PhD thesis, University Hannover
22. El Yaagoubi M, El Maanaoui H, Meier J (2020) Lifetime prediction of carbon black filled elastomers based on the probability distribution of particles. Polymer 208:122973
23. Boller F, Geisler H, Schuster RH (1991) Kautschuk Gummi Kunststoffe 44:1166
24. Lake GJ, Lindley PB, Thomas AG (1962) Fundamental study of the fatigue of rubber, In: Proceedings of the second international conference, London IRI, London, p 428
25. Klüppel M (2003) The role of disorder in filler reinforcement of elastomers on various length scales. Adv Polym Sci 164:1–86

26. Klüppel M, Meier J, Dämgen M (2005) Modelling of stress-softening and filler-induced hysteresis of elastomer materials. In: Constitutive models for rubber IV, Balkema, pp 171–176
27. Lorenz H, Klüppel M (2012) Micro-structure based modelling of arbitrary deformation histories of filler-reinforced elastomers. J Mech Phys Solids 60:1842–1861
28. Plagge J, Klüppel M (2020) Micromechanics of stress softening and hysteresis of filler reinforced elastomers with applications to thermo-oxidative aging. Polymers 12:1350
29. Carslaw HS, Jaeger CJ (2003) Conduction of heat in solids.2nd edn. Clarendon Press, Oxford
30. Fritzsche J, Klüppel M (2011) Simulation of internal sample temperature of rubber components during cyclic deformation. Kautschuk Gummi Kunststoffe 64(6):44–51
31. Hänggi P, Talkner P, Borkovec M (1990) Reaction rate theory: fifty years after Kramer. Rev Mod Phys 62:251
32. Wunde M, Klüppel M (2021) Review on the role of phase morphology and energy dissipation around the crack tip during fatigue crack propagation of filler-reinforced elastomer blends. Adv Polym Sci 286:245–272
33. Stoček R, Stěnička M, Kipscholl R (2019) Heat build-up characterization under realistic load. In: Constitutive models for rubber XI, Balkema, pp 157–162
34. Product information from Lanxess, Germany

Adv Polym Sci (2023) 289: 49–70
https://doi.org/10.1007/12_2021_110

Published online: 30 December 2021

Novel Approach on Analyzing Mechanochemical Wear Mechanism of Rubber Vulcanizates in Molecular Scale

Shinya Nakano, Tomofumi Zushi, and Toshio Tada

Contents

Abstract Abrasion, in other words wear, is one of the most critical degradation processes for rubber compound, and the wear resistance is the determinant factor of the tread performance and service lifetime of vehicle tires. In the present study, wear mechanism at molecular scale is investigated by quantum mechanics-based atomistic simulation with focusing on the wear phenomena of rubber compounds caused by

S. Nakano (✉), T. Zushi, and T. Tada
Sumitomo Rubber Industries Ltd., Kobe, Japan
e-mail: shinya-nakano.az@srigroup.co.jp

mechanochemical reaction. By using quantum chemical calculation (QCC) techniques, we found that the weakest bond strength of carbon–carbon single bond of saturated rubber is higher than that of unsaturated diene rubber, meaning that saturated rubber is higher resistant to the mechanical scission and thus generates less polymer radicals. In addition, it is clarified that an alkoxy radical that is formed from the reaction of the polymer radical with oxygen promotes the polymer dissociation which would enhance the wear degradation. These QCC results suggest that hydrogenation of diene rubber is the straight strategy to strengthen the wear resistance against the mechanochemical degradation because the decrease in carbon–carbon double bond suppresses the polymer oxidation and the subsequent chain scission of polymer. To confirm the simulation predictions, the effect of hydrogenation on the wear performance is studied experimentally. It is shown that, as the QCC predicted, hydrogenated styrene-butadiene rubber is less sensitive to the mechanochemical wear when compared with conventional non-hydrogenated rubber. Gel permeation chromatography and near edge X-ray absorption fine structure spectroscopy are also carried out for structural analysis of the wear particle. The experimental results show that the hydrogenated rubber is less reactive with oxygen, which is in line with QCC results, in relevant to the result that the wear rate is greatly decreased. This paper, for the first time, demonstrates the successful development of new class of polymer with higher wear resistance, hydrogenated styrene-butadiene rubber, inspired by QCC results.

Keywords Hydrogenated styrene-butadiene rubber · Mechanochemical reaction · Mechanochemical wear · Oxidation · Quantum chemical calculation

1 Introduction

1.1 Wear of Tires

Along with the recent technological progress in the automotive industry such as autonomous and ride-sharing, the requirements for better tire performance have been considerably increased [1]. The long-life tire, which tolerates to the damage and degradation during its usage, is one of the most desired performance properties for the vehicles because it can provide the longer driving distance with less maintenance. However, in general, tire performance is gradually degraded due to the abrasion of tire surface and degradation of rubber itself, which are unavoidable phenomena in the course of tire usage [2].

For example, the shallow tread grooves of worn tires could easily induce aquaplaning, which arises from the difficulty of drainage of a water film from the contact pitch. The increase in elastic modulus of the tire tread, which is one of the typical rubber property changes caused by the oxidation of polymer molecules and/or bleed out of plasticizer from the tire surface, induces the reduction of a real contact area between the pavement and the tread surface, leading to the deterioration of the grip performance. Yet the higher elastic modulus promotes the worsening of the wear resistance. This is because the tire tread compound becomes more brittle as the modulus increases. Therefore, it is crucial for producing long-life tire to improve the abrasion resistance from the rubber degradation point of view.

Besides the tire performance requirements, the improvement of the wear resistance is of growing importance from environmental view. It has recently been indicated that tire wear particles, which are generated during the wear down process, cause the pollution of the environment as microplastics [3]. In addition, the wear particles might be harmful to not only the environment but also human health [4]. If one could learn more about the wear degradation mechanism in molecular level, it might be possible to reduce tire wear, which would be beneficial from both an industrial and ecological point of view.

However, in spite of these concerns, the impact of the rubber degradation on wear process has not been fully understood although there are some studies regarding the thermal aging effect on the abrasion performance for nitrile butadiene rubber which is widely used in industries as O-ring, oil seal, and so on [5, 6]. This is because that the rubber degradation is a complicated process involving various mechanisms, e.g., mechanical failure, thermal degradation, and chemical reaction, preventing the comprehensive understanding of degradation behavior in experiments.

1.2 Mechanochemical Wear of Tire Rubber

The detachment of rubber particles from a tire tread surface during cornering and braking is generally denoted as abrasion or wear, and the wear resistance must be improved to produce the long-life tire. Rubber wear can be mainly categorized into three mechanisms: abrasive wear, fatigue wear, and smearing wear. The abrasive wear and fatigue wear originate from the mechanical failure, and the smearing wear, which is defined as mechanochemical wear in this study, is induced by mechanochemical reaction [7, 8]. In general, all the wear processes occur simultaneously in practice.

The abrasive wear occurs on substrates consist of very sharp edges on the surface, which cut the rubber fragments away from the rubber surface. The abrasive wear phenomena can be divided in accordance to the size of the asperity. The wear over the asperity in the size of micrometer range is well known as intrinsic wear, which generates very small rubber particles of a micrometer scale. This dimension is found to be independent of the applied normal force and thus it essentially depends on the intrinsic characteristic size of the asperities of the abrading surface [9, 10]. The wear over the asperity in the size of millimeter to centimeter range is well known as Cut and Chip wear [11, 12]. In the case of tires, Cut and Chip wear refers to the detachment or breakage of rubber material from treads when riding on a rough road surface (e.g., gravel roads, roots, stalks). This is more likely associated with off-the-road tires, light truck and SUV tires. The fatigue wear is caused by crack propagation on blunt substrates [7]. The periodic wave pattern perpendicular to the sliding direction, known as abrasion pattern, can be observed, in particular, on the surface of worn rubber samples which undergone the fatigue wear. The fatigue wear is related to the tear mechanism and the wear rate can be predicted by the tearing energy [9, 13]. The mechanochemical wear is attributed to local large deformations and high temperature at the contact patch [14]. Large deformation triggers molecular chain scission and generates polymer free radicals. Then, the free radicals easily react with oxygen molecules at high temperature; these processes are called as the mechanochemical reaction. As the molecular-level degradation proceeds at the surface, a sticky layer, composed of degraded polymers, forms on the rubber surface. The formed sticky layer is finally torn off due to the subsequent contact with the substrate. For example, the mechanochemical wear of rubber is often observed in natural rubber based compounds due to its thermal instability despite mechanical strength.

For the investigation of mechanochemical wear mechanism, Gent and Pulford [10] performed fundamental studies with a blade equipped to their handmade wear tester. The mechanism of the wear on asphalt pavement or concrete is significantly important for tire industries as a practical case. According to the Persson's theory [15, 16], at an asperity contact region, the rubber surface is subject to high temperature, so-called the flash temperature, which is a result of the frictional heat build-up. Molecular scission potentially occurs due to the large deformation and high temperatures. Therefore, sticky layer would be formed on the rubber surface under tire operational conditions on the asphalt and concrete. The formation of the sticky layer would affect the surface rubber properties, resulting in changes of wear behavior.

In both scientific and tire development views, the mechanisms of the abrasive wear and the fatigue wear have been relatively well investigated [8, 17–19]. On the other hand, however, there are fewer reports on the mechanochemical wear mechanism. This is attributed to difficulty of consideration of the molecular-level degradation under the mechanical stress in experiments, e.g., chemical bond scission, bond recombination, oxidation, generation of by-products, and other physical-chemical phenomena in rubber, which would be critical factor for determining the mechanochemical degradation rate. The molecular structure change is reported to influence the wear characteristics during the abrasion. For example, the effect of the rubber reacted with oxygen molecules against the wear behavior was investigated [10, 20]. They have concluded that the wear rate in an inert gas atmosphere is much less than that in the presence of oxygen. From these former researches, it is clarified that oxidation of rubber molecules is related to the mechanochemical wear.

Thus, in order to elucidate the detailed mechanism of the wear performance change in response to the mechanochemical degradation process, the molecular-level mechanochemical reaction mechanism should be clarified, which is relevant to the wear process. For this purpose, it is highly desirable to develop the novel approach which can investigate the molecular-level degradation and its effect on the wear resistance.

1.3 Multiscale Simulation

Tire needs to meet versatile requirements for customer's safety and environment friendliness. Design of tire structure and compounds is a time-consuming process because performance depends on a lot of factors, which are complicatedly linked to each other. Tire industry is one of pioneers on introducing the CAE in the R&D stage to shorten the development time. In the early stage of the CAE in the tire industries, finite element analysis (FEA) is often implemented to optimize a whole tire structure in a macroscopic scale. In the early 2000s, FEA began to be applied in material science, focusing on the effect of phase structure of polymer blend system on mechanical properties in terms of stress concentration over micrometer scale, for example. Due to the drastic advance in the computational technology in the past decade, both the time and spatial scales of a simulation model are greatly enlarged. In fact, high performance super computer facilities are built up by means of government funding, enables the tire industries to simulate more complicated model involving fracture mechanics on a massive scale. Coarse-grained molecular dynamic (MD) simulation and quantum chemical simulation are more commonly applied for the system which consists of polymer molecules. In addition to R&D stage, MD simulation is also used to elucidate the fundamental mechanism of rubber dynamics in molecular scale. Hagita et al., for example, applied Kremer Grest model to study the effect of filler particles dispersion on tensile property of filled polymer network [21]. Smith et al. employed MD simulation on carbon black filled elastomer and elucidated jamming effect of filler particles plays a crucial role in the mechanical response [22]. There are a number of studies that the molecular simulation contributed to answer the invisible and undetectable problems only with experiments in more detail. The molecular simulation is increasingly essential to complement experimental results to reach plausible answers.

Recently, the main interest was drawn to failure behavior of elastomers, as more practical but complicated problems, including fatigue crack propagation behavior and so on [23–25]. Nevertheless, the durability of compounds is the most important performance commonly for tire manufacturers. It is not easy to apply directly the molecular simulation on the durability issues of rubber compound because rubber undergoes strains much higher than the linear dynamic viscoelasticity. Moreover, it was revealed that chemical reaction is involved in the failure behavior of rubber in relatively mild severity. In order to add the chemical structures into the coarse-grained MD model, all-atom MD simulation is a possible solution. Substantially, it is not possible to implement all-atom MD simulation for the analysis covering molecular reaction due to the limitation on the computer capacity.

The concept of multiscale simulation emerged in the background. Figure 1 illustrates an overview of the concept of multiscale simulation. The ordinate and the abscissa represent time and length scales, respectively. Here, the multiscale simulation is defined as a combination of different simulation platforms to fully cover the modeling scale ranging from electronic structure of a molecule to a whole tire. The smallest scale including electrons being involved with chemical reaction are

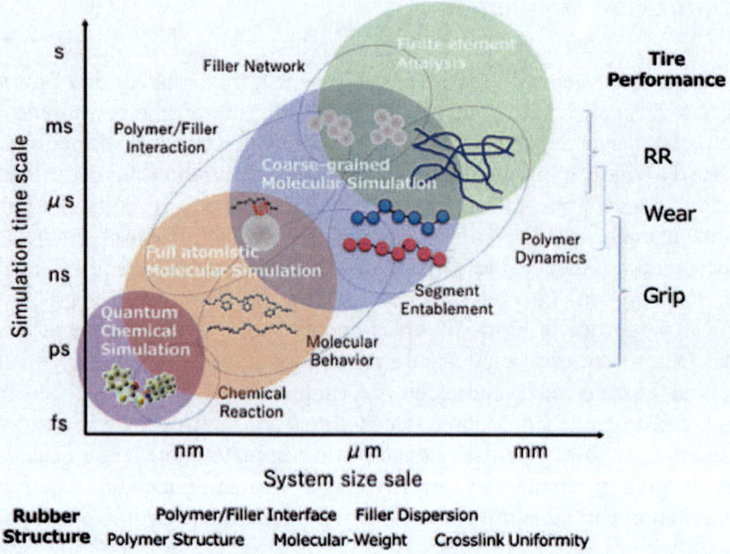

Fig. 1 Concept of multiscale simulation

analyzed by the quantum chemical calculations (QCC). The basis of QCC is quantum mechanics and it can provide electronic structure of a molecule which is responsible for the smallest scale behavior such as chemical reaction while the FEA is responsible for the largest scale. Various types of MD simulations exist for the intermediate scale between the QCC and the FEA. All-atom MD and coarse-grained MD simulations are often employed to analyze the dynamics of rubber molecules and its network structures. The former can simulate the time evolution of atoms comprising the rubber molecules. The spatial scale of the all-atom MD is not so large enough to consider a complex structure such as inhomogeneity in cross-linking density or distribution of filler particles, for example. The latter is a suitable simulation to deal with a model in larger scale containing entangled polymer network to analyze the segmental motion including the micro-Brownian motion around glass transition temperature, rubbery plateau, and terminal flow regions. However, there is no longer real atomistic structure in the coarse-grained MD simulation.

It is widely accepted that the tire performance is determined by hierarchical structure [26]. The relationship between smallest and intermediate scale still remains unclear. Establishment of reliable model is a common issue for major tire manufacturers. The multiscale simulation is the one of the powerful tools to elucidate the wear mechanism of rubber.

1.4 Mechanochemical Analysis with QCC

Molecular simulation is a powerful tool to investigate the mechanisms of mechanical properties of rubber in molecular scale. As noted in the previous section, there are many studies on the mechanical properties of rubber network with MD simulations. Fujimoto et al. [27] performed all-atom MD simulation on glassy polymers, and their results suggest that brittle fracture of brittle polymer is attributed to the less flexible backbone with fewer entablements. Naito [28] employed coarse-grained MD simulation and showed that the void formation at the silica surface is a key factor to determine the molecular-level breakdown in rubber. Recently, it was shown that, by coarse-grained MD simulation and numerical calculation, viscoelastic properties of rubber are dependent on the polymer dissociation rate and suggested that bond breakage of polymer is a part of factors causing the Mullins effect [29, 30]. Classical MD simulation is yet to describe the pictures of mechanical failure process. Beside the mechanical failure, it is evident that chemical degradation must be additional factor on wear. Obviously, formation of sticky layer on the rubber surface during wear process at certain condition strongly suggests that chemical reaction accompanies mechanical failure during the wear process. The coupling of mechanical and chemical phenomena on a molecular scale is defined as mechanochemical reaction. The mechanochemical reaction should also be an influential factor on rubber wear. Thus, for the exploration of mechanochemical wear, the other approach, which can simulate the electronic states of materials, is necessary. In order to investigate the effect of chemical reaction on wear behavior, QCC is one of the most effective methods.

In general, the length scale of QCC is not enough to analyze the effect of network density distribution which plays a significantly important role in the stress concentration and causes crack initiation and propagation. Macroscopic failure caused by stress concentration would be explained by the QCC, based on the concept of the multiscale simulation. Recently, Euchler et al. [31] reported the cavitation behavior of rubber, which is a flaw into the rubber matrix and responsible for crack initiation and growth of elastomer. Schneider et al. [32] observed that the ellipsoidal cavities oriented toward the crack opening direction, around the crack tip or rubber with small angle X-ray scattering technique, scanning around the crack tip of rubber sheet under stretch. These results suggest that the stress concentration at the crack tip is a dominant factor because it becomes unstable to fall into catastrophic rupture. Based on the idea, the cavitation is closely related with the wear through crack initiation of rubber and oxidation reaction. The length scale of the cavitation is covered by QCC.

Therefore, QCC has been applicable for molecular degradation process accompanying with chemical reaction. Density functional theory (DFT) is one of the most commonly used methods for electronic structure calculation of molecules. The oxidation process and Criegee mechanism, which is well-known ozone degradation process of polymer, are well investigated by DFT [33, 34]. DFT can also be employed to evaluate the synergy effect between mechanical force and chemical reaction. Beyer [35] introduced the concept of constrained geometry which simulates external force for mechanochemistry, and some authors used the approach to investigate the bond rupture characteristics under the strain [36, 37]. For example, Iozzi et al. [36] characterized the mechanochemistry of the sulfur–sulfur bonding under external stretching force by using the DFT approach. In addition to DFT calculation, first-principles MD is also one of the possible approaches to evaluate the mechanochemical effect on molecular reactivity. In ref. [38], first-principles MD was performed to investigate the cis-trans isomerization of cis-polyacetylene under tensile stress.

However, despite these progress of QCC approach, the application of QCC to mechanochemical wear of tire is still ambitious, limiting the basic understanding of the wear process and design of polymer or tire materials which can mitigate the degradation. In this study, using the QCC techniques, mechanochemical wear processes, i.e., polymer bond breaking, the radical reaction with oxygen followed by additional bond breaking, are analyzed. The strategy for improving the mechanochemical wear resistance is considered from material design point of view basis on the QCC results. Our QCC results suggest that hydrogenation of diene rubber is the simple way to strengthen the wear resistance, and it is experimentally validated. Our novel approach inspired by QCC demonstrates the successful development of new class of polymer, hydrogenated styrene-butadiene rubber (SBR).

2 Simulations and Experiments

There is limited QCC research regarding the mechanochemical wear process, and thus the atomistic origin of the wear process has not been fully understood. It is essential to clarify the cause of wear in order to design the new class of tire materials for practical application. In this paper, we investigate the origin of the mechanochemical behavior of rubber in a molecular scale by using QCC and attempt to develop novel polymer class rationalized by the simulation results which reduces the mechanochemical degradation process. The novelty of this study is to validate the QCC prediction experimentally and to demonstrate the QCC-motivated development strategy for new class of polymer. This chapter describes the simulation approach to investigate the mechanochemical wear process and experimental methods used for the validation and demonstration.

2.1 *Quantum Chemical Simulations for Analyzing Mechanical Behavior*

It is widely accepted that oxygen or ozone molecules are playing an important role in the mechanochemical reaction. It is reported that the oxidation of rubber molecules is involved in the mechanochemical wear [10]. Degradation of rubber molecule through mechanochemical reaction would proceed as a result of oxidation of rubber molecule following chain scission by mechanical force. Therefore, for analyzing the atomistic degradation origin, this study focuses on both the strength against mechanical chain scission and the reactivity of polymer free radical with oxygen molecules. To obtain practical guidance for developing the new polymer class with better mechanochemical wear resistance, effect of chemical structure of polymer on the mechanochemical behavior is also investigated.

The mechanical strength of rubber can be evaluated by the bond dissociation energy of the carbon–carbon single bonds, which is calculated by density functional theory (DFT). All the DFT calculations in this study were performed using the Becke-three-parameter Lee-Yang-Parr functional [39] in conjunction with the 6-311 + G(d,p) basis set with Gaussian16 software [40]. The basis set superposition error was corrected by the counterpoise method for the calculation of the dissociation energy. In this study, bond dissociation energy (D_0) is defined as the energy difference between the parent molecules (for example, $A-B$) and the product due to the homolytic bond breaking into radicals A and B (A· and B·).

$$D_0 = E_{\text{opt}}(A\cdot) + E_{\text{opt}}(B\cdot) - E_{\text{opt}}(A - B) \tag{1}$$

where E_{opt} is the energy of the optimized molecule calculated by DFT calculation. The reactivity of polymer with oxygen is demonstrated by conducting molecular dynamics (MD) simulations based on density-functional tight-binding (DFTB) method [41]. The third order of DFTB with the 3ob Slater-Koster parameters was used, and the MD (canonical ensemble) simulations were run at 300 K. Dissociated SBR model, i.e., unstable polymer radical, is employed as the initial structure to accelerate the reaction with oxygen because DFTB-MD simulation is a time-consuming technique, as compared with classical MD. Oxygen molecule is placed at the distance of 0.3 nm from the one of the carbon radicals of the polymer, as shown in Fig. 5. BIOVIA Materials Studio [42] was used to perform the DFTB calculations and to generate the graphical results.

Figure 2 shows an example of the model polymer used in the QCC. The 3-mer system consisting of a pair of butadiene and styrene repeating units is modelled, i.e., the smallest representative fraction of a SBR. In this work, no sulfur crosslink is considered in the model polymer. The degradation of sulfur crosslinks potentially

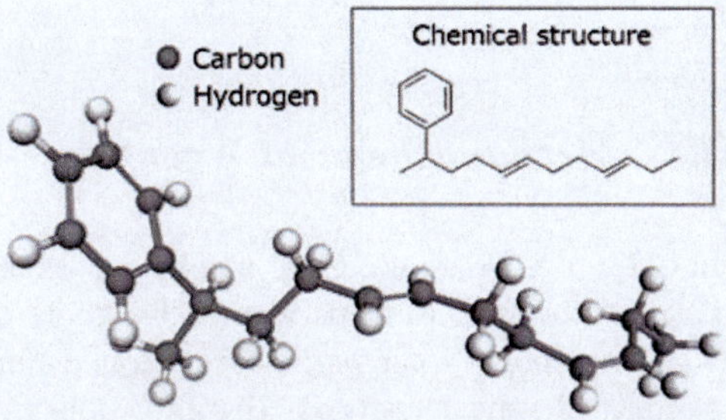

Fig. 2 SBR model used in QCC. No sulfur crosslink is considered for the simplification of analysis. Inset shows the chemical structure of SBR model

affects the mechanochemical process and thus the wear performance. For example [43], Cuneen found that sulfur reacted with oxygen, and this process promotes the oxidation degradation of rubber. On the other hand, it is reported that polymer backbone scission is predominant factor to determine the non-linear viscoelastic properties of rubber. The details of the impact of sulfur crosslink degradation on rubber properties are still controversial. In this work, in order to focus on the investigation of fundamental chemical polymer, only polymer degradation is analyzed. The detailed investigation of the impacts of sulfur crosslinks degradation on wear process is our future study subject, which will be attacked by our methodology presented here.

2.2 Material Preparation

Silica filled rubber vulcanizates were prepared using SBR or hydrogenated SBR (H-SBR), both of which are supplied by JSR corporation, Japan. As shown later, the QCC results suggest that hydrogenation of diene rubber is the straight approach to strengthen the wear resistance through the mechanochemical degradation. Hence, H-SBR is also prepared for evaluating the effect of hydrogenation on mechanochemical wear. The hydrogenation ratio of H-SBR was optimized for rubber compounding in terms of Mooney viscosity and hardness after vulcanization. The mixing process consisted of two steps: (1) 100 parts per hundred rubber (phr) of SBR or H-SBR were mixed with 2.0 phr of stearic acid (NOF Corporation, Japan) and 2.0 phr of zinc oxide (Mitsui Mining & Smelting Co., Ltd., Japan) within a Banbury mixer at 140°C for 3 min. For the filled samples, 60 phr of silica (Ultrasil VN3, Evonik Industries AG, Germany) and 5.2 phr of silane coupling agent (bis(triethoxysilylpropyl) disulfide, Evonik Industries AG, Germany) were added. (2) SBR or H-SBR compound was mixed with 1.6 phr of sulfur (Oil Treated Sulfur Powder, Tsurumi Chemical Industry Co., Ltd., Japan) and 4.0 phr of accelerators (N-tert-butyl-2-benzothiazolesulfenamide, Sanshin Chemical Industry Co., Ltd., Japan and 1,3-Diphenylguanidine, Sumitomo Chemical Co., Ltd., Japan) by a two-roll mixing mill at 80°C. In addition to the vulcanized rubber sheets for mechanical tests, wheel shaped samples were also prepared for wear test using a heating press at 170°C. The diameter and width of the rubber wheel were 80 mm and 18 mm, respectively.

2.3 Mechanochemical Wear Test

In order to validate the QCC results and examine the wear mechanism at a contact patch, wear tests were performed by home-made wear testing machines with the wheel samples. Wear particles generated during the wear tests were collected to conduct chemical analysis. Figure 3 illustrates the wear test machine. A wheel

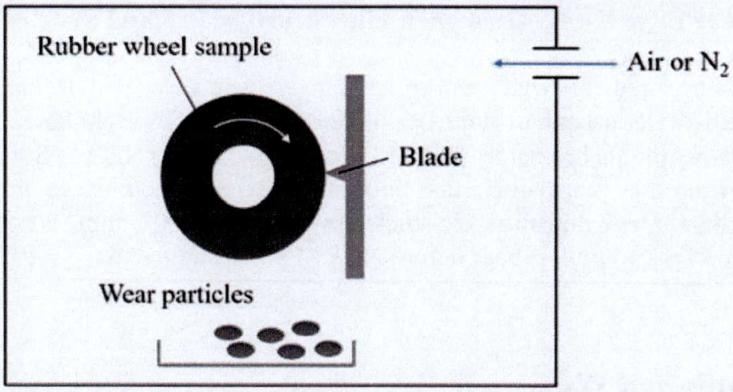

Fig. 3 The wear test machine employed in the study. The test machine is installed in an environmental chamber, capable of conducting wear tests in the air or nitrogen gas atmosphere

shaped sample was pushed on a blade with load of 6.5 N and revolved at a constant speed of 60 rpm, which corresponded to a sliding speed of 0.9 km/h. We referred the similar investigation of Gent and Pulford that the wear rate changes in inert gas and this change would be correlated with chemical reaction [10]. The wear test machine was installed in an environmental chamber, which has a transparent acrylic window, and the atmosphere in the chamber was replaced by high purity nitrogen gas. The nitrogen concentration in the chamber was monitored by Nitrogen Concentration Meter (AJX-N2B, AS ONE Corporation, Japan) and was adjusted to be higher than 98%. Wear particles during the wear test were collected with an aluminum container set beneath the blade. At the end of the wear test, the weight loss of the rubber wheel was measured to evaluate the wear rate.

2.4 *Molecular Structure Analysis*

2.4.1 Gel Permeation Chromatography (GPC)

GPC measurements were carried out to determine the molecular weight distribution of the wear particles. The collected wear particles and bulk rubber, as a reference, were dissolved in tetrahydrofuran (THF) immediately after the collection of the particles because the solubility of wear particles in solvent is likely to decrease with increasing interval time between the particle formation and the solubility test [44]. The extracts from the wear particles and bulk rubber in THF were used for the GPC analysis. The used columns were two Shodex GPC LF-804 (8.0 mm I.D. × 300 mm) and one Waters Styragel Column HR 0.5 (50 Å, 5 μm, 7.8 mm × 300 mm) (Showa Denko, Japan). Injection volume was 20 μL and column temperature was 40°C.

2.4.2 Near Edge X-ray Absorption Fine Structure (NEXAFS)

NEXAFS measurements were performed to investigate wear particles on an atomic scale. The data were acquired at the beamline BL17 of SAGA Light Source in Japan, using a grating monochromator. The C K-edge (275–320 eV) NEXAFS signals were observed from the wear particle and bulk rubber sample mounted in an ultra-high vacuum chamber. A thin film, the thickness of which is 0.2 mm, is prepared for rubber particles and bulk rubber before NEXAFS measurements.

3 Results and Discussion

3.1 Quantum Mechanics Simulations for Investigating Effect of Molecular Structure on Mechanical Strength and Reactivity with Oxygen

Degradation of rubber molecules throughout the mechanochemical reaction would proceed as a result of the oxidation of rubber molecules, following the chain scission by the mechanical force. The resistance against mechanical chain scission is evaluated by calculating the bond dissociation energy (D_0) of a carbon–carbon single bond of polymer which is weaker than a double bond. The calculated D_0 of each bond in SBR model is summarized in the second column in Table 1. D_0 is estimated at 4 and 6 different bond positions for SBR and H-SBR (to be discussed later), respectively. The definition of the bond positions is given in Table 1. For D_0 of SBR, the weakest carbon–carbon single bond, i.e., bond with less tolerant to the mechanical scission, is found between a pair of butadiene repeating units (bond labeled with B3). This result can be interpreted as the resonance effect of the double bonds of butadiene fragments. The radicals formed after bond breaking at B3 would be

Table 1 Bond dissociation energy D_0 for SBR and H-SBR model, which is calculated at the B3LYP/6-311 + G(d,p) level

Bond	D_0 (kcal/mol)	D_0 (kcal/mol)
Carbon–carbon single covalent bond	SBR 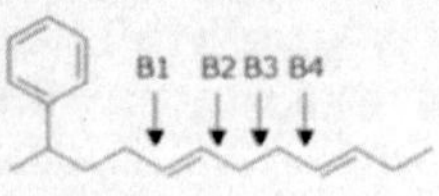	H-SBR 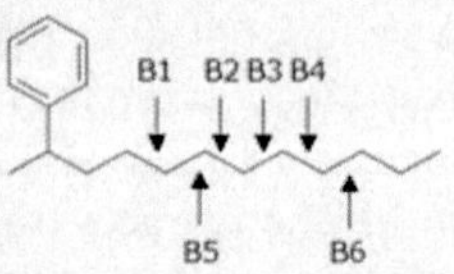
B1	90.6	77.1
B2	90.9	76.5
B3	46.0	75.5
B4	88.6	75.1
B5	–	77.0
B6	–	75.1

stabilized by the electron delocalization, which leads to the lower energy and lower D_0 as compared with the other bonds by electron resonance effect. This D_0 analysis suggests that that hydrogenation of the butadiene fragment is favorable way to improve the wear resistance because it suppresses the resonance effect.

To confirm this finding, D_0 for H-SBR is evaluated, as is given in the third column in Table 1. The H-SBR in this study is modelled by adding hydrogen–carbon bonds at carbon–carbon double bonds of butadiene units (see figure inset in Table 1). For the H-SBR model, D_0 shows nearly an identical energy level without particularly weaker bonds independently of the bond position. As compared with D_0 of the SBR model, the lowest D_0 for the H-SBR model is higher than that of the SBR model, meaning that the H-SBR model is stronger against the mechanical chain scission than the SBR model. The potential curve for a carbon–carbon covalent bond as a function of the bond length is also investigated. Figure 4 shows the potential curves of the weakest bond (B3) for the SBR and H-SBR models. The bonding energy potential curve for the H-SBR model is steeper and lower minimum energy than that of the SBR model, confirming that the strength of the single bond neighboring a double bond becomes weaker. This QCC result suggests that H-SBR is highly resistant to mechanical scission as compared with conventional SBR.

In general, oxidation is considered as one of the possible molecular degradation origins, which is enhanced in combination with heat, ultraviolet, and so on. It is a plausible hypothesis that the oxygen is considered as an essential part of the wear behavior throughout the mechanochemical reaction.

DFTB-MD technique is implemented to investigate the reactivity of polymer free radicals with the oxygen molecules. Effect of oxidation of a polymer chain on molecular degradation (or molecular chain scission) is studied in such a way that oxygen molecules are approaching to the radical of the polymer chain, generated after a chain scission. Figure 5a, b shows the snapshots of the MD trajectory on the

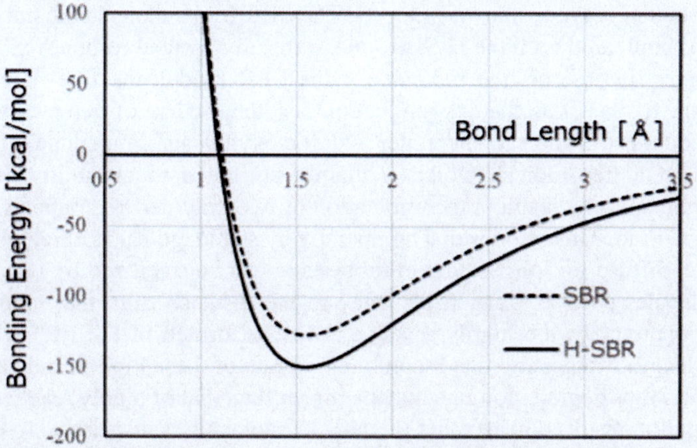

Fig. 4 The potential curve, shown for the weakest bond (B3 in Table 1) as a function of the bond length. Solid and dashed lines are for H-SBR and SBR, respectively

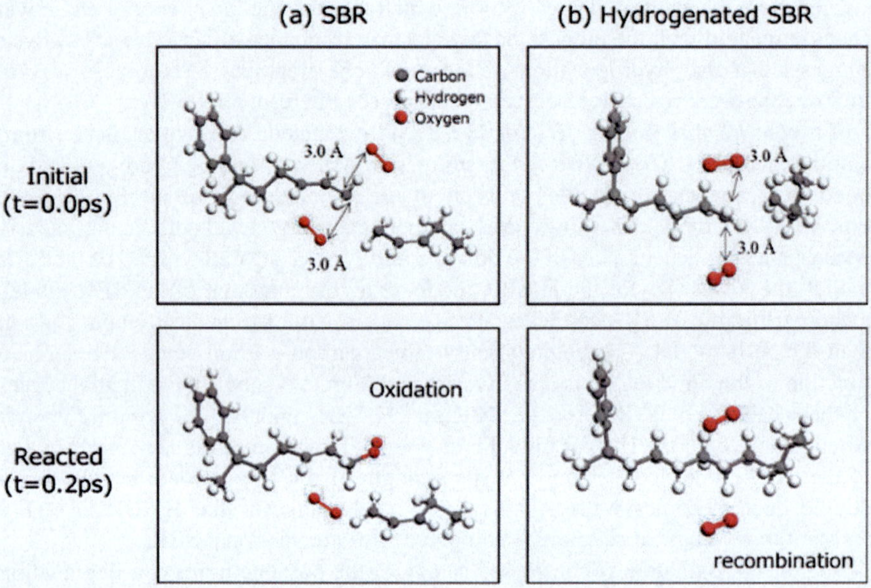

Fig. 5 Snapshot image of QCC for the reactivity with oxygen for (**a**) SBR and (**b**) H-SBR. The trajectories are obtained by conducting DFTB-MD simulations. Initial and reacted structures are shown

reaction process with oxygen for the SBR and H-SBR models, respectively. As shown in Fig. 5a, polymer free radicals of the SBR model tend to react with oxygen molecule in the course of the MD simulation. On the other hand, the polymer free radical of the H-SBR model tends to recombine with the other free radical end rather than reacting with the oxygen molecules, owing to the stronger bond strength and steeper potential curve as shown in Fig. 4. The SBR forms more stable polymer free radicals, as compared with the H-SBR, due to the allyl radical resonance. Thus, due to the longer lifetime of free radicals for the SBR model, the free radicals exist enough time to react with the oxygen. In contrast, the lifetime of free radicals for the H-SBR model is too short to encounter with the oxygen and immediately bonds the other end of a free radical, which is mainly attributed to instability of H-SBR polymer radical. As a result, a recombination of two polymer free radicals instantaneously occurred. This implies that comparatively stable radical (formed in the SBR model) can diffuse for longer time in rubber and can be stabilized by reacting with other molecules such as oxygen far away in the distance from the bond scission occurred, contrary to the highly reactive radicals (formed in H-SBR) with lower diffusion length, suggesting that recombination rate of these kind of radicals would be lower and thus degradation rate may be higher than that of highly reactive radical. Our calculation result indicates that the H-SBR molecule is insensitive to oxidation as compared to the non-hydrogenated SBR molecule.

As seen in Fig. 5, SBR is more susceptible to oxidation as compared with H-SBR, suggesting that the oxidation products of polymer, such as hydroperoxide and alkoxy radical, would be more easily formed in SBR rubber during the mechanochemical process. It is known that an alkoxy radical undergoes the decomposition into a carbonyl and alkyl species via carbon–carbon bond breaking, meaning the high possibility of enhancement of the polymer degradation. Since alkoxy radicals would be frequently formed inside SBR rubber due to the mechanochemical process as demonstrated in Fig. 5, we investigate the effect of the alkoxy radical formation on the bond breaking, and D_0 for the degraded SBR with alkoxy radical is calculated. The lowest D_0 is evaluated as 17.4 kcal/mol, which is much lower than that of SBR and H-SBR. This means that alkoxy radicals, which are formed by mechanochemical reaction, also promote the bond breaking of a polymer. It is possible that this additional bond scission promotes the wear degradation of tire rubber.

From the analyses of the D_0, the reactivity of free radical with oxygen, and the stability of alkoxy radical, it is concluded that hydrogenation of diene rubber is the practical method to improve the wear resistance against the mechanochemical degradation because the decrease in carbon–carbon double bond suppresses the atomistic degradation processes, i.e., mechanical scission, polymer oxidation, and the subsequent chain scission of polymer. This finding is validated experimentally in the following sections.

3.2 *Effect of Oxidation on Rubber Wear Behavior*

As described above, it is considered that oxidation of polymer is a crucial factor to determine the mechanochemical wear. In this section, the wear experimental pilot tests with different oxygen levels to verify the effect of oxidation qualitatively on the mechanochemical degradation of SBR and H-SBR are discussed. In order to reduce the oxygen level, the air in the chamber, in which the wear test machine is installed, is replaced by the nitrogen gas before the wear tests. Figure 6a shows the results of wear tests for the SBR and H-SBR compounds in the air and nitrogen gas atmosphere. For the SBR compound, the wear rate in the nitrogen gas atmosphere was nearly half value of the corresponding wear test in the air. Here, wear rate is defined as the volume loss per unit time. The volume loss is calculated as weight loss divided by specific gravity of rubber compound. Since the attribute of mechanochemical reaction can be negligible in the nitrogen gas atmosphere, the wear mechanism observed in the nitrogen gas atmosphere would be mainly attributed to the fatigue wear, dominated by the mechanical crack propagation. A ridge pattern was observed on the worn surface of the rubber wheel after testing in the nitrogen gas atmosphere. The presence of ridge patterns also indicates that the observed wear, in the nitrogen gas atmosphere, proceeded mainly through the mechanical crack propagation. Therefore, the difference of the wear rates between the tests in the air and in nitrogen gas atmosphere can be explained by the mechanochemical reactions. Here, it is revealed that nearly half value of wear rate in the air is due to the mechanochemical wear.

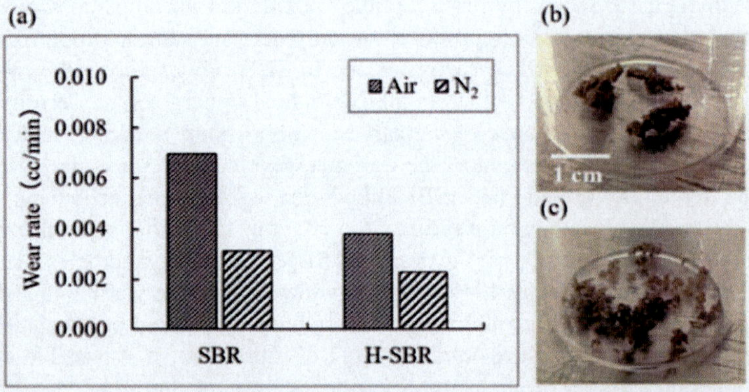

Fig. 6 Wear rate of rubber wheel in the air and in nitrogen gas atmosphere for the SBR and H-SBR compounds (**a**). Pictures of wear particles of the SBR compound in the air (**b**) and nitrogen gas atmosphere (**c**), respectively. Wear particles in the air were sticky, and wear particles in nitrogen gas were powdery. The mean size of wear particles in the air is larger than that in nitrogen gas atmosphere

The attribute of the mechanochemical reaction can also be found in the difference of appearance of the wear particles. Figure 6b, c shows the pictures of the wear particles in the air and nitrogen gas atmosphere. The size of the wear particles in air is larger than that in the nitrogen gas atmosphere. The difference of the sizes can be explained by a growing process of the wear particles. The wear particles produced in the air are sticky and tend to merge into larger particles during the growing process, whereas the primary particle size produced in the nitrogen gas atmosphere is nearly constant due to a lack of stickiness. The stickiness of the surface is attributed to the degraded rubber molecules, mainly due to the molecular chain scission by the oxidation, indicating a higher attribute of the mechanochemical reaction in the air.

3.3 Effect of Molecular Structure on Mechanochemical Wear

From the QCC, it is suggested that hydrogenation of diene is practical and simple way to improve the mechanochemical degradation. To validate this strategy, the relationship between the molecular structure and wear mechanism is investigated experimentally. Specifically, the wear rate in the different atmospheres is compared between the SBR and H-SBR compounds. The results for the H-SBR compound are also given in Fig. 6a. The attribute of mechanochemical reaction can be evaluated as the difference of wear rate in the air and nitrogen gas atmosphere. The difference of wear rate for the H-SBR compound was 40% of the SBR compound. A smaller difference of the wear rate corresponds to the less attribute from the mechanochemical reaction to the total wear rate. It is concluded that the wear rate for the H-SBR compound proceeds mainly as crack propagation but not as mechanochemical wear.

Here, we can discuss the effect of molecular structure on the degree of mechanochemical wear in terms of the number of double bonds. It is noted that the total number of double bonds for the H-SBR molecule is much less than that for the SBR molecule.

The wear rate for the H-SBR compound in the air was almost half of the SBR compound while the wear rates in the nitrogen gas atmosphere were on comparable levels. Thus, the difference of the wear rate in the air is mainly attributed to the corresponding level of the mechanochemical wear. This result suggests that the H-SBR compound shows better wear performance due to insensitive to mechanochemical reactions, supporting our QCC results. In the case of the nitrogen gas atmosphere, we expect similar levels of the wear rates for the two cases of the SBR and H-SBR. However, contrary to the prediction, we find slight differences as shown in Fig. 6a. Obviously, this slight difference between the wear rates of the H-SBR and SBR compounds in the nitrogen gas atmosphere is reflecting the different microstructures of the investigated polymers and the resulting impact on the mechanical properties of the final compounds. Generally, the fatigue wear behavior is thought to strongly depend on mechanical properties [8, 17–19]. We note that the value of elongation at break of the H-SBR compound was 1.8 times higher than that of the SBR compound. Therefore, the H-SBR compound would show slight lower fatigue wear rate than the SBR compound.

3.4 *Effect of Oxygen on Molecular Weight Distribution of Wear Particles*

It is reported that the wear particles, generated throughout the mechanochemical wear process, contain lower molecular weight components as compared to the original polymer before the vulcanization [44]. Thus, the amount of low molecular weight components is assumed to depend on the oxygen level. If the molecular chain scission is accelerated in the presence of oxygen, the amount of low molecular weight components in the nitrogen gas atmosphere would be less than that in the air.

Figure 7 shows the GPC chart obtained for the extracts from the original rubber before the wear test and wear particles of the SBR compound. The wear particles are collected after the wear test running for 1 h in the air or nitrogen gas atmosphere. The GPC chart reveals that the molecular distribution for each sample, which are the original and wear particles in the air or nitrogen gas atmosphere, consist of two fractions: the fraction 1 corresponds to the molecular weight 50,000 g/mol and the fraction 2 is around 700 g/mol. Comparing the GPC chart of the bulk rubber with those of the wear particles, we can find that the amounts of fraction 1 significantly increases through the wear process, while the amounts of fraction 2 is almost identical in all the samples. Therefore, fraction 1 and fraction 2 would be mainly attributed to low molecular weight polymers generated by the wear process and the bulk rubber component such as a process oil and accelerator, respectively. The

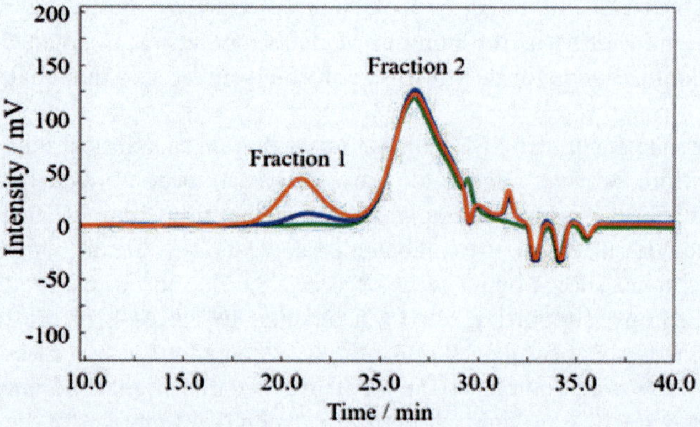

Fig. 7 GPC chart obtained for bulk rubber (green line), wear particles in the air (orange line), and wear particles in nitrogen gas atmosphere (blue line) of the SBR compound. Fraction 1 corresponds to low molecular weight SBR and fraction 2 corresponds to low molecular weight additives such as process oil

Table 2 The area ratio (A_N/A_A) of fraction 1 in GPC chart obtained from wear particles of the SBR and H-SBR compounds. The lower the ratio, the larger the difference of the amount of low molecular weight component measured from wear particles between in the air and in nitrogen gas atmosphere

	SBR compound	H-SBR compound
A_N/A_A	0.248	0.466

amount of each molecular fraction can be quantitatively evaluated by the area of the GPC chart.

The area of the fraction 1 for wear particle in the nitrogen gas atmosphere was smaller than that for the wear particle in the air. This result means, at a higher oxygen level, the lower molecular fraction is larger than that in the nitrogen gas atmosphere. Therefore, the result indicates that oxygen accelerates the degradation of rubber molecule during the wear process, as we assume.

In order to compare the effect of double bond on the mechanochemical reaction between the SBR and H-SBR, the areas of the GPC charts in the air and nitrogen gas atmosphere are evaluated. Table 2 shows the A_N/A_A of the fraction 1 obtained from the wear particles of the SBR compound and H-SBR compound, where the quantities, A_N and A_A, represent areas of the fraction 1 in the nitrogen gas atmosphere and in the air, respectively. Higher A_N/A_A ratios means higher resistance against the mechanochemical reaction. The A_N/A_A ratio for the H-SBR compound was higher than that for the SBR compound. This result indicates that H-SBR molecules are less influenced by the oxygen on the degradation during the wear process. The results about the lower wear rate for the H-SBR compared to the SBR compound (Fig. 6a) and, correspondingly, the lower molecular fraction for the H-SBR compound as shown in Table 2, indicate that the formation of lower molecular components increases the wear rate. Moreover, the decrease in the amount of carbon–carbon double bonds of the polymer molecules would lead to the reduction of the degradation, resulting in the improvement of mechanochemical wear, or vice versa. These experiments are in line with the QCC results in Sect. 3.1, and our strategy, hydrogenate of diene, is shown to be a useful method to strengthen the polymer against the mechanochemical wear.

3.5 *Effect of Molecular Structure on Reactivity with Oxygen*

In order to investigate the degradation mechanism during the wear process in a molecular scale, NEXAFS was carried out for the wear particles of the SBR and H-SBR compounds using a synchrotron radiation source. Figure 8 shows the changes in the number of specific functional groups in the SBR and H-SBR compounds after the formation of the wear particles. We focus on the number of carbonyl groups, which represents the degree of the oxidation and reflects the number of the molecular chain scissions. The ordinate represents the number of the functional groups, such as carbon–carbon double bond and so on. A positive value means an increasing number of the functional groups. As shown in Fig. 8, the number of carbon–carbon double bonds in the SBR compound decreased during the wear process. It is suggested that the oxidation of the carbon–carbon double bonds through an ozonide formation may result in the reduction in the number of the double bonds. In contrast, the carbon–carbon double bonds in the H-SBR compound slightly increased. The difference between the H-SBR and SBR compounds could be explained by the difference of free radical reaction pathways. In the case of the SBR it is easy to react with the oxygen molecules after a formation of free radicals as mentioned above, on the basis of the molecular simulation. Contrarily, in the case of the H-SBR compound, free radical formation after a molecular scission is likely to induce a recombination or disproportionation which causes an increase of the double bonds as shown in Eq. 2.

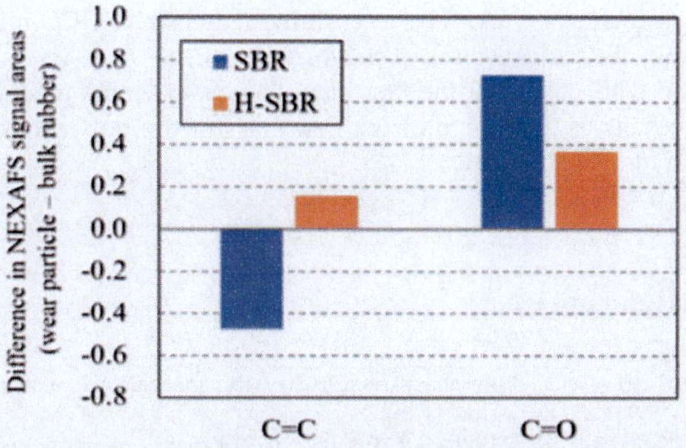

Fig. 8 Difference in NEXAFS signals area corresponding to C=C bond and C=O bond calculated as the signal area of bulk rubber subtracted from that of wear particles

$$-CH_2 - CH_2\cdot + \cdot CH_2 - CH_2- \rightarrow -CH_2 - CH_3 + CH_2 = CH- \quad (2)$$

In terms of the oxidation, the increase in the number of the carbonyl groups after the wear process for the H-SBR is smaller than that for the SBR compound. This result directly reveals that the H-SBR molecule is less reactive with oxygen than the SBR molecules. These experimental results are consistent with the prediction of the QCC.

4 Conclusion

In order to improve the wear performance by reducing the mechanochemical reaction, which occurs at a tire tread surface during a wear process, the effect of oxygen on the mechanochemical reaction and the effect of molecular structure on the mechanochemical wear were studied by performing the QCC and wear experiments.

Our QCC approach suggests that hydrogenation of diene rubber is the straight strategy to strengthen the wear resistance against the mechanochemical degradation because the decrease in carbon–carbon double bond suppresses the polymer oxidation and the subsequent chain scission of polymer. The QCC prediction is validated by comprehensive range of experimental results. GPC analysis and NEXAFS results for the wear particles generated in the different oxygen levels reveal that the oxygen molecules accelerate the degradation of rubber molecule during the wear process. The wear test for the SBR compound performed in the air and nitrogen gas atmosphere indicates that the mechanochemical wear dominates more than half of the total wear rate in the air. The effect of oxygen on the mechanochemical reactions was much smaller for the H-SBR molecule, whose carbon–carbon double bonds are highly hydrogenated. The wear rate of the H-SBR compound is about half of the wear rate of the non-hydrogenated SBR compound. When designing tire tread compounds with polymer molecules having less amount of carbon–carbon double bonds, a further upgrading of wear performance is expected through the control of mechanochemical reactions. Through our novel approach on analyzing the mechanochemical wear, we successfully demonstrate QCC-motivated development strategy for the new class of polymer.

This paper, for the first time, demonstrates the practical application of the QCC results to the design of the new class of tire polymer with higher wear resistance, hydrogenated SBR.

References

1. Nagase T, Löwenhaupt B (2020) Mechanochemical analysis, Tire technology. Int Annu Rev:82–84
2. Wright KRS, Botha TR, Els PS (2019) Effects of age and wear on the stiffness and friction properties of an SUV tyre. J Terramechanics 84:21–30

3. Alexandrova O, Kaloush KE, Allen JO (2007) Impact of asphalt rubber friction course overlays on tire wear emissions and air quality models for Phoenix, Arizona, airshed. In: Environmental issues (2011, pp. 98–106). Transportation research record; no. 2011
4. Gualtieri M, Mantecca P, Cetta F, Camatini M (2008) Organic compounds in tire particle induce reactive oxygen species and heat-shock proteins in the human alveolar cell line A549. Environ Int 34:437–442
5. Dong CL, Yuan CQ, Bai XQ, Yan XP, Peng Z (2015) Tribological properties of aged nitrile butadiene rubber under dry sliding conditions. Wear 322–323:226–237
6. Jiang B, Jia X, Wang Z, Wang T, Guo F, Wang Y (2019) Influence of thermal aging in oil on the friction and Wear properties of nitrile butadiene rubber. Tribol Lett 67:86
7. Muhr AH, Roberts AD (1992) Rubber abrasion and wear. Wear 158:213–228
8. Persson BNJ (2009) Theory of powdery rubber wear. J Phys Condens Matter 21:485001–485008
9. Huang M, Guibert M, Thevenet J, Fayolle C, Chaussée T, Guy L, Vanel L, Loubet J-L, Sotta P (2018) A new test method to simulate low-severity wear conditions experienced by rubber tire materials. Wear 410–411:72–82
10. Gent AN, Pulford CTR (1983) Mechanisms of rubber abrasion. J Appl Polym Sci 28:943–960
11. Stoček R, Heinrich G, Schulze A, Wunde M, Klüppel M, Vatterott C, Tschimmel J, Lacayo-Pineda J, Kipscholl R (2020) Chip & cut wear of truck tire treads: comparison between laboratory and real tire testing. Kautschuk Gummi Kunststoffe 73:51–55
12. Stoček R, Heinrich G, Kipscholl R, Kratina O (2021) Cut & chip wear of rubbers in a range from low up to high severity conditions. Appl Surf Sci Adv 6:100152
13. Southern E, Thomas AG (1979) Studies of rubber abrasion. Rubber Chem Technol 52:1008–1018
14. Ahagon A, Kirino Y (2006) Aging of black filled rubber under deformation. Rubber Chem Technol 79:641–652
15. Persson BNJ (2006) Rubber friction: role of the flash temperature. J Phys Condens Matter 18: 7789–7823
16. Lorenz B, Persson BNJ, Fortunato G, Giustiniano M, Baldoni F (2013) Rubber friction for tire tread compound on road surfaces. J Phys Condens Matter 25:095007–095014
17. Thomas AG, J. (1974) Factors influencing the strength of rubbers. Polym Sci, Symp 48:145–157
18. Fukahori Y, Yamazaki H (1994) Mechanism of rubber abrasion. Part I: abrasion pattern formation in natural rubber vulcanizate. Wear 171:195–202
19. Emami A, Khaleghian S (2019) Investigation of tribological behavior of styrene-butadiene rubber compound on asphalt-like surfaces. Tribol Int 136:487–495
20. Schallamach A (1968) Abrasion, fatigue, and smearing of rubber. J Appl Polym Sci 12:281–293
21. Hagita K, Morita H, Doi M, Takano H (2016) Coarse-grained molecular dynamics simulation on filled polymer nanocomposites under uniaxial elongation. Macromolecules 49:1972–1983
22. Smith M, Simmons S (2019) Poisson ratio mismatch drives low-strain reinforcement in elastomeric nanocomposites. Soft Matter 15:656–670
23. Ghosh P, Stocek R, Gehde M, Mukhopadhyay R, Krishnakumar R (2014) Investigation of fatigue crack growth characteristics of NR/BR blend based tyre tread compounds. Int J Fract 188:9–21
24. Lake GJ, Lindley PB (1965) Cut growth and fatigue of rubbers. II. Experiments on a noncrystallizing rubber. Rubber Chem Technol 38:301–313
25. Stoček R, Ghosh P, Machů A, Chanda J, Mukhopadhyay R (2020) Fatigue crack growth vs. chip and cut wear of NR and NR/SBR blend-based rubber compounds. In: Heinrich G, Kipscholl R, Stoček R (eds) Fatigue crack growth in rubber materials. Advances in polymer science, vol 286. Springer, Cham
26. Takenaka M (2013) Analysis of structures of rubber-filler systems with combined scattering methods. Polym J 45:10–19

27. Fujimoto K, Tang Z, Shinoda W, Okazaki S (2019) All-atom molecular dynamics study of impact fracture of glassy polymers. I: molecular mechanism of brittleness of PMMA and ductility of PC. Polymer 178:121570
28. Naito M (2016) Simulation for tire materials. Innovative molecular simulation and analysis technologies for designing tire materials. Tire Technol Int:26–28
29. Wan H, Ke G, Li S, Zhang L, Wu X, Wang X, Liu J (2019) Chemical bond scission and physical slippage in the Mullins effect and fatigue behavior of elastomers. Macromolecules 52: 4209–4221
30. Huber G, Vilgis T (1999) Universal properties of filled rubbers: mechanisms for reinforcement on different length scales. Kautschuk Gummi Kunststoffe 52:102–107
31. Euchler E, Bernhardt R, Schneider K, Heinrich G, Tada T, Wießner S, Stommel M (2021) Cavitation in rubber vulcanizates subjected to constrained tensile deformation. Adv Polym Sci 286:203–224
32. Euchler E, Bernhardt R, Wilde F, Schneider K, Tada T, Heinrich G, Wießner S, Stomme M (2021) First-time investigations on cavitation in rubber subjected to constrained tension using in situ synchrotron X-ray microtomography (SRμCT). Adv Eng Mater:2001347
33. Dibble TS (1999) A quantum chemical study of the C-C bond fission pathways of alkoxy radicals formed following OH addition to isoprene. J Phys Chem A 103:8559–8565
34. Zhang D, Zhang R (2002) Mechanism of OH formation from ozonolysis of isoprene: a quantum-chemical study. J Am Chem Soc 124:2692–2703
35. Beyer MK (2000) The mechanical strength of a covalent bond calculated by density functional theory. J Phys Chem 112:7307–7312
36. Iozzi MF, Helgaker T, Uggerud E (2011) Influence of external force on properties and reactivity of disulfide bonds. J Phys Chem 115:2308–2315
37. Klein IM, Husic CC, Kovács DP, Choquette NJ, Robb MJ (2020) Validation of the CoGEF method as a predictive tool for polymer mechanochemistry. J Am Chem Soc 142:16364–16381
38. Röhrig UF, Frank I (2001) First-principles molecular dynamics study of a polymer under tensile stress. J Phys Chem 152:8670–8674
39. Beck AD (1993) Density-functional thermochemistry. III. The role of exact exchange. J Chem Phys 98:5648
40. Frisch MJ, Trucks GW, Schlegel HB, Scuseria GE, Robb MA, Cheeseman JR, Scalmani G, Barone V, Petersson GA, Nakatsuji H, Li X, Car-icato M, Marenich AV, Bloino J, Janesko BG, Gomperts R, Mennucci B, Hratchian HP, Ortiz JV, Izmaylov AF, Sonnenberg JL, Williams-Young D, Ding F, Lipparini F, Egidi F, Goings J, Peng B, Petrone A, Henderson T, Ranasinghe D, Zakrzewski VG, Gao J, Rega N, Zheng G, Liang W, Hada M, Eha-ra M, Toyota K, Fukuda R, Hasegawa J, Ishida M, Nakajima T, Honda Y, Kitao O, Nakai H, Vreven T, Throssell K, Montgomery Jr JA, Peralta JE, Ogliaro F, Bearpark MJ, Heyd JJ, Brothers EN, Kudin KN, Staroverov VN, Keith TA, Kobayashi R, Normand J, Raghavachari K, Rendell AP, Bu-rant JC, Iyengar SS, Tomasi J, Cossi M, Millam JM, Klene M, Adamo C, Cammi R, Ochterski JW, Martin RL, Morokuma K, Farkas O, Foresman JB, Fox DJ (2016) Gaussian˜16 Revision C.01. Gaussian Inc., Wallingford
41. Frauenheim T, Seifert G, Elstner M, Niehaus T, Köhler C, Amkreutz M, Sternberg M, Hajnal Z, Di CA, Suhai S (2002) Atomistic simulations of complex materials: ground-state and excited-state properties. J Phys Condens Matter 14:3015
42. Ver. 2018, BIOVIA Materials Studio®, San Diego, CA, USA
43. Cunneen JI (1968) Oxidative aging of natural rubber. Rubber Chem Technol 41:182–208
44. Wu G (2016) The mechanisms of rubber abrasion. PhD thesis, Queen Mary University of London

Adv Polym Sci (2023) 289: 71–102
https://doi.org/10.1007/12_2022_118

Published online: 31 March 2022

Tire-Abrasion Particles in the Environment

R. Gieré and V. Dietze

Contents

Abstract Driving a vehicle on a road is inevitably associated with the generation of tire-abrasion particles. Annually, nearly six million tons of this material are released

R. Gieré (✉)
Department of Earth and Environmental Science, University of Pennsylvania, Philadelphia, PA, USA
e-mail: giere@upenn.edu

V. Dietze
Air Quality Department, Research Center Human Biometeorology, German Meteorological Service, Freiburg, Germany

into the environment globally. Most of the tire debris is encrusted to variable extent with other particles, which are derived from brake and road wear, as well as from soil. First emitted into the atmosphere or deposited onto the road surface, these tire-abrasion particles are subsequently transported by wind and water to different environmental compartments, whereby road runoff plays a crucial role in the distribution. Once in the environment, these particles interact with air, light, and water, which triggers degradation and concomitant release of crucial tire components, such as zinc and various organic tire additives. The tire-abrasion particles and the leachates derived from them during their degradation in the environment can cause various biological responses in aquatic and terrestrial biota, as determined by numerous ecotoxicological studies. Challenges exist with the identification and quantification of the abundance of tire-abrasion particles in the various environmental compartments, the determination of their toxicity, and with the reduction of both their emission and immission into the environment.

Keyword Characterization · Emission factors · Immissions to environment · PAH · Remediation strategies · Road wear · Toxicity

1 Introduction

Driving a vehicle on a road is inevitably associated with frictional contact between the tires and the road surface, especially during acceleration, braking, and cornering. This friction causes abrasion of the tire tread, i.e., the outermost layer of a tire, thus generating tire-abrasion, or tire-wear particles. Even though these particles are small, depending on the exact definition generally less than several hundred micrometers (μm) in size [1], much larger pieces of tire tread can also break off during driving, but the latter will not be discussed here.

Tires consist of natural and synthetic rubber; various fillers to maximize performance and lifespan, e.g., carbon black (CB) and amorphous silica (SiO_2) with grain sizes smaller than ~500 nm [2–4]; reinforcing textile fibers (e.g., nylon, polyester, rayon) and steel wires; extenders or softeners (e.g., petroleum process oils); vulcanizing agents and accelerators (e.g., zinc oxide, stearic acid, organic sulfur compounds); and various antioxidants and preservatives [5, 6]. The amounts of these materials vary widely among manufacturers and types of tires [7], with an example shown in Fig. 1.

Tire-abrasion particles are derived from the tire tread, i.e., the outermost layer, which directly interacts with the road surface. Within the context of tire abrasion, therefore, the composition of the tread is more relevant than that of the entire tire. On the other hand, the tire-tread composition is also highly variable and often unknown across different tire brands and models [8], but the predominance of rubber polymers and fillers in the tread (Fig. 2) implies that tire abrasion initially generates particles that are mostly composed of rubber polymers and fillers. Conversely, foreign

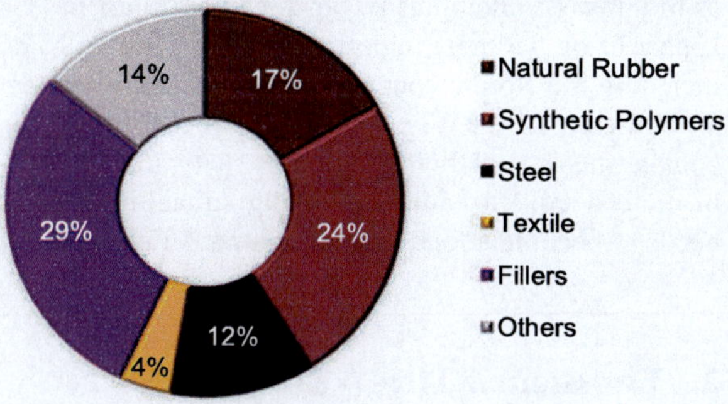

Fig. 1 Composition of reference passenger car tire in the European Union. Data (in wt%) from OECD [5]

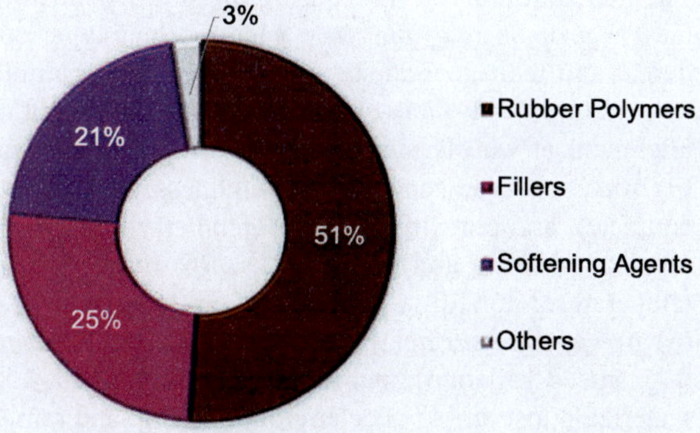

Fig. 2 Example of a tire-tread composition. Data (in wt%) from Winther and Slentø [9]

materials, such as particles eroded from the road surface or from the vehicle brakes, may have been incorporated into the tread during driving [10]. Therefore, after emission and specifically after having been deposited on, and resuspended from the road surface, most of these tire-abrasion particles are encrusted and/or mixed with other materials, especially road-wear particles [6, 10–16]. Consistent with the literature [8, 17–20], the term *tire- and road-wear particles* (TRWPs) will be used here for particles that are encrusted and mixed with foreign materials (brake wear, road wear, and dust, including soil particles), whereas original emissions of pure tire-tread material are termed *tire-core particles* [15]. It is of note, however, that tire-core particles are rarely found in the environment [21], including in air samples that were collected along various types of roads [15].

The abundance of rubber polymers in pure tire tread (Fig. 2) and the typical sizes of TRWPs found in the environment entail that the TRWPs can be classified as *microplastics* (MPs), according to both the commonly used terminology (<5 mm in size) and to the more recently recommended size range, i.e., from 1 μm up to 1 mm [22].

Specifically, TRWPs are classified as *primary* MPs, i.e., plastic particles that have been generated within the size range of MPs and then directly released into the environment [23]. Recent research has shown that TRWPs contribute considerably to the flow of microplastics into the environment [1, 15, 24, 25]. Indeed, TRWPs appear to be the most important type of primary MPs emitted in several countries, including the Nordic countries, for which excellent data are available [26–29].

Emission of TRWPs, their subsequent immission into various environmental compartments, and their potential toxicity are therefore subjects that must be studied in more detail. In addition, finding suitable measures to reduce TRWP pollution needs to become a focus of environmental research.

2 Emission of Tire-Wear Particles

The generation of TRWPs during driving depends on numerous factors, including the (1) *properties of the tires*, such as their type, composition, size, shape, tread depth, and their general state, influenced, for example, by their age, pressure, and temperature; (2) *vehicle characteristics* (e.g., weight, load distribution, number and alignment of wheels, suspension, steering geometry, general state of maintenance); (3) *road surface*, such as its roughness or texture, its material (e.g., bitumen, concrete), and porosity; (4) *road geometry and topography* (uphill, downhill, flat, number of bends and intersections); (5) *ambient environmental parameters* (e.g., temperature, humidity, and road conditions, such as dry, wet, or snow-covered); (6) *tire-to-road-surface contact* (e.g., rolling, sliding, slipping); (7) *driving style* (e.g., speed variation); and (8) *traffic flow* [11, 13, 15, 17, 30–32]. Most of the tire material is lost during acceleration, braking, and cornering, and therefore, emission of TRWPs will be greatest near intersections, roundabouts, and bends [31].

2.1 Emission Estimates

The amounts of TRWPs generated during driving of a vehicle over a certain period of time are typically estimated by one of two approaches: (1) using an emission factor (EF), commonly reported as weight loss (in milligram) per vehicle-kilometer driven (mg/veh-km), and multiplied by the total distance driven; or (2) determining the weight loss per tire, which is then multiplied by the number of tires consumed [1]. The EF is also known as the wear factor [11].

EFs can be derived experimentally, either in the laboratory with test tires or test vehicles and road simulators, or in field settings, such as road tunnels and test strips [17, 33, 34]. Published EF values vary widely between different studies [18], but on average, the EF for passenger cars is ~100 mg/veh-km (Fig. 3), in good agreement with early estimates [40]. This result implies that driving a car for 10 km would typically release about 1 g of TRWP into the environment. As shown in Fig. 2, this mass of abraded tread only partially consists of rubber polymers, because it also contains other materials, including CB and silica filler, and oils [40]. Moreover, larger and heavier vehicles feature distinctly higher EF values than smaller and

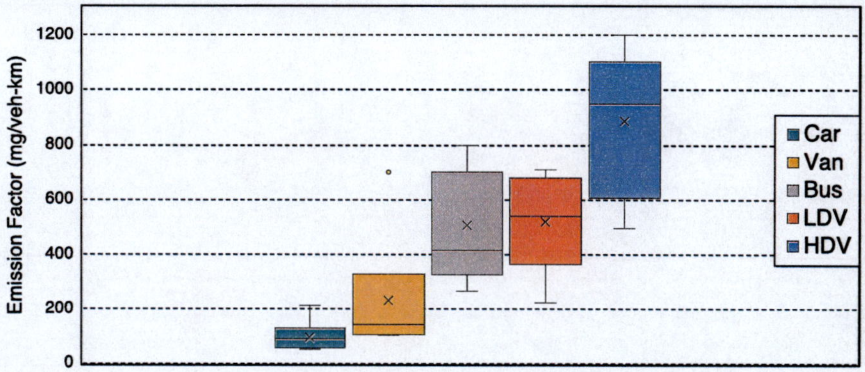

Fig. 3 TRWP emission factors (EFs) for various types of vehicles. EF values shown in milligram per vehicle-kilometer (mg/veh-km). Data from various sources [1, 16, 31, 33–39]. *LDV* light-duty vehicle, *HDV* heavy-duty vehicle

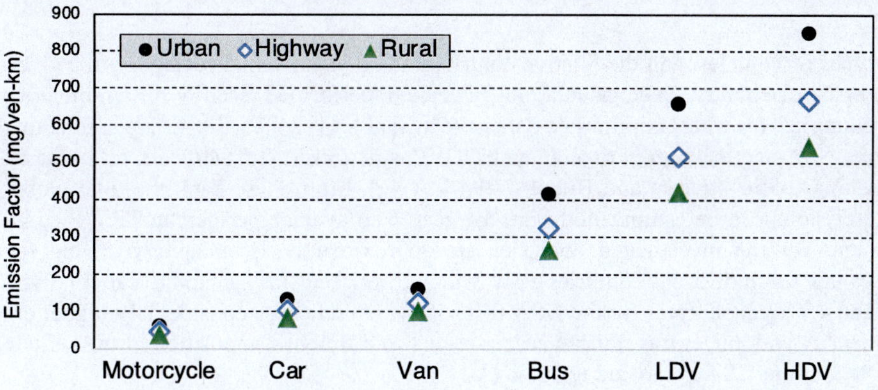

Fig. 4 TRWP emission factors (EFs) for various types of vehicles and roads. EF values shown in milligram per vehicle-kilometer (mg/veh-km). Data from Verschoor et al. [16]. *LDV* light-duty vehicle, *HDV* heavy-duty vehicle

lighter vehicles (Fig. 3), as also documented in both experimental and on-road studies [41, 42]. This general trend is maintained across different types of roads, whereby EFs on rural roads are lower than those on urban roads for all vehicle categories (Fig. 4).

2.2 *Global TRWP Emissions*

The two approaches explained above can be used to estimate the total amounts of TRWPs emitted per vehicle category, the total amounts of TRWPs emitted by all

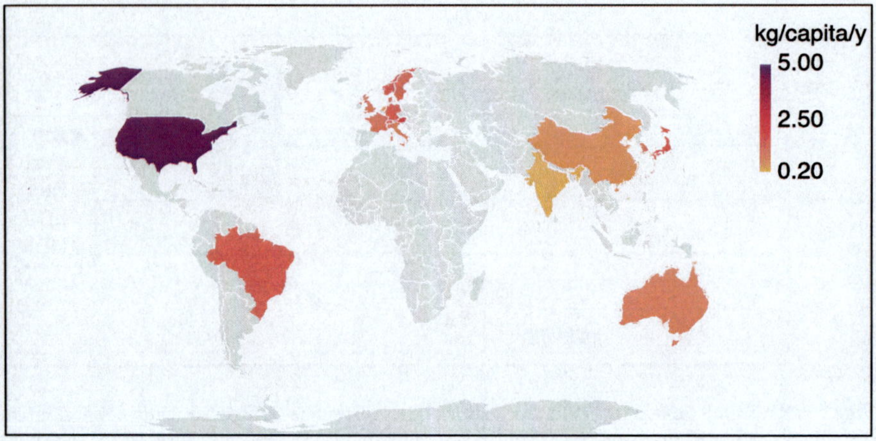

Fig. 5 Global TRWP emissions in kilograms per capita per year. Data from Kole et al. [24], augmented by recent data for Korea [38], Switzerland [43], and Austria [25]

types of vehicles, and the relative contributions of each vehicle category in a given region or country. Such calculations have been performed recently for a number of countries, for which appropriate data were available [1, 24, 25, 38, 43], revealing that the USA had the highest total annual TRWP emissions (>1.5 Mt/year), followed by China (~756 kt/year) and Brazil (~294 kt/year). From these data, the total global emissions can be estimated at ~3.6 Mt/year. The average per-person TRWP emissions for the investigated countries are approximately 1 kg/capita/year, but the values for individual countries vary widely, between 0.2 kg/capita/year for India and 4.7 kg/capita/year for the USA (Fig. 5), which reflects a considerably higher car density and longer travel distances, as well as a different composition of the vehicle fleet in the USA compared to India [1].

The analysis of Kole et al. [24] included 14 countries, which account for approximately 3.6 billion inhabitants and nearly 1.1 billion vehicles; when assuming that the distance driven by the number of vehicles considered in their estimates is representative of all vehicles in the world, the authors calculated that the global TRWP emissions amount to ~5.9 Mt/year, equivalent to 0.80 kg/capita/year globally.

Considerable variation among different countries is also observed for the proportions between the TRWP emissions from passenger cars and the total TRWP emissions from all vehicles. The percentage of car-to-total TRWP emissions ranges from 18% (India) to 71% (the Netherlands), with an average value of 37% (Fig. 6). These data highlight differences in the use of heavier vehicles (see data for LDV + HDV in Fig. 6), which could also reflect transportation via other means, such as railways.

The current increase in the share of electric vehicles (EVs), which due to the battery weight are heavier than conventional vehicles, will lead to a general increase in TRWP and other non-exhaust particulate matter (PM) emissions, both in absolute

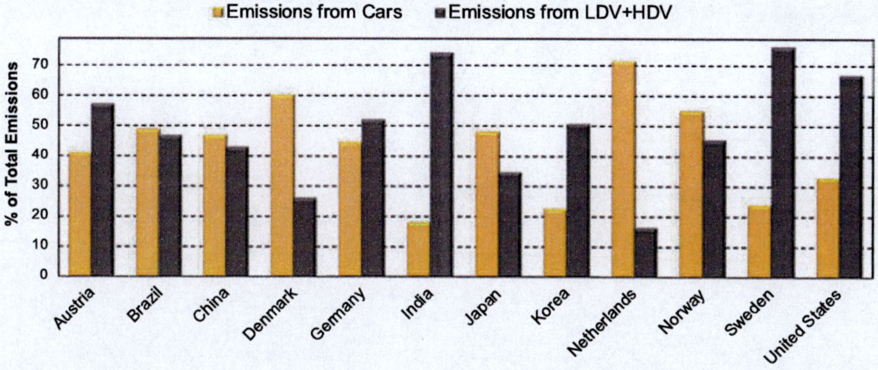

Fig. 6 TRWP emissions from passenger cars and LDV + HDV as percentage of total TRWP emissions. Data from Lee [38], Kole et al. [24], and Prenner et al. [25]. *LDV* light-duty vehicle, *HDV* heavy-duty vehicle

terms and relative to exhaust emissions [15, 18, 44–46]. In addition, it must also be stressed that even though EVs do not generate exhaust emissions, traffic will continue to release ultrafine (<100 nm across) particles (mostly CB), but these are derived from tire components and generated by the tire–pavement interaction rather than through combustion [47].

3 Characteristics of Tire-Wear Particles

The TRWPs exhibit a number of physical and chemical features that need to be known in order to correctly identify these types of particles in the environment. Here we compile the most important characteristics of TRWP, as retrieved from the literature and our own research.

3.1 Physical Features

Typical TRWPs display an elongate, cylindrical, and in some cases "sausage-like" shape (Fig. 7) and most commonly consist of tire-core particles, which are to various extents coated by, or encrusted with, particles from other sources (Fig. 8), especially road-wear particles [10, 12–15, 48]. The extent of coating of the tire-core particles depends on various factors, including driving speed, fleet composition, traffic density or volume, and traffic flow [15].

The TRWPs with diameters >10 μm (i.e., super-coarse particles) and their coatings are most easily identified and characterized by using a combination of transmitted light microscopy (TLM) and scanning electron microscopy (SEM) with

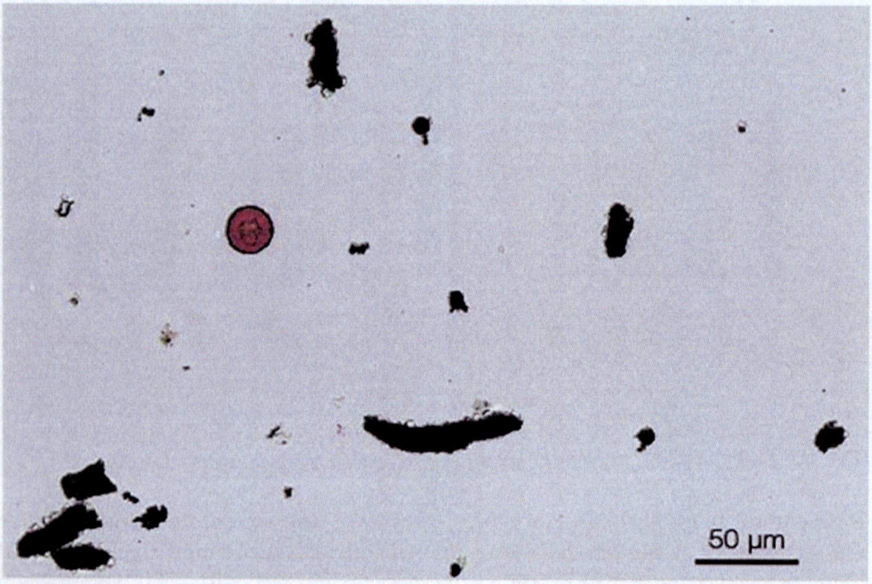

Fig. 7 TLM image of ambient aerosol collected from a kerbside location in Germany. The TRWPs are the elongate and cylindrical, opaque particles

Fig. 8 SEM image in backscatter electron mode of an individual TRWP, which is mixed with, and encrusted by, foreign material. Bright angular fragments comprise road-wear and soil particles. Photo: Dr. Frank Sommer (used with permission)

integrated energy-dispersive X-ray spectroscopy (SEM-EDX), because the technique not only returns morphological parameters (size, shape, topography) but also the chemical composition of the individual components [2, 15, 48]. On the other hand, characterization by SEM-EDX of TRWPs in the PM_{10} fraction (i.e., particulate matter with diameters that are generally 10 μm and smaller) of atmospheric PM is challenging, because they display a very complex chemical composition [49].

TRWPs have also been studied by focused-ion beam SEM methods, which revealed that they exhibit considerable porosity [50], and by various types of mass spectrometry, which allowed for additional characterization of organic components [19, 21, 48, 51–54].

The size distribution of TRWPs is relatively difficult to assess [36], as some studies report sizes of airborne particles, whereas others focus on the TRWPs deposited on the road. From early studies it is known that most of the TRWP mass is released as particles that range in size from ~5–7 to ~100 μm, with some very coarse particles (~500 to ~1,100 μm) found in roadway gutters and soil samples [40, 55, 56]. These relatively large TRWPs have also been referred to as sedimentary particles, highlighting their tendency to settle on or close to roadways [55]. In a more recent experimental study, Kreider et al. [12] documented that the TRWPs emitted and collected during outdoor driving on asphalt-based road surfaces had a unimodal size distribution by volume that was ranging between 4 μm and 280 μm, with the mode centered at ~50 μm (determined by laser diffraction); the corresponding size distribution by number of particles was also unimodal, with a similar range (4–265 μm), but the mode located at ~25 μm. Using SEM, Sommer et al. [15] observed sizes up to 200 μm for super-coarse TRWPs collected from the ambient air along roads, whereas Kovochich et al. [48], studying TRWPs collected in road simulator experiments, reported a range 6–120 μm, with average sizes of 34 μm and 49 μm by number and volume distribution, respectively. Kovochich et al. [54] have also described TRWPs from road-dust samples, where the particle size (determined by TLM and SEM) ranged between 6 and 649 μm, with averages of 158 μm and 224 μm by number and volume, respectively. Overall, these recent studies are quite consistent, returning TRWP sizes in the coarse and super-coarse modes.

Some studies, however, reported the occurrence of TRWPs with smaller dimensions, including those in the fine ($PM_{2.5}$, i.e., $\leq$2.5 μm across) and the ultrafine ranges [42, 57, 58]. Such small particles are emitted primarily under extreme driving conditions, i.e., racing start, full-stop braking, and extreme cornering [11, 59]. As documented by transmission electron microscopy (TEM), the ultrafine particles studied so far consisted mostly of tire-filler materials, especially the reinforcing CB filler, but also some softening agents, such as mineral oils, and condensates from thermomechanically volatilized tire components [47, 58, 59]. Indeed, considerable amounts of CB are emitted during tire abrasion, as suggested by a recent material-flow analysis, which revealed that about 5,500 t/year of CB are emitted along with 21,100 t/year of plastic in Austria [25]. These authors, however, concluded that nearly all CB remained embedded in the TRWPs, and that only small amounts were released as nanoparticles.

Recently, Klöckner et al. [53] reported that the density of TRWPs, occurring in road dust from a highway tunnel, was mostly between 1.3 and 1.7 g/cm^3. This result is in good agreement with the estimated density of 1.26 g/cm^3, calculated for encrusted airborne TRWPs, where the total volume of the attached non-tire particles accounted for 6–10 vol% of the whole TRWP [15]. Due to the variable extent of encrustation observed for TRWPs and because of the presence of various types of non-tire particles in the encrustations [15], the density of TRWPs may vary considerably. According to a recent review, the TRWP density may be as high as 2.2 g/cm^3 [17].

3.2 Chemical Composition

In addition to natural and synthetic rubber, TRWPs may also contain considerable amounts of ZnO [10, 13, 19, 52], which is added as a vulcanizing agent during tire manufacture [60]. Therefore, TRWPs represent an important source of Zn to the environment, whereby the input from TRWPs in urban-suburban watersheds may rival or even exceed the Zn emissions into the atmosphere derived from other sources, such as waste incineration and fossil fuel combustion [10, 61, 62].

Zinc, however, is not necessarily characteristic of TRWPs occurring in the environment, because it may not be easily detectable due to the encrustations of the TRWPs by road- and brake-wear particles as well as other materials [15]. It is due to these encrustations that several other metals have been documented to occur at high concentrations in TRWPs, including Al, Ca, Mg, Fe, Ti, V, Pb, K, and Na, which are usually present only at low levels in the original tire tread [10, 12, 15, 62]. This observation also holds for the metalloid silicon, even though tires that are based on silica filler are relatively rich in Si to begin with. Clearly these elevated concentrations of Si and metals are due to the particles that are attached to the TRWPs, but originated from other sources, such as the road surface, brake wear, or the natural environment (e.g., soils and dust). The addition of non-tire material to tire-core particles leads to a relative depletion of characteristic tire elements, such as Zn and S. Nevertheless, Zn has been used as a chemical marker for TRWPs found in the environment [10, 19, 63], where they represent an important source of Zn [61]. Other components used for tire manufacturing, including styrene-butadiene rubber and some benzothiazole compounds, have also been used successfully as chemical tracers for TRWPs occurring in the environment [6, 52, 64–71]. Some of the tracers used, however, are not unique for tires because they may be emitted from other sources as well, which complicates the assessments of the TRWP abundance in the environment [72].

TRWPs further contain polycyclic aromatic hydrocarbons (PAH), mostly associated with CB and oils in the tire [73–77]. Kreider et al. [12] reported that TRWPs collected during outdoor driving on asphalt-based road surfaces in France contained significantly higher amounts of PAH than the cryogenically ground pure tread particles of the same silica-based tires, suggesting that considerable amounts of the

PAH present in the outdoor samples were derived from non-tire sources (e.g., from asphalt or exhaust emissions). Due to health concerns, however, the use of extender oils with very high contents of certain PAHs has been discontinued in the European Union [11], and therefore, the results of Kreider et al. [12] may not be representative of other regions on Earth, where similar laws do not exist.

4 Immission of Tire- and Road-Wear Particles

Recent analyses and models of the material flows for particles generated by tire abrasion revealed that TRWP immissions have to be considered and investigated in several distinct environmental compartments: (1) *road surface*; (2) *atmosphere*; (3) *road runoff and wastewater;* (4) *soil and vegetation*; (5) *surface water and groundwater*; and (6) *sediments* [1, 15, 18, 23, 25, 43, 68, 78–80]. Of these compartments, road surface, atmosphere, vegetation, and wastewater mainly act as transient hosts for TRWPs, whereas soil and surface water represent long-term sinks, with road runoff playing a central role in the TRWP transport (Fig. 9). Road runoff, especially in urban areas, has been shown to be a major transport path for pollutants, including TRWPs, to surface waters [78, 81–86].

Compared to other MPs in the environment, our understanding of both occurrence and fate of TRWPs in the environment is limited [87], primarily because of analytical challenges, and especially also, because standardized sampling, analysis, and monitoring procedures for these particles are largely missing [8, 72, 88]. Our

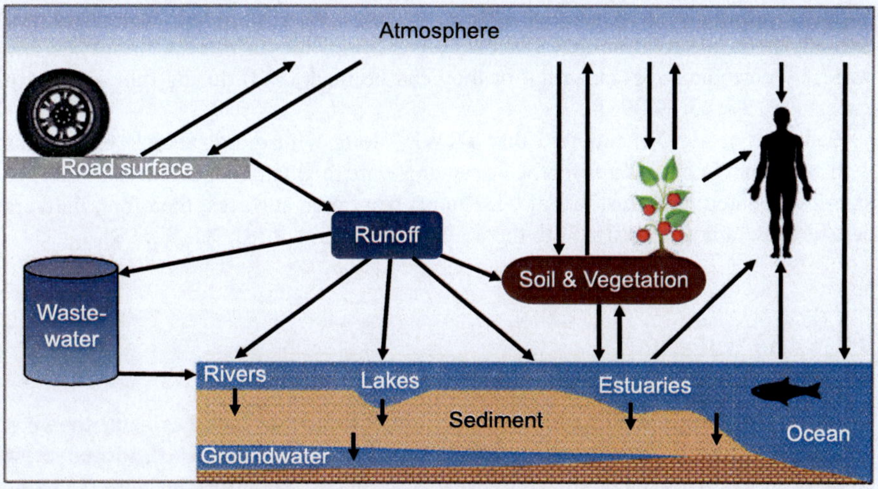

Fig. 9 Schematic diagram showing the TRWP pathways into the various environmental compartments, highlighting the central role of road runoff in the transport of TRWP into soil and surface water

knowledge about environmental concentrations of TRWPs, however, is improving as a result of an increasing number of recent publications on this topic [17]. Nevertheless, many researchers deliver quantitative TRWP data obtained for a specific location, under certain conditions, and with a particular analytical technique, and subsequently, based on a number of assumptions and selected tire markers, deduce explicit values for TRWP abundances and mass flows between different environmental compartments, which are at best rough estimates and in many cases inconsistent with other assessments [72, 89]. For this reason, there is very little reliable quantitative information about the TRWP amounts distributed between different environmental compartments and their dispersion in nature [67, 90]. Consequently, it is nearly impossible to compare the datasets from different investigations and to provide a valuable compilation of all data on this subject. Therefore, we will avoid discussing details of published mass flows below.

4.1 Road Surface

It has been estimated that 90–95% by mass of the total TRWP emissions consist of relatively large, non-airborne particles [17], which thus are first deposited on the road surface [55, 56], where they become part of the road dust (Fig. 9). A wide range of TRWP concentrations (0.7–210 g/kg dry weight) has been reported for road dust, with the highest values detected inside tunnels [68, 69, 80, 91]. Moreover, the TRWP mass in road dust appears to be dominated by particles derived from cars compared to those from trucks, at least for samples collected in the German city of Oldenburg [92]. New data from Seoul further indicate that TRPW concentrations are higher in industrial areas than in residential areas [93]. TRWPs can be removed partially from the road surface by resuspension into the air due to turbulence and wind, by periodic street cleaning, or they can be flushed off during rain and storm events [27, 37, 43, 94–96].

Pengchai et al. [77] reported that TRWP, along with diesel-vehicle exhaust as well as asphalt or bitumen from the pavement, were the major contributors of PAH to size-fractionated road dust in Tokyo. Runoff from road surfaces, therefore, delivers considerable amounts of PAH to the aquatic environment [81–85, 97, 98].

4.2 Atmosphere

The share of non-exhaust particles in the atmospheric PM load has been shown a while ago to be comparable to that of exhaust particles [63, 99, 100]. Moreover, it continues to increase, as predicted by Rexeis & Hausberger [101], because (1) stringent standards on exhaust emissions and the transition to EVs will both lead to a continuous reduction of exhaust emissions, and (2) non-exhaust PM emissions are largely unregulated [15, 18, 44, 46, 102, 103].

Typically, the larger size fractions of TRWPs will be deposited on the road surface, whereas the smaller particles become airborne, thus contributing to the PM load in the atmosphere. Under typical driving conditions, cars are estimated to emit $<10\%$ by mass of their total tire wear as PM_{10} [55, 80, 94, 104, 105]. Roadside ambient air contains TRWPs in concentrations of up to 11 μg/m^3, based on highest measured concentrations of tire markers [80].

The size of individual TRWPs occurring in ambient air varies from ultrafine to super-coarse, but the TRWP mass concentrations in the smallest fractions are generally low [15, 21, 49, 63, 70, 105, 106].

Even though the fraction of TRWPs emitted into the ambient air is small compared to the total emissions, these particles can be transported over long distances via atmospheric processes. Moreover, this fraction is inhalable and thus needs to be characterized in detail and quantified.

4.3 *Runoff and Wastewater*

Because road dust exhibits elevated concentrations of organic and inorganic pollutants, including TRWPs, the road-runoff water (i.e., stormwater) needs to be treated before it reaches roadside soils or the aquatic environment [96, 107]. Concentrations of TRWP in road runoff are highly variable depending on factors, such as location, traffic characteristics, properties of the road surface (e.g., presence or absence of open asphalt concrete), rain intensity, and time of measurement relative to the start of a heavy rainfall [1, 90], and the concentrations may be as high as ~180 mg/L, whereby they are typically higher in the first flush compared to the later stages of a particular heavy rain event [80]. Installation of special roadside drainage systems helps in reducing the amounts of TRWP that reach adjacent soils or surface waters via stormwater. In rural areas, the road effluents may be managed in road-runoff treatment plants, which use either a natural (e.g., sedimentation ponds, filter basins) or technical (e.g., gully pots, engineered filters) approach [43, 108]. In many urban areas, road runoff is collected by a sewer system and transported either alone (separate sewer system) or mixed with sanitary sewage (combined sewer system) into wastewater-treatment plants, where it is purified before being discharged into surface waters [43, 104]. Wastewater-treatment facilities retain a major part of the TRWPs [78, 104]. In wastewater-treatment plants, the TRWPs accumulate in the sewage sludge [43], where concentrations of up to 43 g/kg have been observed [80]. If sewage sludge is used as a fertilizer in agricultural applications, considerable amounts of TRWP will thus be transferred to farmland soils.

4.4 Soil and Vegetation

Soil is a dominant sink for TRWPs, which reach this environmental compartment via the atmosphere due to wind and vehicle-induced turbulence or via road runoff, including splash (Fig. 9). Because many roads are not connected to drainage or sewer systems, the stormwater ends up on the roadside land surface, where it eventually infiltrates the soil, thus depositing TRWPs and other contaminants, such as brake-wear particles [6, 17, 104]. This diffuse infiltration, combined with TRWP deposition directly from the atmosphere, generates a polluted zone of up to several meters width along either side of a road [109, 110].

The concentration of TRWPs in roadside soils may be as high as 158 g/kg dry weight, but decreases rapidly with increasing distance from the edge of the road as well as with increasing depth beneath the surface [17, 55, 65, 80, 109]. Of note is that even soils located far away from roads may be contaminated by TRWP, especially if sewage sludge is applied as agricultural fertilizer.

Vegetation is known to improve air quality by influencing particle-pollutant deposition and dispersion, especially in urban environments [111, 112], but the underlying processes are so far not fully understood [113, 114]. Airborne particles are deposited when they pass surfaces, both natural and artificial, whereby vegetation has generally a larger surface, thus increasing the chance for deposition due to the higher friction, which reduces wind speed [115]. The deposition rate of atmospheric particles, including TRWPs, on vegetation and thus the filtering efficiency of vegetation, also depends on the type of plants and the properties of their surfaces (e.g., leaves), including their windspeed-reducing capabilities, as well as on particle size and meteorological parameters. In most cases, however, vegetation only temporarily reduces the atmospheric PM load, and the particles will eventually be rinsed off the leaves by precipitation, followed by deposition on the ground.

4.5 Surface Water and Groundwater

Like soil, surface water is a dominant sink for TRWPs [90]. These particles reach the riverine, lacustrine, and marine environments (Fig. 9) either directly or indirectly via other environmental compartments [18]. The dispersion of TRWPs in surface water depends on various properties of the particles, such as their size, density, shape, and degree of encrustation by road-wear particles, as well as on the characteristics of the water in which they are transported or received (e.g., its flow rate and salinity). For example, the difference in density between TRWPs and water along with the size of the TRWPs are the major parameters influencing settling velocities, which thus determine how TRWPs are distributed within the water column, how long they will be floating, and which of the various aquatic compartments (Fig. 9) they will reach before eventually accumulating in the bottom sediments. Therefore, rivers and estuaries, for example, may act either as TRWP pathways to the ocean or as deposition sites.

Estimated amounts of TRWPs entering surface waters vary considerably, depending on methods and assumptions (see above). In the Netherlands, for example, roughly 10% of the emitted TRWP mass have been inferred to reach surface waters [16]. On the other hand, Unice et al. [87] estimated that about 20% of the TRWPs generated in the Seine watershed in France (1.8 kg/capita/year) are transported to the aquatic environment, whereby only about 2% reached the estuary, which points to appreciable retention and/or degradation of TRWPs prior to their export to the ocean.

Road runoff has been recognized as a major transport pathway for TRWPs to surface waters [78, 86], but only a few studies reported actual TRWP concentrations in surface water. These absolute TRWP concentrations in surface water, derived from the concentration of various tire markers, are generally low, ranging between 0.09 and 6.4 mg/L [80]. Concentrations of TRWP, just like those of other MPs, have also been reported as number of particles per liter, because they have been determined directly by counting the number of visually identified (via TLM and/or SEM) particles present rather than calculated from the measured concentration of tire markers [116, 117]. Tire-derived chemical components and their transformation products appear to be ubiquitous in urban watersheds [118], and some of these are potentially toxic to aquatic organisms (see below). The aqueous concentrations of TRWP-derived components, however, can also be reduced via their possible adsorption onto sediment or soil particles [119].

In urban areas and close to highly frequented roads, groundwater can become contaminated by traffic-derived pollutants, including various metals [120]. When MPs, including TRWPs, are transported through soil, especially due to actions of ground-dwelling organisms (e.g., earthworms), there is a possibility that they might also reach the groundwater [121], although this scenario has been called into question [6]. In addition, there are only very few studies on occurrence and dispersion of MPs in groundwater [90], and it is unknown whether these particles, where present, were delivered directly from the roadside soils or through sewerage [122]. On the other hand, certain TRWP components can be leached out of the particles during their degradation in soils and therefore, can contaminate the groundwater in dissolved form, thus deteriorating its quality [6].

In a recent report, it was estimated that 270 kt of TRWPs reach the marine environment annually, accounting for approximately 28 wt% of all primary MPs, the highest proportion of all MP sources [123]. The same percentage, second only to textile-derived primary MPs, was also calculated by Boucher & Friot [23], but their absolute value was considerably higher (~420 kt), even though their model only considered synthetic rubber; when including natural rubber, their model returned an absolute value of ~920 kt/year, which accounted for ~46% of all primary microplastics. These estimated absolute values for TRWP immissions into the marine environment are consistent with those of Kole et al. [1, 24] who concluded that 5–10% of the global TRWP emissions, i.e., ~300–600 kt/year, reach the oceans.

4.6 Sediments

Eventually, TRWPs and other MPs in the aquatic environment will settle at the bottom of water bodies or along riverbeds, where they accumulate in sediments together with other materials, such as minerals (e.g., clay minerals, quartz), rock fragments, and organic debris (e.g., leaves). Therefore, TRWP and MP concentrations in sediments are typically higher than those in the corresponding water [6, 116]. Consequently, sediments are important sinks for those TRWPs that have reached the aquatic environment.

Most of the quantitative data for TRWP concentrations in sediments have been obtained from the analysis of certain tire markers, i.e., Zn, benzothiazole and its derivatives. Published TRWP concentrations show a considerable range for various settings and watersheds, varying between 0.3 and 155 g/kg of dry sediment, whereby the highest values were observed near busy traffic sites, such as bridges [1, 6, 20, 80].

Concentrations of TRWPs in sediments, just like those of other types of MPs, have also been reported in different units, i.e., as number of particles per square meter or per kilogram, because they have been determined directly by counting the visually identified (via TLM and/or SEM) number of particles present [116, 117]. These authors described variations in TRWP concentrations between different types of sediments, namely those deposited in intertidal vs. subtidal settings, in the Charleston Harbor estuary, South Carolina (USA), where TRWPs and blue fibers were the two most abundant types of MPs present in the tributaries. Moreover, TRWP concentrations in each of the two studied sediment types were highest in samples collected close to high-traffic areas [117].

The presence of the mentioned tire markers in sediments highlights the importance of investigating the potential toxic effects of these components on benthic organisms (see below).

5 Toxicity

The occurrence of TRWPs in the environment is a major concern, because of their chemical composition, including that of the encrustations, and because the particles are present in a size range that can be easily ingested and digested by various organisms. TRWPs have been estimated to contribute 5–10% of the total global MPs ending up in the oceans [1]. Moreover, it has been estimated that the contribution by weight of TRWPs in the ambient air is <7% of $PM_{2.5}$ and $< 11\%$ of PM_{10} [1, 11, 17, 63, 70, 105], which suggests that, even though TRWPs emitted into the atmosphere are coarser than PM from exhaust emissions, they may still represent a human health risk as a result of inhalation [124]. The potential toxicity of TRWPs at environmentally relevant concentrations, therefore, must be carefully assessed, especially because they contain relatively high concentrations of some components of concern, including Zn, butadiene, benzothiazoles, and PAHs, which may be released from or leached out of the TRWPs during their physical, chemical (including photo-oxidation), and microbial degradation in the environment (i.e., weathering) or within the bodies of organisms [6, 10, 17, 80, 89, 119, 125–127]. A recent study emphasized the environmental risks of TRWPs and several TRWP-associated chemical substances in runoff water, surface water, and sediments [128].

Several researchers reported on the bioavailability of both TRWPs and some compounds leached from them [17, 80, 88, 129]. Aquatic organisms may be exposed to TRWPs through ingestion and ventilation. Indeed, TRWPs and other MPs have been found in the digestive tracts and gills of various organisms, including various fish species, living in benthic, epibenthic, and pelagic marine environments and estuaries [130, 131]. Some of these and other organisms are important commercial seafood species, indicating that TRWPs may reach humans via consumption of polluted seafood [1]. On the other hand, shrimps have been shown to be able to depurate both ingested and ventilated TRWPs after having been exposed to particle-free water [130].

5.1 TRWP Components of Concern

As highlighted above, Zn is an important chemical element in tires. After emission in the form of TRWPs, Zn becomes a significant pollutant in road dust, runoff water, air, soil, water, and sediments, as it can be released to the environment along with other metals [10, 127, 132–137]. Zinc can be toxic to aquatic organisms at elevated concentrations, and its adverse biological effects include increased mortality, behavioral changes, and decreased invertebrate diversity and abundance [138]. A recent study has shown, however, that only a small fraction of the Zn contained in TRWPs was bioavailable when freshwater benthic macroinvertebrates were exposed to sediment + TRWP mixtures [139].

Butadiene and styrene-butadiene are the two main synthetic rubber polymers used for tire manufacturing. Butadiene has recently been evaluated as carcinogenic to humans [140]. Low levels of butadiene are typically present in ambient air within urban and suburban settings, and the substance is one of the hazardous air pollutants that pose the greatest potential health threat in urban areas [141].

Some benzothiazole compounds, used as vulcanization accelerators (e.g., mercaptobenzothiazole), are present in tires at relatively high concentrations [66, 126, 142]. At the same time, these tire components are biologically active, exhibiting various health effects, and some benzothiazoles are biocides, potential aquatic toxins, or potentially mutagenic [64]. Benzothiazole compounds and other tire-associated organic chemicals and their transformation products have been found at elevated concentrations in road runoff, effluents from wastewater-treatment plants, and urban rivers [66, 142–144]. In addition, the presence of benzothiazole compounds in the inhalable fraction (PM_{10}) of urban air suggests that they are derived from TRWPs and that human exposure to these chemicals has been underestimated, because it is not limited to occupational settings only [64].

TRWPs further contain elevated concentrations of PAHs [12, 13, 75], which thus are also released to the environment during particle degradation [127]. These chemical substances are among the most concerning organic pollutants, as they are toxic, potentially carcinogenic, and ubiquitous in the environment [145]. In urban environments, PAH concentrations are elevated in dusts deposited on impervious surfaces, including road surfaces and roofs [75, 146–148]. As road runoff, triggered by heavy rain events, is the primary pathway for the transport of road dust, including TRWPs, to the environment (Fig. 9), considerable amounts of PAHs can be transported along with the TRWPs into local receiving waters [97, 98], thus representing an elevated risk to aquatic life [81–85]. Sources of PAH in road runoff include diesel- and gasoline-vehicle exhaust, TRWPs, road-wear particles (e.g., asphalt, bitumen), and lubricating oils [77, 149, 150]. Road runoff represents an intermittent flux of contaminants to the environment, because strong rain storms are typically relatively short, producing high contaminant concentrations, which are time-dependent, both over short time spans and seasons [151, 152]. The quantities of road runoff as well as its potential toxicity are thus difficult to model.

5.2 *Toxic Effects*

Most ecotoxicological studies have been carried out with TRWP leachates rather than directly with TRWPs or with particles generated by grinding tire or tire-tread material, and they investigated the effects primarily on various aquatic organisms, with only a few studies on terrestrial biota [17, 80, 88, 89, 129, 153, 154]. When tire or tire-tread material was pulverized, the resulting leachates were generally the most toxic, presumably because of the larger specific surface area of the smaller particles [80]. There is, however, a large variability of the toxicity data obtained in different laboratories and with different experimental protocols, tire materials, leaching solutions, toxicant concentrations, and organisms [6, 80, 88]. Studies have further shown that some tire-derived leachates can also exhibit mutagenic, teratogenic, and estrogenic activities [80, 155].

The observed acute and chronic toxicity of TRWP leachates is thought to be caused by Zn and organic compounds, especially benzothiazole and its derivatives, released from the tire rubber [125–127, 129, 155–160]. Leachates are very complex cocktails of chemicals and therefore, experiments performed with leachates cannot easily identify which compound is the most potent toxicant, and interpretations are primarily based on the abundance of specific compounds in the mixture [126]. In a recent study, however, a transformation product derived from a widely used tire-rubber antioxidant (6PPD) was reported to occur at toxic concentrations in creeks impacted by road runoff, and it was linked to the acute mortality phenomenon observed annually when adult coho salmon return to spawn in freshwaters [144].

Investigations on ingestion of TRWPs by various benthic macroinvertebrates and by fish from both marine and freshwater environments revealed that the ingested particles or mixtures of particles + sediment did not cause significant adverse effects on the tested animals, suggesting that, compared to the effects caused by TRWP leachates, those induced by TRWPs can be relatively mild or even absent under ecologically relevant conditions [130, 139, 161]. The limited toxicity observed for grass shrimp may be due to the rapid clearing of the majority of ingested and ventilated particles [130]. On the other hand, recent experimental studies with another crustacean (*Hyalella azteca*) provided evidence for acute mortality and reduction of long-term growth related to ingested TRWPs, whereby the TRWP-containing aqueous suspensions were generally more toxic than the TRWP leachates [129, 162].

Only a few studies exist on the ecotoxicity of TRWPs in soils. Various biological effects have been observed in soil-vegetation systems, including dose-dependent phytotoxic effects in the tested plants and impacts on the biological activity of the associated soils [80, 163]. Recently, it has further been shown that TRWPs in soil can be ingested by earthworms, causing the particles to degrade inside their bodies, which resulted in bioaccumulation of various heavy metals, oxidative stress, and tissue damage [164].

The toxicity of airborne TRWPs has been investigated by only a few studies. In-vitro exposure of human lung epithelial cells to organic extracts from TRWPs revealed various biological responses, including increase in cell mortality, DNA damage, modification of cell morphology, and formation of reactive oxygen species [156, 165]. In-vivo experiments with mice showed that intra-tracheal instillation of TRWPs caused inflammatory responses for particles <10 μm in the bronchial area, and strong cytotoxic effects and/or acute inflammatory responses for particles <2.5 μm in the alveolar region [166, 167]. Acute pulmonary toxicity of respirable TRWPs, experimentally observed in rats after intra-tracheal instillation, has been suggested to result from the presence of water-soluble Zn and Cu in TRWPs [168]. In contrast to these investigations, no significant TRWP-related effects were observed in another study involving rats during nose-only-inhalation experiments [169]. In a follow-up study, Kreider et al. [170] evaluated the human health risk of TRWPs in air and concluded that TRWPs pose a low risk.

As TRWPs typically consist of tire material that is encrusted with particles from other sources and with different physical and chemical, and thus toxicological properties (e.g., road- and brake-abrasion particles, soil particles; see above), it is not surprising that the reported toxicological effects vary widely. It is further not known whether or not the TRWPs, once they reached the aquatic environment or the bottom sediments, will release some or all of the particles present in their surface encrustments (Fig. 8). Moreover, testing particulate exposure often involves instillation of a TRWP + fluid mixture, and thus, such experiments are actually a combination of particulate and leachate exposure [88]. Thus, investigating the toxicity of such complex, multicomponent particles remains a challenge.

6 Reducing TRWP Pollution

Tire abrasion is one of the more difficult MP sources to control and limit, and thus, it is crucial to improve our knowledge on the extent of TRWP pollution and its impacts on both the environment and human health.

6.1 Reducing Emissions

The general concept for all mitigation measures should be focused on the source, i.e., the priority must be on minimization or avoidance of TRWP emissions into the environment. One approach involves a technical concept to develop tire materials with higher abrasion resistance under all possible driving and road conditions. A first start was made by the development of the so-called self-healing or self-repairing tire materials, whereby commercial rubber is modified into a self-healing material, which is capable of repairing tire damages and tire aging. The goals are (1) to minimize the degradation process, which occurs over time mostly as a result of chemical reaction within the rubber components, (2) to extend the lifetime of a tire, (3) to increase abrasion resistance, and at the same time, (4) to lower fuel consumption [171–174]. Moreover, the addition of devices that capture TRWPs behind the wheels of vehicles appears to be another promising way forward in minimizing TRWP emissions into the ambient air and onto the road [175]. Furthermore, there is an urgent need for substituting certain tire components with more sustainable and more environmentally friendly, less toxic materials to avoid the release of toxic compounds into the environment, which would lower the ecotoxicological impacts as well as the potential human health effects of TRWPs [176].

Another approach involves reduction of the traffic density on roads as much as possible by implementing new policies that promote or mandate alternative transport options, such as public transport and shift of cargo transport from road to railways or ships [79]. Traffic density, however, is not the only important factor influencing TRWP emission. Equally important parameters are traffic speed, mode, and fleet composition, generally described as level of service (LOS), and must also be considered. On-board driver-assistance systems and computer-directed traffic management systems that control the velocity, reduce braking frequency, and facilitate a smooth traffic flow, for example, can reduce TRWP emission, as they aim at avoiding "stop-and-go" driving conditions [15, 46]. Implementation of lower speed limits and optimization of vehicle use and maintenance, including proper fit, inflation and balancing of the tires, would further help in reducing TRWP emissions, with minimum costs and efforts [177]. In addition, the trend toward heavier vehicles should be reversed and, especially in urban areas, the number of SUVs should be reduced [35]. Moreover, according to the latest report from Europe's largest motoring association, which evaluated approximately 100 different tires (summer, winter and all-season tires), consumers already today have the possibility to purchase high-quality tire brands with abrasion characteristics as low as ~90 mg/veh-km, i.e., significantly lower than the average of ~120 mg/veh-km [178].

The approaches highlighted above suggest that, in addition to advanced tire engineering, improved consumer education in regard to tire choice, driving habits, and environmental impacts of driving is crucial to raise public awareness of the issue and to achieve a reduction of TRWP emissions.

6.2 Reducing Immissions

There exist also several approaches to reduce the distribution in the distal environment of the TRWPs that have been emitted. Porous asphalt as a road material, for example, can trap TRWPs in the road surface for later removal. This approach has already been widely implemented in some countries (e.g., Netherlands) and seems to be very promising [1]. Porous asphalt is also known to be a safer road surface for drivers, as it reduces standing water and spray [179].

Sweeping and rinsing activities at "hotspot" areas are recommended, especially a combination of the two techniques, to avoid TRWPs deposited on the road from further comminution caused by vehicle-driving movements [180, 181]. Technical systems for suction and filtering directly along roads represent another possibility for lowering TRWP pollution, but this approach is limited to urban environments, because the traffic-related abrasion particles are distributed over large road areas [180]. In order to reduce the transport of TRWPs from the road surface via runoff to distal environmental compartments, such as soils, surface waters and sediments, better stormwater management and interception (e.g., roadside settling ponds, grass-filled swales, bioretention cells, gully pots), connection of road-drainage systems to wastewater-treatment infrastructure, particularly on bridges and along busy roads, and planting of vegetation along roads to reduce particulate pollution, especially evergreen hedges and plants, are examples of mitigations strategies that must be more widely implemented [26, 111, 113, 114, 180, 182].

Because sewage sludge can contain important amounts of TRWPs [80], agricultural application of sewage sludge as fertilizer should be banned to reduce the amounts of TRWPs transferred to soils. Such bans have been implemented by certain countries (e.g., Switzerland), where sewage sludge is classified as hazardous waste [43]. In addition, procedures for the abatement of the PAHs contained in road dust should be included in the treatment protocols for both road runoff and wastewater [152].

7 Conclusions

The release of nearly six million tons of TRWP into the environment worldwide represents challenges on various fronts. First, tire abrasion is one of the more difficult sources of MPs to control and limit, and innovation is needed to reduce emissions while maintaining essential tire properties and associated safety requirements. We further have to develop and install new measures to lower the immission of these complex particles into the environment, whereby solutions cannot be based solely on engineering interventions, but they must also include other measures, such as driver education aimed at changing driving habits. Moreover, it is crucial that we improve our knowledge on the extent of TRWP pollution by developing standardized sampling and characterization techniques. Finally, because the degradation of TRWPs in the environment releases various components, further research is needed to evaluate the potential toxicity of TRWPs and compounds leached from them in ecologically relevant concentrations, in an effort to better understand the environmental and health effects of this abundant type of MPs.

Acknowledgements We thank the editors and reviewers of our manuscript for invaluable suggestions and constructive criticism, which helped us improve this article.

References

1. Kole PJ, Löhr AJ, Van Belleghem FGAJ, Ragas AMJ (2017) Wear and tear of tyres: a stealthy source of microplastics in the environment. Int J Environ Res Public Health 14:1265
2. Camatini M, Crosta GF, Dolukhanyan T, Sung C, Giuliani G, Corbetta GM, Cencetti S, Regazzoni C (2001) Microcharacterization and identification of tire debris in heterogeneous laboratory and environmental specimens. Mater Charact 46:271–283
3. ETMRA (2012) Reinforcing fillers in the rubber industry. European Tyre & Rubber Manufacturer's Association (ETMRA), Brussels. https://www.etrma.org/wp-content/uploads/2019/09/201201-etrma-fact-sheet-carbon-black-and-silica-2.pdf
4. Norman DT (2001) Rubber grade carbon blacks. Witco Corporation, Concarb Division, Houston. http://www.continentalcarbon.com/pdfs/What_Is_Carbon_Black.pdf
5. OECD (2014) Nanotechnology and tyres: greening industry and transport. OECD Publishing, Paris. https://doi.org/10.1787/9789264209152-en
6. Wagner S, Hüffer T, Klöckner P, Wehrhahn M, Hofmann T, Reemtsma T (2018) Tire wear particles in the aquatic environment – a review on generation, analysis, occurrence, fate and effects. Water Res 139:83–100
7. Gieré R, LaFree ST, Carleton LE, Tishmack JK (2004) Environmental impact of energy recovery from waste tyres. In: Gieré R, Stille P (eds) Energy, waste, and the environment: a geochemical perspective. The Geological Society, London, pp 475–498
8. Rauert C, Rødland ES, Okoffo ED, Reid MJ, Meland S, Thomas KV (2021) Challenges with quantifying tire road wear particles: recognizing the need for further refinement of the ISO technical specification. Environ Sci Technol Lett 8:231–236
9. Winther M, Slentø E (2010) Heavy metal emissions for Danish road transport; NERI technical report n. 780. National Environmental Research Institute, Aarhus. https://www.dmu.dk/Pub/FR780.pdf

10. Adachi K, Tainosho Y (2004) Characterization of heavy metal particles embedded in tire dust. Environ Int 30:1009–1017
11. Grigoratos T, Martini G (2014) Non-exhaust traffic related emissions. Brake and Tyre wear PM. JRC89231. European Commission, Joint Research Centre, Institute of Energy and Transport, Ispra
12. Kreider ML, Panko JM, McAtee BL, Sweet LI, Finley BL (2010) Physical and chemical characterization of tire-related particles: comparison of particles generated using different methodologies. Sci Total Environ 408:652–659
13. Panko J, Kreider ML, Unice KM (2018) Review of tire wear emissions. In: Amato F (ed) Non-exhaust emissions. Academic Press, London, pp 147–160. https://doi.org/10.1016/B978-0-12-811770-5.00007-8
14. Rauterberg-Wulff A, Israel GW, Fesch M, Schlums C (1995) Bestimmung des Beitrages von Reifenabrieb zur Russimmission an stark befahrenen Strassen. VDI Ber 1228:81–92 (in German)
15. Sommer F, Dietze V, Baum A, Sauer J, Gilge S, Maschowski C, Gieré R (2018) Tire abrasion as a major source of microplastics in the environment. Aerosol Air Qual Res 18:2014–2028
16. Verschoor A, de Poorter L, Dröge R, Kuenen J, de Valk E (2016) Emission of microplastics and potential mitigation measures. RIVM report 2016–0026. National Institute for Public Health and the Environment, Bilthoven. https://www.rivm.nl/bibliotheek/rapporten/2016-0026.html
17. Baensch-Baltruschat B, Kocher B, Stock F, Reifferscheid G (2020) Tyre and road wear particles (TRWP) – a review of generation, properties, emissions, human health risk, ecotoxicity, and fate in the environment. Sci Total Environ 733:137823
18. Jekel M (2019) Scientific report on tyre and road wear particles, TRWP, in the aquatic environment. European Tyre & Rubber Manufacturers Association (ETMRA), Brussels. https://www.tyreandroadwear.com/wp-content/uploads/2019/10/FINAL-Scientific-Report-on-Tyre-and-Road-Wear-Particles.pdf
19. Klöckner P, Reemtsma T, Eisentraut P, Braun U, Ruhl AS, Wagner S (2019) Tire and road wear particles in road environment – quantification and assessment of particle dynamics by Zn determination after density separation. Chemosphere 222:714–721
20. Unice KM, Kreider ML, Panko JM (2013) Comparison of tire and road wear particle concentrations in sediment for watersheds in France, Japan, and the United States by quantitative pyrolysis GC/MS analysis. Environ Sci Technol 47:8138–8147
21. Dall'Osto M, Beddows DCS, Gietl JK, Olatunbosun OA, Yang X, Harrison RM (2014) Characteristics of tyre dust in polluted air: studies by single particle mass spectrometry (ATOFMS). Atmos Environ 94:224–230
22. Hartmann NB, Hüffer T, Thompson RC, Hassellöv M, Verschoor A, Daugaard AE, Rist S, Karlsson T, Brennholt N, Cole M, Herrling MP, Hess MC, Ivleva NP, Lusher AL, Wagner M (2019) Are we speaking the same language? Recommendations for a definition and categorization framework for plastic debris. Environ Sci Technol 53:1039–1047
23. Boucher J, Friot D (2017) Primary microplastics in the oceans: a global evaluation of sources. International Union for Conservation of Nature (IUCN), Gland
24. Kole PJ, Löhr AJ, Van Belleghem FGAJ, Ragas AMJ (2019) Wear and tear of tyres in the global environment: size distribution, emissions, pathways and health effects. SETAC Europe 29th annual meeting. Helsinki
25. Prenner S, Allesch A, Staudner M, Rexeis M, Schwingshackl M, Huber-Humer M, Part F (2021) Static modelling of the material flows of micro- and nanoplastic particles caused by the use of vehicle tyres. Environ Pollut. https://doi.org/10.1016/j.envpol.2021.118102
26. Furuseth S, Rødland S (2020) Reducing the release of microplastics from tire wear: Nordic efforts. Report NA2020:909, ISSN 2311–0562. Norwegian Institute of Water Research
27. Järlskog I, Strömvall A-M, Magnusson K, Gustafsson M, Polukarova M, Galfi H, Aronsson M, Andersson-Sköld Y (2020) Occurrence of tire and bitumen wear microplastics on urban streets and in sweepsand and washwater. Sci Total Environ 729:138950

28. Lassen C, Hansen SF, Magnusson K, Hartmann NB, Rehne Jensen P, Nielsen TG, Brinch A (2015) Microplastics: occurrence, effects and sources of releases to the environment in Denmark. Environmental project no. 1793. Danish Environmental Protection Agency. https://mst.dk/service/publikationer/publikationsarkiv/2015/nov/rapport-om-mikroplast
29. Vogelsang C, Lusher AM, Dadkhah ME, Sundvor I, Umar M, Ranneklev SB, Eidsvoll D, Meland S (2020) Microplastics in road dust – characteristics, pathways and measures. Report SNO. 7526–2020. Norwegian Institute for Water Research, Oslo
30. EPA (2014) Brake and tire wear emissions from on-road vehicles in MOVES2014; EPA-420-R-14-013. U.S. Environmental Protection Agency
31. Luhana L, Sokhi R, Warner L, Mao H, Boulter P, McCrae I, Wright JP, Osborn D (2004) Measurement of non-exhaust particulate matter. Version 2.0. PARTICULATES – Characterisation of exhaust particulate emissions from road vehicles. Deliverable 8. European Commission, Directorate General Transport and Environment
32. Tian Z, Dietze V, Sommer F, Baum A, Kaminski U, Sauer J, Maschowski C, Stille P, Cen K, Gieré R (2017) Coarse-particle passive-sampler measurements and single-particle analysis by transmitted light microscopy at highly frequented motorways. Aerosol Air Qual Res 17:1939–1953
33. ADAC (2019) Verschleiss und Reifenabrieb (Tire wear and abrasion). Cited in Prenner et al 2021: Allgemeiner Deutscher Automobil-Club (ADAC), available upon personal request (in German)
34. Grigoratos T, Gustafsson M, Eriksson O, Martini G (2018) Experimental investigation of tread wear and particle emission from tyres with different treadwear marking. Atmos Environ 182: 200–212
35. Gebbe M, Hartung L, Berthold O (1997) Quantifizierung des Reifenabriebs von Kraftfahrzeugen in Berlin, Teil II: Endbericht. TU Berlin, ISS-Fahrzeugtechnik, im Auftrag der Senatsverwaltung für Stadtentwicklung, Umweltschutz und Technologie, Berlin, pp 77
36. Gustafsson M (2002) Non-exhaust particles in the road environment: a literature review. PIARC 2002 XIth International Winter Road Congress 28–31 January 2002-Sapporo (Japan). Swedish National Road and Transport Research Institute. https://www.diva-portal.org/smash/get/diva2:673724/FULLTEXT02
37. Hillenbrand T, Toussaint D, Böhm E, Fuchs S, Scherer U, Rudolphi A, Hoffmann M (2005) Einträge von Kupfer, Zink und Blei in Gewässer und Böden – Analyse der Emissionspfade und möglicher Emissionsminderungsmaßnahmen. UBA-Texte 19/05. Umweltbundesamt, Desslau. https://www.umweltbundesamt.de/sites/default/files/medien/publikation/long/2936.pdf
38. Lee H, Ju M, Kim Y (2020) Estimation of emission of tire wear particles (TWPs) in Korea. Waste Manag 108:154–159
39. UNECE (2013) Particulate matter emissions by tyres. Informal document GRPE-65–20, transmitted by the expert from the Russian Federation. United Nations Economic Commission for Europe, Working Party on Pollution and Energy (GRPE), Geneva. https://unece.org/DAM/trans/doc/2013/wp29grpe/GRPE-65-20e.pdf
40. Dannis M (1974) Rubber dust from the normal wear of tires. Rubber Chem Technol 47:1011–1037
41. Aatmeeyata, Kaul DS, Sharma M (2009) Traffic generated non-exhaust particulate emissions from concrete pavement: a mass and particle size study for two-wheelers and small cars. Atmos Environ 43:5691–5697
42. Oroumiyeh F, Zhu Y (2021) Brake and tire particles measured from on-road vehicles: effects of vehicle mass and braking intensity. Atmo Environ X 12:100121
43. Sieber R, Kawecki D, Nowack B (2020) Dynamic probabilistic material flow analysis of rubber release from tires into the environment. Environ Pollut 258:113573
44. Amato F, Karanasiou A, Moreno T, Alastuey A, Orza JAG, Lumbreras J, Borged R, Boldoe E, Linares C, Querol X (2012) Emission factors from road dust resuspension in a Mediterranean freeway. Atmos Environ 61:580–587

45. Timmers VRJH, Achten PAJ (2016) Non-exhaust PM emissions from electric vehicles. Atmos Environ 134:10–17
46. Vanherle K, Lopez-Aparicio S, Grythe H, Lükewille A, Unterstaller A, Mayeres I (2021) Transport non-exhaust PM emissions. Eionet report ETC/ATNI 2020/5. European Topic Centre on Air pollution, Transport, Noise and Industrial Pollution; European Environment Agency, Kjeller
47. Dahl A, Gharibi A, Swietlicki E, Gudmundsson A, Bohgard M, Ljungman A, Blomqvist G, Gustafsson M (2006) Traffic-generated emissions of ultrafine particles from pavement-tire interface. Atmos Environ 40:1314
48. Kovochich M, Liong M, Parker JA, Oh SC, Lee JP, Xi L, Kreider ML, Unice KM (2021) Chemical mapping of tire and road wear particles for single particle analysis. Sci Total Environ 757:144085
49. Weinbruch S, Worringen A, Ebert M, Scheuvens D, Kandler K, Pfeffer U, Bruckmann P (2014) A quantitative estimation of the exhaust, abrasion and resuspension components of particulate traffic emissions using electron microscopy. Atmos Environ 99:175–182
50. Milani M, Pucillo FP, Ballerini M, Camatini M, Gualtieri M, Martino S (2004) First evidence of Tyre debris characterization at the nanoscale by focused ion beam. Mater Charact 52:283–288
51. Eisentraut P, Dümichen E, Ruhl AS, Jekel M, Albrecht M, Gehde M, Braun U (2018) Two birds with one stone—fast and simultaneous analysis of microplastics: microparticles derived from thermoplastics and tire wear. Environ Sci Technol Lett 5:608–613
52. Klöckner P, Seiwert B, Eisentraut P, Braun U, Reemtsma T, Wagner S (2020) Characterization of tire and road wear particles from road runoff indicates highly dynamic particle properties. Water Res 185:116262
53. Klöckner P, Seiwert B, Weyrauch S, Escher BI, Reemtsma T, Wagner S (2021) Comprehensive characterization of tire and road wear particles in highway tunnel road dust by use of size and density fractionation. Chemosphere 279:130530
54. Kovochich M, Parker JA, Oh SC, Lee JP, Wagner S, Reemtsma T, Unice KM (2021) Characterization of individual tire and road Wear particles in environmental road dust, tunnel dust, and sediment. Environ Sci Technol Lett 8:1057–1064
55. Cadle SH, Williams RL (1978) Gas and particle emissions from automobile tires in laboratory and field studies. J Air Pollut Control Assoc 28:502–507
56. Pierson WR, Brachaczek WW (1974) Airborne particulate debris from rubber tires. Rubber Chem Technol 47:1275–1299
57. Beji A, Deboudt K, Khardi S, Muresan B, Lumière L (2021) Determinants of rear-of-wheel and tire-road wear particle emissions by light-duty vehicles using on-road and test track experiments. Atmos Pollut Res 12:278–291
58. Park I, Kim H, Lee S (2018) Characteristics of tire wear particles generated in a laboratory simulation of tire/road contact conditions. J Aerosol Sci 124:30–40
59. Mathissen M, Scheer V, Vogt R, Benter T (2011) Investigation on the potential generation of ultrafine particles from the tire–road interface. Atmos Environ 45:6172–6179
60. JATMA (2002) Tire industry of Japan. www.jatma.or.jp
61. Councell TB, Duckenfield KU, Landa ER, Callender E (2004) Tire-wear particles as a source of zinc to the environment. Environ Sci Technol 38:4206–4214
62. Hjortenkrans DST, Bergbäck BG, Häggerud AV (2007) Metal emissions from brake linings and tires: case studies of Stockholm, Sweden 1995/1998 and 2005. Environ Sci Technol 41: 5224–5230
63. Harrison RM, Jones AM, Gietl J, Yin J, Green DC (2012) Estimation of the contributions of brake dust, tire wear, and resuspension to nonexhaust traffic particles derived from atmospheric measurements. Environ Sci Technol 46:6523–6529
64. Avagyan R, Sadiktsis I, Bergvall C, Westerholm R (2014) Tire tread wear particles in ambient air—a previously unknown source of human exposure to the biocide 2-mercaptobenzothiazole. Environ Sci Pollut Res 21:11580–11586

65. Fauser P, Tjell JC, Mosbaek H, Pilegaard K (2002) Tire-tread and bitumen particle concentration in aerosol and soil samples. Pet Sci Technol 20:127–141
66. Kloepfer A, Jekel M, Reemtsma T (2005) Occurrence, sources, and fate of benzothiazoles in municipal wastewater treatment plants. Environ Sci Technol 39:3792–3798
67. Knight LJ, Parker-Jurd FNF, Al-Sid-Cheikh M, Thompson RC (2020) Tyre wear particles: an abundant yet widely unreported microplastic? Environ Sci Pollut Res 27:18345–18354
68. Kumata H, Sanada Y, Takada H, Ueno T (2000) Historical trends of N-cyclohexyl-2-benzothiazolamine, 2-(4-morpholinyl)benzothiazole, and other anthropogenic contaminants in the urban reservoir sediment core. Environ Sci Technol 34:246–253
69. Kumata H, Yamada J, Masuda K, Takada H, Sato Y, Sakurai T, Fujiwara K (2002) Benzothiazolamines as Tire-derived molecular markers: Sorptive behavior in street runoff and application to source apportioning. Environ Sci Technol 36:702–708
70. Panko JM, Hitchcock KM, Fuller GW, Green D (2019) Evaluation of tire wear contribution to PM2.5 in urban environments. Atmos 10:99
71. Piscitello A, Bianco C, Casasso A, Sethi R (2021) Non-exhaust traffic emissions: sources, characterization, and mitigation measures. Sci Total Environ 766:144440
72. Dalmau E, Augsburg K, Wenzel F, Ivanov V (2020) Tire particle emissions: demand on reliable characterization. Tire Sci Technol TSTCA 48:107–122
73. Aatmeeyata, Sharma M (2010) Polycyclic aromatic hydrocarbons, elemental and organic carbon emissions from tire-wear. Sci Total Environ 408:4563–4568
74. Alves CA, Vicente AMP, Calvo AI, Baumgardner D, Amato F, Querol X, Pio C, Gustafsson M (2020) Physical and chemical properties of non-exhaust particles generated from wear between pavements and tyres. Atmos Environ 224:117252
75. Boonyatumanond R, Murakami M, Wattayakorn G, Togo A, Takada H (2007) Sources of polycyclic aromatic hydrocarbons (PAHs) in street dust in a tropical Asian mega-city, Bangkok, Thailand. Sci Total Environ 384:420–432
76. Hwang H-M, Fiala MJ, Wade TL, Park D (2019) Review of pollutants in urban road dust: Part II. Organic contaminants from vehicles and road management. Int J Urban Sci 23:445–463
77. Pengchai P, Nakajima F, Furumai H (2005) Estimation of origins of polycyclic aromatic hydrocarbons in size-fractionated road dust in Tokyo with multivariate analysis. Water Sci Technol 51:169–175
78. Parker-Jurd FNF, Napper IE, Abbott GD, Hann S, Thompson RC (2021) Quantifying the release of tyre wear particles to the marine environment via multiple pathways. Mar Pollut Bull 172:112897
79. Sundt P, Syversen F, Skogesal O, Schulze P-E (2016) Primary microplastic pollution: measures and reduction potentials in Norway. Mepex Consult, for the Norwegian Environment Agency, Asker. https://www.miljodirektoratet.no/globalassets/publikasjoner/M545/M545.pdf
80. Wik A, Dave G (2009) Occurrence and effects of tire wear particles in the environment – a critical review and an initial risk assessment. Environ Pollut 157:1–11
81. Boxall ABA, Maltby L (1997) The effects of motorway runoff on freshwater ecosystems: 3. Toxicant confirmation. Arch Environ Contam Toxicol 33:9–16
82. Maltby L, Boxall ABA, Forrow DM, Calow P, Betton CI (1995) The effects of motorway runoff on freshwater ecosystems: 2. Identifying major toxicants. Environ Toxicol Chem 14: 1093–1101
83. Maltby L, Forrow DM, Boxall ABA, Calow P, Betton CI (1995) The effects of motorway runoff on freshwater ecosystems: 1. Field study. Environ Toxicol Chem 14:1079–1092
84. Marsalek J, Rochfort Q, Brownlee B, Mayer T, Servos M (1999) An exploratory study of urban runoff toxicity. Water Sci Technol 39:33–39
85. Schiff K, Bay S, Diehl D (2003) Stormwater toxicity in Chollas Creek and San Diego Bay, California. Environ Monit Assess 81:119–132
86. Sundt P, Schulze P-E, Syversen F (2014) Sources of microplastic-pollution to the marine environment. Mepex Consult, for the Norwegian Environment Agency, Asker. https://www.miljodirektoratet.no/globalassets/publikasjoner/m321/m321.pdf

87. Unice KM, Weeber MP, Abramson MM, Reid RCD, van Gils JAG, Markus AA, Vethaak AD, Panko JM (2019) Characterizing export of land-based microplastics to the estuary – part I: application of integrated geospatial microplastic transport models to assess tire and road wear particles in the Seine watershed. Sci Total Environ 646:1639–1649
88. Halle LL, Palmqvist A, Kampmann K, Khan FR (2020) Ecotoxicology of micronized tire rubber: past, present and future considerations. Sci Total Environ 706:135694
89. Luo Z, Zhou X, Su Y, Wang H, Yu R, Zhou S, Xu EG, Xing B (2021) Environmental occurrence, fate, impact, and potential solution of tire microplastics: similarities and differences with tire wear particles. Sci Total Environ 795:148902
90. Svensson N, Andersson-Sköld Y (2021) Dispersion and fate models for microplastics from tyre and road wear. VTI rapport 1061A (vti.se/publications). Swedish National Road and Transport Research Institute
91. Saito T (1989) Determination of styrene-butadiene and isoprene tire tread rubbers in piled particulate matter. J Anal Appl Pyrolysis 15:227–235
92. Goßmann I, Halbach M, Scholz-Böttcher BM (2021) Car and truck tire wear particles in complex environmental samples – a quantitative comparison with "traditional" microplastic polymer mass loads. Sci Total Environ 773:145667
93. Youn J-S, Kim Y-M, Siddiqui MZ, Watanabe A, Han S, Jeong S, Jung Y-W, Jeon K-J (2021) Quantification of tire wear particles in road dust from industrial and residential areas in Seoul, Korea. Sci Total Environ 784:147177
94. Boulter P (2005) A review of emission factors and models for road vehicle non-exhaust particulate matter. Transport Research Laboratory
95. Gnecco I, Berretta C, Lanza LG, Barbera PL (2005) Storm water pollution in the urban environment of Genoa, Italy. Atmos Res 77:60–73
96. Järlskog I, Strömvall A-M, Magnusson K, Galfi H, Björklund K, Polukarova M, Garção R, Markiewicz A, Aronsson M, Gustafsson M, Norin M, Blom L, Andersson-Sköld Y (2021) Traffic-related microplastic particles, metals, and organic pollutants in an urban area under reconstruction. Sci Total Environ 774:145503
97. Menzie CA, Hoeppner SS, Cura JJ, Freshman JS, LaFrey EN (2002) Urban and suburban storm water runoff as a source of polycyclic aromatic hydrocarbons (PAHs) to Massachusetts estuarine and coastal environments. Estuaries 25:165–176
98. Oros DR, Ross JRM, Spies RB, Mumley T (2007) Polycyclic aromatic hydrocarbon (PAH) contamination in San Francisco Bay: a 10-year retrospective of monitoring in an urbanized estuary. Environ Res 105:101–118
99. Bukowiecki N, Lienemann P, Hill M, Furger M, Richard A, Amato F, Prévôt ASH, Baltensperger U, Buchmann B, Gehrig R (2010) PM10 emission factors for non-exhaust particles generated by road traffic in an urban street canyon and along a freeway in Switzerland. Atmos Environ 44:2330–2340
100. Lenschow P, Abraham HJ, Kutzner K, Lutz M, Preuß JD, Reichenbächer W (2001) Some ideas about the sources of PM10. Atmos Environ 35:S23–S33
101. Rexeis M, Hausberger S (2009) Trend of vehicle emission levels until 2020 – prognosis based on current vehicle measurements and future emission legislation. Atmos Environ 43:4689–4698
102. Amato F, Dimitropoulos A, Farrow K, Oueslati W (2020) Non-exhaust particulate emissions from road transport. OECD, Paris. https://www.oecd.org/environment/non-exhaust-particulate-emissions-from-road-transport-4a4dc6ca-en.htm
103. Jeong C-H, Traub A, Huang A, Hilker N, Wang JM, Herod D, Dabek-Zlotorzynska E, Celo V, Evans GJ (2020) Long-term analysis of PM2.5 from 2004 to 2017 in Toronto: composition, sources, and oxidative potential. Environ Pollut 263:114652
104. Baensch-Baltruschat B, Kocher B, Kochleus C, Stock F, Reifferscheid G (2021) Tyre and road wear particles – a calculation of generation, transport and release to water and soil with special regard to German roads. Sci Total Environ 752:141939

105. Panko JM, Chu J, Kreider ML, Unice KM (2013) Measurement of airborne concentrations of tire and road wear particles in urban and rural areas of France, Japan, and the United States. Atmos Environ 72:192–199
106. Rausch J, Jaramillo-Vogel D, Perseguers S, Schnidrig N, Grobéty B, Yajan P (2022) Automated identification and quantification of tire wear particles (TWP) in airborne dust: SEM/EDX single particle analysis coupled to a machine learning classifier. Sci Total Environ 803:149832
107. Loganathan P, Vigneswaran S, Kandasamy J (2013) Road-deposited sediment pollutants: a critical review of their characteristics, source apportionment, and management. Crit Rev Environ Sci Technol 43:1315–1348
108. Mengistu D, Heistad A, Coutris C (2021) Tire wear particles concentrations in gully pot sediments. Sci Total Environ 769:144785
109. Blok J (2005) Environmental exposure of road borders to zinc. Sci Total Environ 348:173–190
110. Werkenthin M, Kluge B, Wessolek G (2016) Assessment of metal retention in newly constructed highway embankments. Environ Sci Pollut Res 23:23619–23629
111. Litschke T, Kuttler W (2008) On the reduction of urban particle concentration by vegetation – a review. Meteorol Z 17:229–240
112. Wolch JR, Byrne J, Newell JP (2014) Urban green space, public health, and environmental justice: the challenge of making cities 'just green enough'. Landsc Urban Plan 125:234–244
113. Escobedo FJ, Kroeger T, Wagner JE (2011) Urban forests and pollution mitigation: Analyzing ecosystem services and disservices. Environ Pollut 159:2078–2087
114. Tiwary A, Sinnett D, Peachey C, Chalabi Z, Vardoulakis S, Fletcher T, Leonardi G, Grundy C, Azapagic A, Hutchings TR (2009) An integrated tool to assess the role of new planting in PM10 capture and the human health benefits: a case study in London. Environ Pollut 157: 2645–2653
115. Hinds WC (1999) Aerosol technology: properties, behavior, and measurement of airborne particles.2nd edn. Wiley, p 504
116. Gray AD, Wertz H, Leads RR, Weinstein JE (2018) Microplastic in two South Carolina estuaries: occurrence, distribution, and composition. Mar Pollut Bull 128:223–233
117. Leads RR, Weinstein JE (2019) Occurrence of tire wear particles and other microplastics within the tributaries of the Charleston Harbor Estuary, South Carolina, USA. Mar Pollut Bull 145:569–582
118. Johannessen C, Helm P, Metcalfe CD (2021) Detection of selected tire wear compounds in urban receiving waters. Environ Pollut 287:117659
119. Degaffe FS, Turner A (2011) Leaching of zinc from tire wear particles under simulated estuarine conditions. Chemosphere 85:738–743
120. Wang H, Nie L, Xu Y, Du C, Zhang T, Wang Y (2018) Effects of highway-related pollutant on the groundwater quality of turfy swamps in the Changbai Mountain area. Int J Environ Res Public Health 15:1652
121. Xu B, Liu F, Cryder Z, Huang D, Lu Z, He Y, Wang H, Lu Z, Brookes PC, Tang C, Gan J, Xu J (2020) Microplastics in the soil environment: occurrence, risks, interactions and fate – a review. Crit Rev Environ Sci Technol 50:2175–2222
122. Qi R, Jones DL, Li Z, Liu Q, Yan C (2020) Behavior of microplastics and plastic film residues in the soil environment: a critical review. Sci Total Environ 703:134722
123. Eunomia. (2016) Plastics in the marine environment. Eunomia, Bristol. https://www.eunomia.co.uk/reports-tools/plastics-in-the-marine-environment/
124. Brunekreef B, Frosberg B (2005) Epidemiological evidence of effects of coarse airborne particles on health. Eur Respir J 26:309–318
125. Evans JJ (1997) Rubber tire leachates in the aquatic environment. Rev Environ Contam Toxicol 151:67
126. Halsband C, Sørensen L, Booth AM, Herzke D (2020) Car tire crumb rubber: does leaching produce a toxic chemical cocktail in coastal marine systems? Front Environ Sci 8:125

127. Lu F, Su Y, Ji Y, Ji R (2021) Release of zinc and polycyclic aromatic hydrocarbons from tire crumb rubber and toxicity of leachate to Daphnia magna: effects of tire source and photoaging. Bull Environ Contam Toxicol 107:651–656
128. Tamis JE, Koelmans AA, Dröge R, Kaag NHBM, Keur MC, Tromp PC, Jongbloed RH (2021) Environmental risks of car tire microplastic particles and other road runoff pollutants. Microplast Nanoplast 1:10
129. Halle LL, Palmqvist A, Kampmann K, Jensen A, Hansen T, Khan FR (2021) Tire wear particle and leachate exposures from a pristine and road-worn tire to Hyalella azteca: comparison of chemical content and biological effects. Aquat Toxicol 232:105769
130. Leads RR, Burnett KG, Weinstein JE (2019) The effect of microplastic ingestion on survival of the grass shrimp Palaemonetes pugio (Holthuis, 1949) challenged with Vibrio campbellii. Environ Toxicol Chem 38:2233–2242
131. Parker BW, Beckingham BA, Ingram BC, Ballenger JC, Weinstein JE, Sancho G (2020) Microplastic and tire wear particle occurrence in fishes from an urban estuary: influence of feeding characteristics on exposure risk. Mar Pollut Bull 160:111539
132. Apeagyei E, Bank MS, Spengler JD (2011) Distribution of heavy metals in road dust along an urban-rural gradient in Massachusetts. Atmos Environ 45:2310–2323
133. Davis AP, Shokouhian M, Ni S (2001) Loading estimates of lead, copper, cadmium, and zinc in urban runoff from specific sources. Chemosphere 44:997–1009
134. Fergusson JE, Kim ND (1991) Trace elements in street and house dusts: sources and speciation. Sci Total Environ 100:125–150
135. Rogge WF, Hildemann LM, Mazurek MA, Cass GR, Simoneit BRT (1993) Sources of fine organic aerosol. 3. Road dust, tire debris, and organometallic brake lining dust: roads as sources and sinks. Environ Sci Technol 27:1892–1904
136. Smolders E, Degryse F (2002) Fate and effect of zinc from tire debris in soil. Environ Sci Technol 36:3706–3710
137. Zafra CA, Temprano J, Tejero I (2011) Distribution of the concentration of heavy metals associated with the sediment particles accumulated on road surfaces. Environ Technol 32:997–1008
138. CCME (1999) Canadian sediment quality guidelines for the protection of aquatic life. Canadian Council of Ministers of the Environment, Winnipeg. https://ccme.ca/en/res/zinc-canadian-sediment-quality-guidelines-for-the-protection-of-aquatic-life-en.pdf
139. Redondo-Hasselerharm PE, de Ruijter VN, Mintenig SM, Verschoor A, Koelmans AA (2018) Ingestion and chronic effects of car Tire tread particles on freshwater benthic macroinvertebrates. Environ Sci Technol 52:13986–13994
140. IARC (2018) 1,3-Butadiene. International Agency for Research on Cancer. https://monographs.iarc.who.int/wp-content/uploads/2018/06/mono100F-26.pdf
141. EPA (2020) Urban air toxic pollutants. U.S. Environmental Protection Agency. https://www.epa.gov/urban-air-toxics/urban-air-toxic-pollutants
142. Ni H-G, Lu F-H, Luo X-L, Tian H-Y, Zeng EY (2008) Occurrence, phase distribution, and mass loadings of benzothiazoles in riverine runoff of the Pearl River Delta, China. Environ Sci Technol 42:1892–1897
143. Seiwert B, Klöckner P, Wagner S, Reemtsma T (2020) Source-related smart suspect screening in the aqueous environment: search for tire-derived persistent and mobile trace organic contaminants in surface waters. Anal Bioanal Chem 412:4909–4919
144. Tian Z, Zhao H, Peter KT, Gonzalez M, Wetzel J, Wu C, Hu X, Prat J, Mudrock E, Hettinger R, Cortina AE, Biswas RG, Kock FVC, Soong R, Jenne A, Du B, Hou F, He H, Lundeen R, Gilbreath A, Sutton R, Scholz NL, Davis JW, Dodd MC, Simpson A, McIntyre JK, Kolodziej EP (2021) A ubiquitous tire rubber-derived chemical induces acute mortality in coho salmon. Science 371:185–189
145. Honda M, Suzuki N (2020) Toxicities of polycyclic aromatic hydrocarbons for aquatic animals. Int J Environ Res Public Health 17:1363

146. Lau S-L, Stenstrom MK (2005) Metals and PAHs adsorbed to street particles. Water Res 39: 4083–4092
147. Murakami M, Nakajima F, Furumai H (2004) Modelling of runoff behaviour of particle-bound polycyclic aromatic hydrocarbons (PAHs) from roads and roofs. Water Res 38:4475–4483
148. Zhao H, Yin C, Chen M, Wang W, Chris J, Shan B (2009) Size distribution and diffuse pollution impacts of PAHs in street dust in urban streams in the Yangtze River Delta. J Environ Sci 21:162–167
149. Aryal R, Furumai H, Nakajima F, Beecham S (2013) Variation in PAH patterns in road runoff. Water Sci Technol 67:2699–2705
150. Takada H, Onda T, Harada M, Ogura N (1991) Distribution and sources of polycyclic aromatic hydrocarbons (PAHs) in street dust from the Tokyo Metropolitan area. Sci Total Environ 107: 45–69
151. Zhang W, Ye Y, Tong Y, Ou L, Hu D, Wang X (2011) Modeling time-dependent toxicity to aquatic organisms from pulsed exposure of PAHs in urban road runoff. Environ Pollut 159: 503–508
152. Zheng Y, Lin Z, Li H, Ge Y, Zhang W, Ye Y, Wang X (2014) Assessing the polycyclic aromatic hydrocarbon (PAH) pollution of urban stormwater runoff: a dynamic modeling approach. Sci Total Environ 481:554–563
153. Capolupo M, Sørensen L, Jayasena KDR, Booth AM, Fabbri E (2020) Chemical composition and ecotoxicity of plastic and car tire rubber leachates to aquatic organisms. Water Res 169: 115270
154. Wik A, Dave G (2006) Acute toxicity of leachates of tire wear material to Daphnia magna—variability and toxic components. Chemosphere 64:1777–1784
155. Chibwe L, Parrott JL, Shires K, Khan H, Clarence S, Lavalle C, Sullivan C, O'Brien AM, De Silva AO, Muir DCG, Rochman CM (2021) A deep dive into the complex chemical mixture and toxicity of tire wear particle leachate in fathead minnow. Environ Toxicol Chem. https://doi.org/10.1002/etc.5140
156. Gualtieri M, Andrioletti M, Mantecca P, Vismara C, Camatini M (2005) Impact of tire debris on in vitro and in vivo systems. Part Fibre Toxicol 2:1
157. Gualtieri M, Andrioletti M, Vismara C, Milani M, Camatini M (2005) Toxicity of tire debris leachates. Environ Int 31:723–730
158. Turner A, Rice L (2010) Toxicity of tire wear particle leachate to the marine macroalga, Ulva lactuca. Environ Pollut 158:3650–3654
159. Wagner M, Lambert S (2018) Freshwater microplastics: emerging environmental contaminants? Springer Nature
160. Wik A, Nilsson E, Källqvist T, Tobiesen A, Dave G (2009) Toxicity assessment of sequential leachates of tire powder using a battery of toxicity tests and toxicity identification evaluations. Chemosphere 77:922–927
161. Panko JM, Kreider ML, McAtee BL, Marwood C (2013) Chronic toxicity of tire and road wear particles to water- and sediment-dwelling organisms. Ecotoxicology 22:13–21
162. Khan FR, Halle LL, Palmqvist A (2019) Acute and long-term toxicity of micronized car tire wear particles to Hyalella azteca. Aquat Toxicol 213:105216
163. Šourková M, Adamcová D, Vaverková MD (2021) The influence of microplastics from ground tyres on the acute, subchronical toxicity and microbial respiration of soil. Environments 8:128
164. Sheng Y, Liu Y, Wang K, Cizdziel JV, Wu Y, Zhou Y (2021) Ecotoxicological effects of micronized car tire wear particles and their heavy metals on the earthworm (Eisenia fetida) in soil. Sci Total Environ 793:148613
165. Gualtieri M, Mantecca P, Cetta F, Camatini M (2008) Organic compounds in tire particle induce reactive oxygen species and heat-shock proteins in the human alveolar cell line A549. Environ Int 34:437–442

166. Mantecca P, Farina F, Moschini E, Gallinotti D, Gualtieri M, Rohr A, Sancini G, Palestini P, Camatini M (2010) Comparative acute lung inflammation induced by atmospheric PM and size-fractionated tire particles. Toxicol Lett 198:244–254
167. Mantecca P, Sancini G, Moschini E, Farina F, Gualtieri M, Rohr A, Miserocchi G, Palestini P, Camatini M (2009) Lung toxicity induced by intratracheal instillation of size-fractionated tire particles. Toxicol Lett 189:206–214
168. Gottipolu RR, Landa ER, Schladweiler MC, McGee JK, Ledbetter AD, Richards JH, Wallenborn GJ, Kodavanti UP (2008) Cardiopulmonary responses of intratracheally instilled tire particles and constituent metal components. Inhal Toxicol 20:473–484
169. Kreider ML, Doyle-Eisele M, Russell RG, McDonald JD, Panko JM (2012) Evaluation of potential for toxicity from subacute inhalation of tire and road wear particles in rats. Inhal Toxicol 24:907–917
170. Kreider ML, Unice KM, Panko JM (2020) Human health risk assessment of tire and road wear particles (TRWP) in air. Hum Ecol Risk Assess Int J 26:2567–2585
171. Araujo-Morera J, Hernández Santana M, Verdejo R, López-Manchado MA (2019) Giving a second opportunity to tire waste: an alternative path for the development of sustainable self-healing styrene–butadiene rubber compounds overcoming the magic triangle of tires. Polymers 11:2122
172. Das A, Sallat A, Böhme F, Suckow M, Basu D, Wießner S, Stöckelhuber KW, Voit B, Heinrich G (2015) Ionic modification turns commercial rubber into a self-healing material. ACS Appl Mater Interfaces 7:20623–20630
173. Le HH, Böhme F, Sallat A, Wießner S, auf der Landwehr M, Reuter U, Stöckelhuber K-W, Heinrich G, Radusch H-J, Das A (2017) Triggering the self-healing properties of modified bromobutyl rubber by intrinsically electrical heating. Macromol Mater Eng 302:1600385
174. Le HH, Hait S, Das A, Wiessner S, Stoeckelhuber KW, Boehme F, Uta R, Naskar K, Heinrich G, Radusch HJ (2017) Self-healing properties of carbon nanotube filled natural rubber/bromobutyl rubber blends. Express Polym Lett 11:230–242
175. Dong J, Huang H, Pei J, Xu Y, Cao J (2021) A methodology for capturing tire wear particles: computational particle fluid dynamics modelling and experimental verification. Powder Technol 384:176–185
176. Ayar M, Dalkiran A, Kale U, Nagy A, Karakoc TH (2021) Investigation of the substitutability of rubber compounds with environmentally friendly materials. Sustainability 13:5251
177. Klüppel M (2021) Wear and abrasion of tires. In: Kobayashi S, Müllen K (eds) Encyclopedia of polymeric nanomaterials. Springer, Berlin
178. ADAC (2022) Dem Mikroplastik auf der Spur: Weniger Reifenabrieb ist möglich. ADAC, Germany. https://www.adac.de/rund-ums-fahrzeug/ausstattung-technik-zubehoer/reifen/reifenkauf/reifenabrieb-mikroplastik; https://assets.adac.de/image/upload/v1639663105/ADAC-eV/KOR/Text/PDF/Tyre_wear_particles_in_the_environment_zkmd3a.pdf, https://assets.adac.de/image/upload/v1639663105/ADAC-eV/KOR/Text/PDF/Tyre_wear_particles_in_the_environment_zkmd1639663103a.pdf
179. Ilić I, Orešković M, Gavran D, Pančić I (2017) The use of porous asphalt for the improvement of the grading plan geometry and drainage of pavement surfaces on urban roads. In: Dell'Acqua G, Wegman F (eds) Transport infrastructure and systems. CRC Press, London, pp 443–450
180. Liebmann B (2015) Mikroplastik in der Umwelt, report REP-0550. Umweltbundesamt, Vienna
181. Querol X, Amato F, Robusté F, Holman C, Harrison RM (2018) Chapter 11 – non-technological measures on road traffic to abate urban air pollution. In: Amato F (ed) Non-exhaust emissions. Academic Press, pp 229–260
182. Smyth K, Drake J, Li Y, Rochman C, Van Seters T, Passeport E (2021) Bioretention cells remove microplastics from urban stormwater. Water Res 191:116785

Adv Polym Sci (2023) 289: 103–120
https://doi.org/10.1007/12_2022_120

Published online: 6 May 2022

Multiphysical Modeling and Simulation of Thermal Damage of Elastomers: State of the Art and Developments Towards Cyber-Physical Systems

R. Behnke and M. Kaliske

Contents

Abstract While different subtopics of degradation of rubber under strain, temperature, wear, tire-road abrasion or fatigue are discussed in detail from an experimental and numerical point of view in other chapters, this chapter mainly aims at the numerical aspects of the model setup from the material to the structural scale. The objective of this chapter is to establish the link between experimental testing on the material scale and predictive simulation on the structural scale. To demonstrate the general procedure in the context of the finite element method (FEM), thermal damage or thermal aging as one phenomenon of material alteration over time is addressed by developing a corresponding continuum mechanical model. The phenomenological model is derived at the material scale and integrated at the structural scale by solving the thermo-mechanically coupled problem for the example of a rubber-cord structure in the form of an automobile tire. With the structural model at hand, simulations of the entire lifetime become feasible. Future trends in this context are introduced by focusing on the further extension of the pure simulative approach to cyber-physical systems (CPS), e.g. for the holistic assessment of the impact of tire

R. Behnke and M. Kaliske (✉)
Institut für Statik und Dynamik der Tragwerke, TU Dresden, Fakultät Bauingenieurwesen, Dresden, Germany
e-mail: ronny.behnke@tu-dresden.de; michael.kaliske@tu-dresden.de

wear particles (TWP) on the environment (definition of system boundaries, material flows, and interactions).

Keywords Continuum mechanics · Cyber-physical system · Finite element method · Thermal aging · Thermal damage · Tire

1 Introduction

With steadily growing computational capacities, solutions of arbitrary, coupled boundary value problems in mechanics have been made accessible and allow a more and more detailed numerical study of the related processes at different length and time scales. The successful numerical representation of the system under investigation is based on the correct formulation of the underlying phenomena and evolution processes in terms of several (sub)models, their coupling and integration on the level of the discretized system (in space and time) and efficient numerical solution methods to solve the resulting system of equations. This mathematical and computational assembling requires expert knowledge from different domains and research areas covering experimental testing and measurement (observation and quantification of the phenomena), computational mechanics (setup of corresponding numerical models), and mathematics (solution schemes).

Nowadays, we are facing the challenge that several well-developed submodels of complex phenomena as well as huge data sets from material characterization are available, but a manual and separate evaluation seems not meaningful to reveal the added value of their holistic exploration. In addition to the afore-mentioned disciplines (metrology, continuum mechanics, mathematics), the means of informatics enable to establish a new digital link between real objects, numerical models, and information (data). Resulting so-called cyber-physical systems (CPS) provide an additional value (insight into relations, extraction of correlated information, control, and optimization) by linking already available system parts and information entities. However, their reliability and accuracy will strongly depend on the underlying components in terms of developed submodels, data quality, and established cyber-physical links (infrastructure).

In the following, aspects of modeling (physical models) and simulating (underlying numerical solution techniques) rubber components at different length and time scales are discussed in the framework of the finite element method (FEM). As an exemplary case, the numerical representation of thermal damage of vulcanized rubber is addressed by focusing on the displacement and temperature field in a coupled manner. As structural example of an industrial rubber component, a cord-rubber structure in the form of an automobile tire is selected and will serve as illustrative example in the following to illustrate potential development steps towards a CPS.

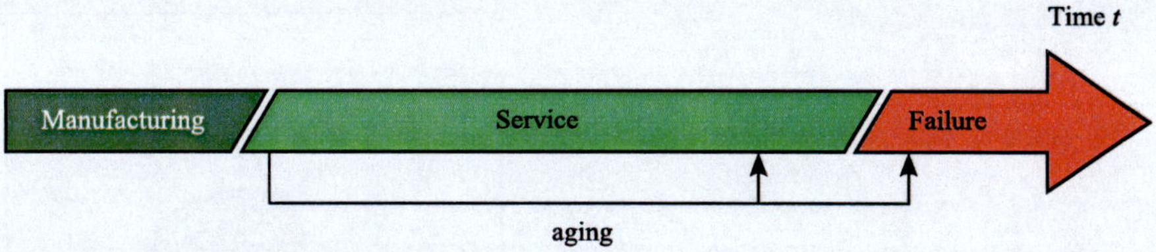

Fig. 1 Phases of the life cycle of a rubber product [3]

In the industrial context, lifetime predictions play an important role in guaranteeing safety and serviceability of rubber products [1, 2]. Covering, analyzing, and optimizing the whole service life (Fig. 1) of such a product involves more and more the digital representation of the whole life cycle starting from manufacturing, service to failure. Here, it becomes obvious that the different phases and states of a rubber product undergoing this life cycle might not yet be captured within one physical model. Nowadays, different submodels for all of these states are already available, but they are of heterogeneous nature from an informational and physical point of view. Interfaces or links between these submodels rarely exist in such a manner that an automated exchange of data and simulation results is possible. The state of the art is that data and partial results of the simulation are mainly transferred manually from one submodel to the other by manually adapting the solution domains and their discretization. The same holds for interfaces between numerical simulation tools and testing facilities, where the planning of tests and the exploration of experimental results on the material and structural scale are still governed by human interaction.

2 Modeling of Thermal Damage

The holistic analysis and exploration of the structural behavior of car tires is of great importance due to safety, driving behavior, and comfort aspects related to their use in different operating situations. Furthermore, optimal design and service have a significant impact on their durability as well as environmental aspects in terms of their environmental sustainability (interactions with the environment). The latter relates not only to the resources and energy used in production, but also to the drive energy required and wear products, e.g. tire wear particles (TWP), during the service life. In turn, it becomes obvious that optimizing one aspect, e.g. minimizing TWP, affects all related system components such as the choice of initial raw materials, curing conditions during manufacturing or operational parameters of the tire in service.

In Fig. 2, a classification of different length scales of an automobile tire is depicted. On the mesoscale, the tire might show finite defects, which are already present since the manufacturing or evolve during the service life.

At the microscale, the molecular components of the rubber matrix are composed of polymer chains, filler particles, and sulfur bridges (vulcanized state of the rubber).

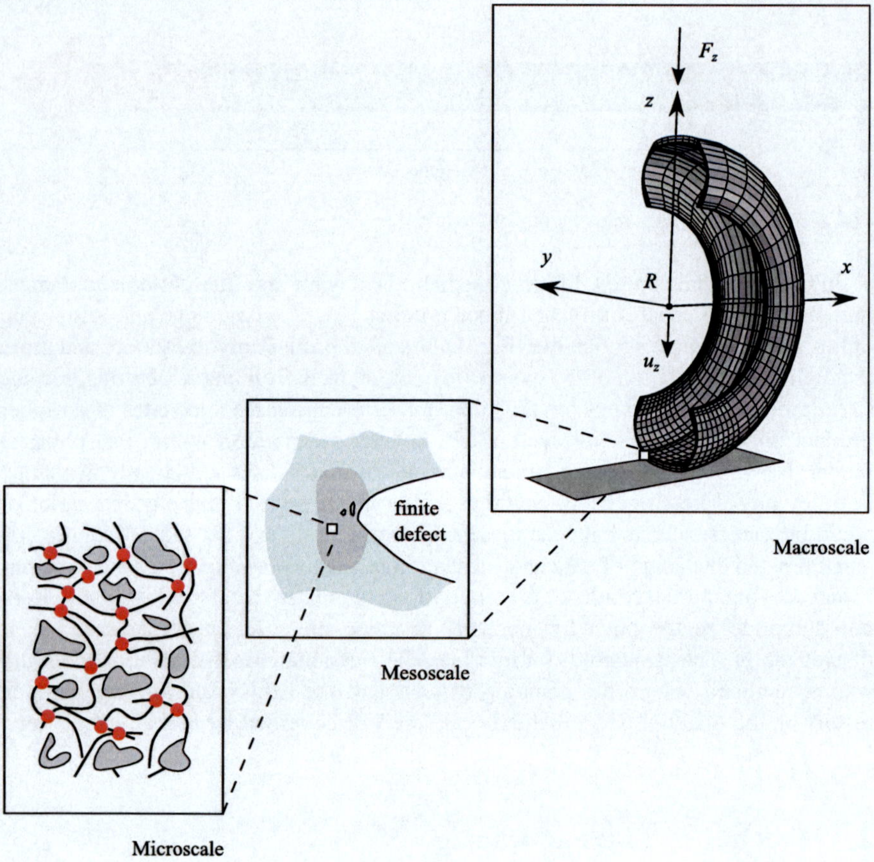

Fig. 2 Automobile tire resolved at different length scales

The network of polymer chains is also subjected to alterations at the microscale during operation of the tire, which will finally influence the characteristics of the composite on the meso- (crack propagation characteristics) and macroscale (driving behavior, tire mass loss due to tire wear).

While a numerical representation of the rubber-cord composite over all the length scales seems promising with respect to highlighting the relations and coupled phenomena, such an ansatz is not yet state of the art. In other words, the prediction of structural features of the tire purely by the chemical formulation of its ingredients and geometrical dimensions is not yet feasible. However, homogenization techniques of different form have already been used to bridge scales and to transfer information from one scale to the other one in terms of mechanical properties, alteration of the material or failure. Hence, several submodels are already available at these scales. However, their automated coupling via interfaces is not yet well established. An overview of different length and time scales is provided in Fig. 3.

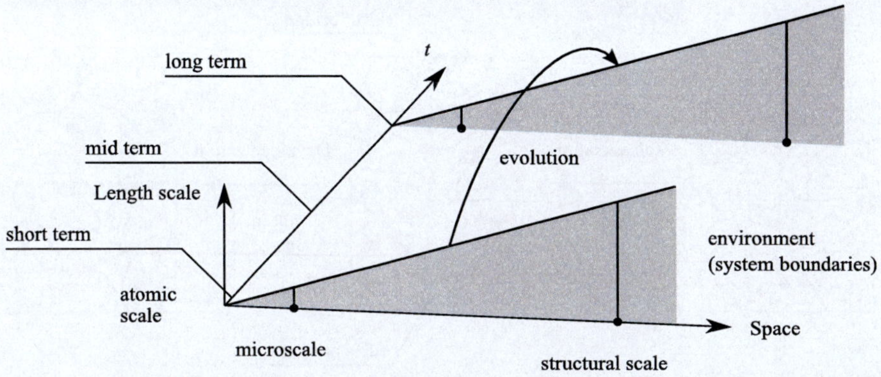

Fig. 3 Evolution of a system: length scales and time scales to be covered by numerical submodels [3]

As a first approach towards the representation of the phases of the life cycle of rubber products, the representation of the molding and vulcanization of the rubber matrix [4, 5] and the interaction with the cords [6, 7] during curing (temperature rise) seem promising to bridge the gap between unvulcanized and vulcanized rubber material within the numerical simulation.

During the vulcanization process, sulfur bridges are created as part of the cross-linked polymer network, which is formed at the higher vulcanization temperature over the duration of an optimum vulcanization time. While a moderate temperature increase under operating conditions leads to a post-vulcanization of the rubber with the creation of further sulfur bridges, the process turns to the destruction of sulfur bridges (sulfur-sulfur (S-S) bonds or carbon-sulfur (C-S) bonds) after exceeding a critical (service) temperature and exposure time. Hence, thermal degradation or thermal damage occurs during the service life and can be generally explained by devulcanization processes [8]. Thermal damage is then linked with a stiffness loss on the material as well as on the structural scale. An illustrative representation of vulcanization and devulcanization processes is given in Fig. 4.

It is worth mentioning that thermal damage occurs in combination with several other phenomena of material alteration during the service life of a tire and, in consequence, thermal damage partially contributes to tire failure [9].

Under operating conditions, the tire is exposed to varying intensities of humidity, salt water, oil, chemical or biological contamination, radiation, and temperature [10]. In consequence, other forms of material alteration (change of properties with increasing time) are chemical aging [11], e.g. by oxygen diffusion [12, 13], or aging by humidity. Finally, material alterations also occur due to mechanical loading, in the case of elastomers often coupled with temperature due to thermo-mechanically coupling effects in case of repetitive loading [14] leading to fatigue [15]. For an overview, the reader is referred, e.g., to [16].

In the following, a numerical representation of thermal damage is discussed as originally published in detail in [17]. This continuum mechanical approach captures

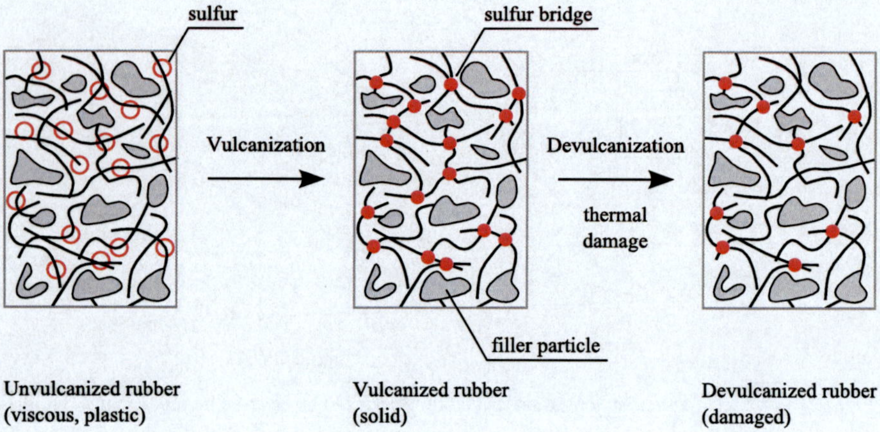

Fig. 4 Vulcanization (manufacturing) and devulcanization (thermal damage during the service life) of the rubber matrix

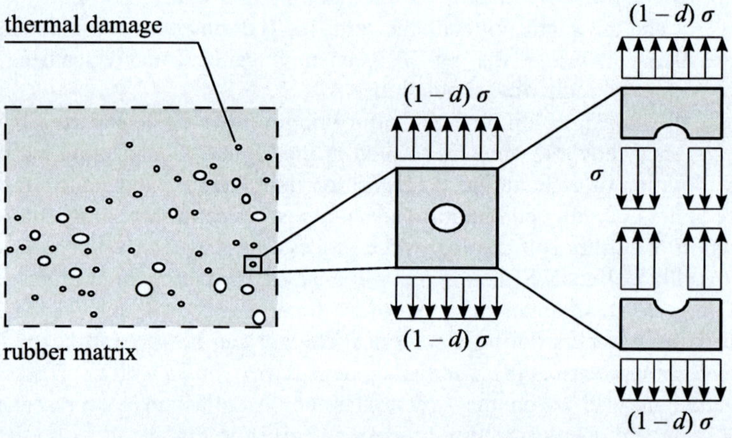

Fig. 5 Thermal damage modeled at the macroscale via a temperature- and time-dependent CDM approach [17]

the material effects of the micro- and mesoscale in a smeared form on the macroscale and, therefore, allows an efficient numerical simulation of larger structures, e.g. a whole tire structure, by avoiding time-consuming homogenization steps between the different length scales. However, the concept behind the phenomenological approach is inspired by the underlying physical processes at these scales. An illustration of the mesodefects, which will increase in number and size during the service life, e.g. due to thermally induced network chain scission, is given in Fig. 5. Here, the effective stress at the macroscale is given by $(1 - d)\,\sigma$, where d is a thermal damage indicator in terms of a scalar damage variable with the property

$$0 \leq d(\Theta, t) \leq 1 \tag{1}$$

and σ stands for the true stress at the mesoscale in the thermally damaged rubber matrix. In case of $d = 0$, no thermal damage is present. In case of $d = 1$, the material has lost its entire stiffness and complete material failure would occur. This concept of continuum damage mechanics (CDM) is well known in the context of stress- or strain-induced damage [18–20], where due to external loading, the material exceeds the critical strain energy above which softening (damage) occurs in continuous or discontinuous form. In case of thermal damage, the damage variable depends not a priori on the stress- or strain field, but on the temperature of the material point considered and the exposure time t.

The evolution of d and, in consequence, the evolution of thermal damage of the rubber matrix are described by the so-called evolution equations prescribing the rate of damage as a function of different influence quantities. In the present case, influence quantities will be the temperature and the damage kinetics (rate of damage increase),

$$\dot{d}(\Theta, t) = \frac{1}{\tau}[d^{\infty}(\Theta) - d(\Theta, t)] \quad \text{if } [d^{\infty}(\Theta) - d(\Theta, t)] \geq 0, \tag{2}$$

$$\dot{d}(\Theta, t) = 0 \quad \text{if } [d^{\infty}(\Theta) - d(\Theta, t)] < 0. \tag{3}$$

The damage kinetics is represented by the material-specific characteristic time τ (in the following example in the range of about 10 h). From Eqs. (2) and (3), it can be observed that thermal damage can only monotonically increase if a difference between a final temperature-dependent long-term damage $d^{\infty}(\Theta)$ and the current damage value $d(\Theta, t)$ is present.

The long-term thermal damage $d^{\infty}(\Theta)$ is an envelope function describing the final thermal damage state of the material, which would be reached if the material is exposed to a constant temperature for an infinite time. The relation between temperature and the long-term damage value can be given in mathematical form via ansatz functions of the form

$$d^{\infty}(\Theta) = \frac{\Theta - \Theta_{\text{crit}}}{\Theta_{\text{max}} - \Theta_{\text{crit}}}, \tag{4}$$

$$d^{\infty}(\Theta) = \sin\left(\frac{\pi}{2}\left[\frac{\Theta - \Theta_{\text{crit}}}{\Theta_{\text{max}} - \Theta_{\text{crit}}}\right]^2\right), \tag{5}$$

$$d^{\infty}(\Theta) = \frac{1}{2}\left[1 + \tanh\left(6\,\frac{\Theta - \Theta_{\text{crit}}}{\Theta_{\text{max}} - \Theta_{\text{crit}}} - 3\right)\right], \tag{6}$$

$$d^{\infty}(\Theta) = 1 - \exp\left(-5\,\frac{\Theta - \Theta_{\text{crit}}}{\Theta_{\text{max}} - \Theta_{\text{crit}}}\right), \tag{7}$$

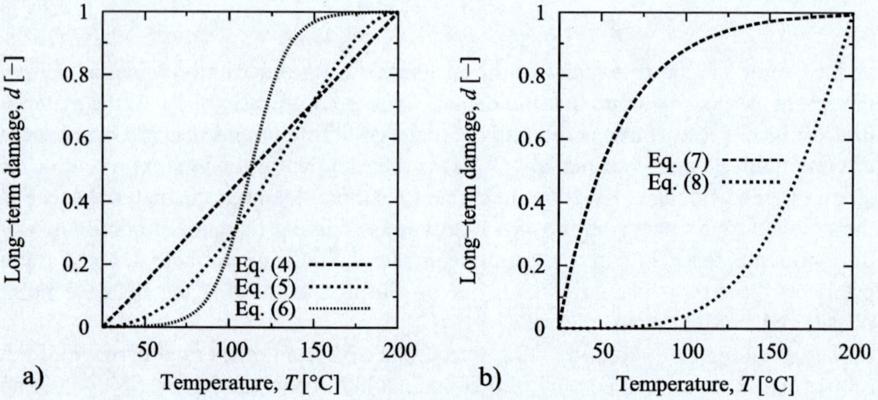

Fig. 6 Long-term thermal damage functions $d^{\infty}(\Theta)$ plotted for the parameters $\Theta_{\text{crit}} = 298$ K, $\Theta_{\max} = 473$ K: (a) Eqs. (4), (5), and (6), (b) Eqs. (7) and (8), see [17]

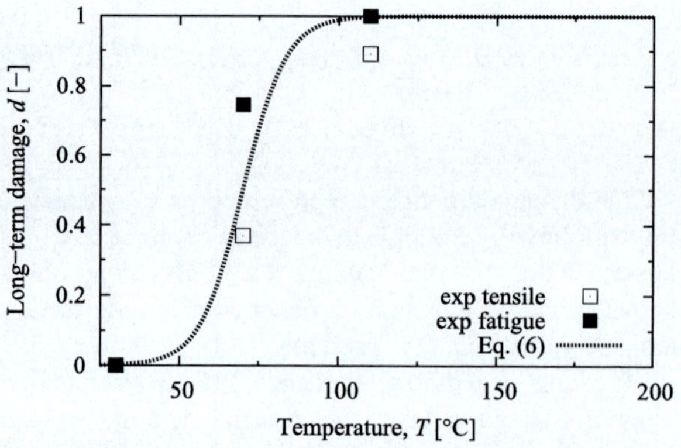

Fig. 7 Identification of the long-term damage envelope function from experimental tensile and fatigue tests on oven-aged rubber specimens [21] as described in detail in [17], $\Theta_{\text{crit}} = 303$ K, $\Theta_{\max} = 383$ K, characteristic time $\tau = 36{,}000$ s

$$d^{\infty}(\Theta) = 1 + \sin\left(\frac{\pi}{2}\left[3 - \left(\frac{\Theta - \Theta_{\text{crit}}}{\Theta_{\max} - \Theta_{\text{crit}}}\right)^2\right]\right), \tag{8}$$

where only five possible function types are exemplarily given, see, e.g., [17]. In Fig. 6, these function types are plotted for a specific set of governing damage parameters Θ_{crit} (critical temperature of thermal damage onset) and $\Theta_{\max}$ (temperature associated with complete thermal damage).

In Fig. 7, the long-term damage function given in Eq. (6) was fitted to experimental results obtained from tests on rubber samples aged at constant temperature in the oven. A detailed description of the control points obtained from the aging experiments is provided in [17].

With the long-term damage function at hand, the time dependence of the thermal damage is added to the evolution equation of d by the characteristic time measure τ, which controls the increase of damage from a current value $d(\Theta, t)$ to the final

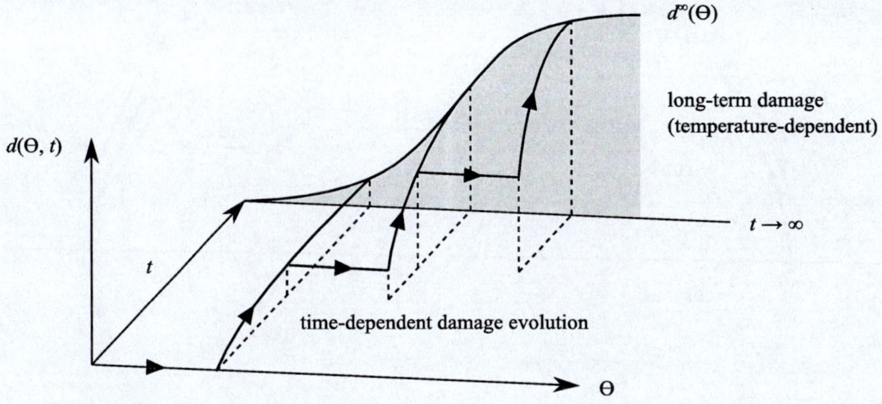

Fig. 8 Evolution of thermal damage for a stepwise increased temperature to the corresponding long-term damage value specified by $d^\infty(\Theta)$ [17]

damage value $d^\infty(\Theta)$ for this temperature. This process is illustrated in Fig. 8 for a stepwise increased temperature of a rubber material undergoing thermal damage over time.

Finally, the evolution equations in the form of Eqs. (2) and (3) are obtained in their strong form at the material scale (material point level). Via the FEM, the coupled temperature-displacement boundary value problem is solved by a space and time discretization of the real object under investigation (automobile tire). The finite element (FE) discretized cross-section of the tire is depicted in Fig. 9. Here, rubber matrix and cords are discretized by FE and a special treatment of the rolling operation of the tire is accomplished by using the so-called arbitrary Lagrangian-Eulerian (ALE) formulation of the problem to describe in a numerically efficient manner the rotation of the tire during rolling at minimal discretization effort. A general description of the ALE method is available in [22].

The computation of thermal damage requires the solution of the balance of energy (temperature field) in combination with the balance of linear and angular momentum (displacement field) since a temperature rise [23] occurs due to dissipation of mechanical input energy during rolling (rolling resistance) as a consequence of inelastic material features of the rubber matrix and other phenomena of dissipative nature (e.g., interaction of rubber matrix and cords [24], crack propagation and evolution of thermal damage itself). Different solution techniques for the coupled problem in the context of tire analysis are available in [25–28]. Also friction in the contact zone associated with the phenomenon of wear [29, 30] and emission of tire wear particles (TWP) additionally contributes to the increase of the temperature and, hence, to the increase of thermal damage. Different approaches to model the friction between tire and rough surfaces (road) [31], even at different length scales, are discussed in [32, 33]. In the present example, the temperature field of the tire cross-section shows a much slower evolution (time scale) than the displacement state of the tire undergoing rotation cycles. In consequence, a sequential solution of the

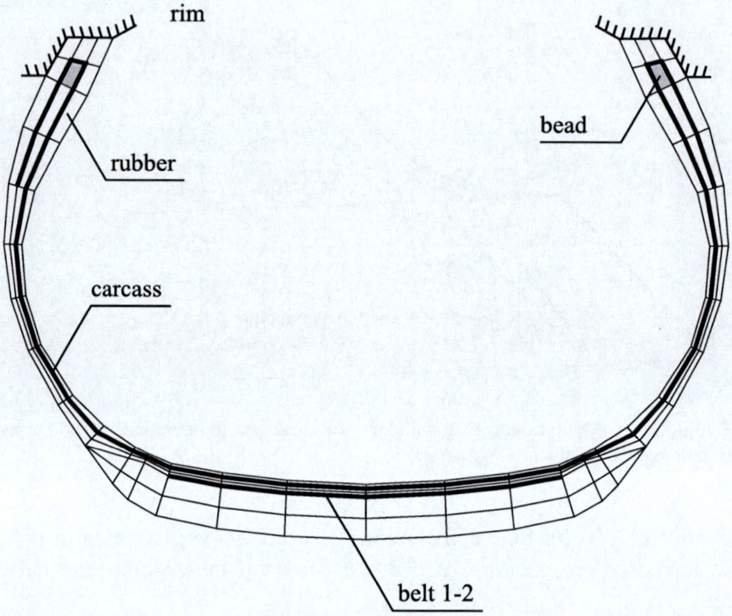

Fig. 9 FE discretized cross-section of a simple automobile tire (passenger car tire 175 SR 14), see [6]

coupled system of equations has been implemented in combination with the ALE formulation of the problem. For a detailed description, the reader is referred to [25].

For the tire depicted with its FE discretized cross-section in Fig. 9, the temperature and, hence, the thermal damage have been computed using the long-term damage envelope function as plotted in Fig. 7. In Fig. 10, different temperature and thermal damage states are shown as a function of time.

The tire 175 SR 14 is made of a viscoelastic tread rubber compound (rubber), two layers of linear elastic steel reinforcement cords (belt 1/2) and a polyethylene PET 1670 × 2370 carcass as well as a linear elastic steel bead wire. For the matrix and cord materials, temperature-dependent mechanical and thermal features are assumed, where the temperature of the tire at time $t = 0$ is $T = 25°\text{C}$. A detailed description including material models and model parameters is available in [6]. The inflation pressure of the tire corresponds to 0.70 bar and represents a severe load case of an underinflated tire, which will give rise to additional mechanical energy loss during rolling and, in consequence, heat-build up in the tire leading to thermal damage. A vertical force of 3300 N is applied to the rim center R, see Fig. 2. For the present investigation, the tire is subjected to a constant translational and rotational speed corresponding to a traveling velocity of 80 km/h in x-direction. Free rolling is assumed on a rigid, plane surface. More details are provided in [6, 17].

The thermal damage induces softening of the whole cord-rubber structure. The stiffness loss on the structural scale [34] due to thermal damage is highlighted in

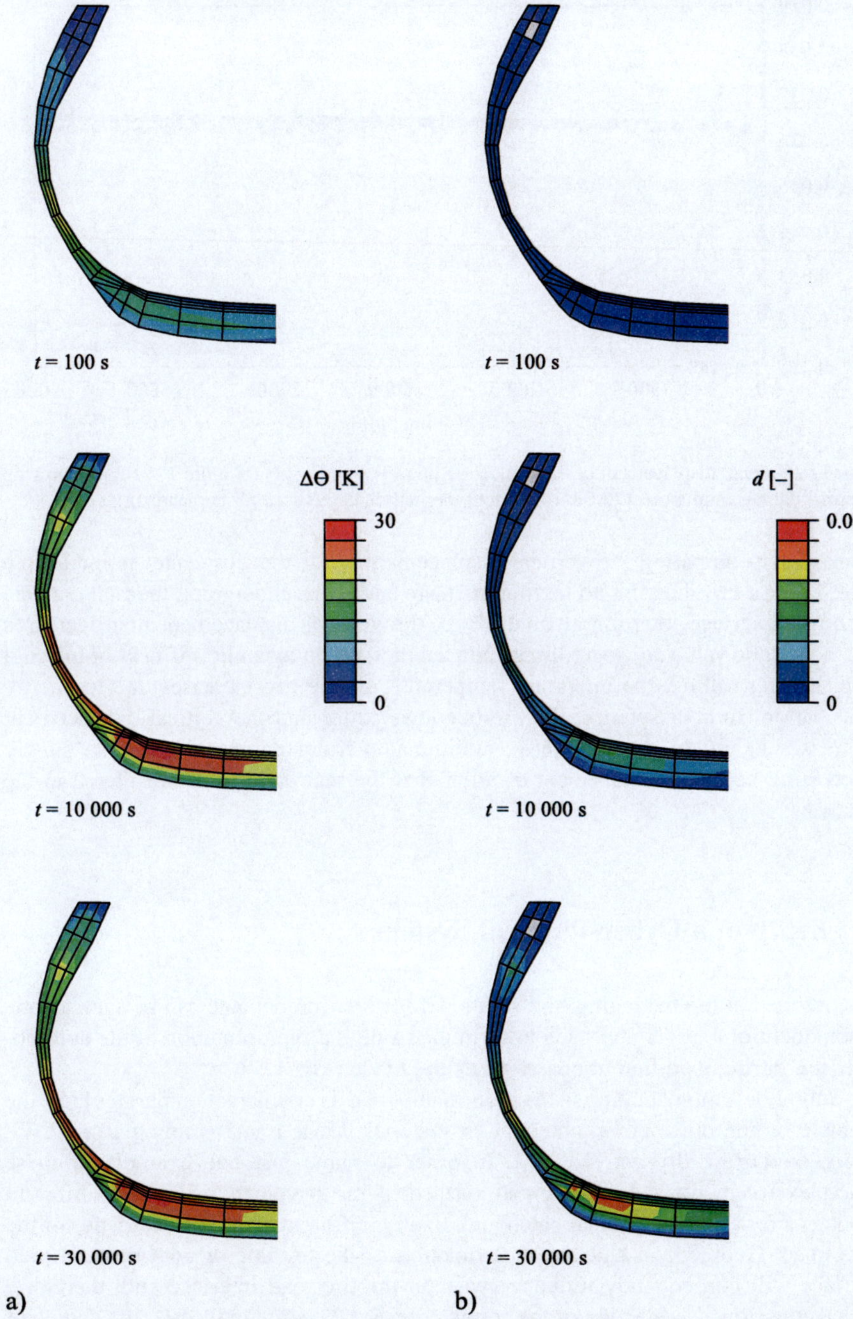

Fig. 10 Cross-sectional distribution (half of the automobile tire 175 SR 14) during rolling for different instants of time: a) temperature, b) thermal damage, see [17]

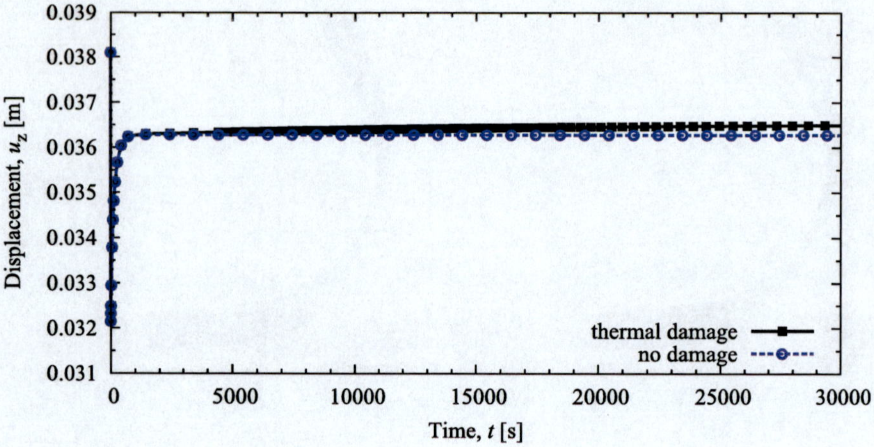

Fig. 11 Vertical displacement of the rim center point R (see Fig. 2) for a tire 175 SR 14 showing thermal damage and a tire 175 SR 14 without thermal damage (used for comparison), see [17]

Fig. 11 by comparing the vertical displacement u_z of the rim center point R (see Fig. 2) for a tire showing no thermal damage and a tire undergoing thermal damage during its service. Starting at time $t = 0$, the vertical displacement first decreases from its static value due to a rolling-induced increase in dynamic stiffness of the cold tire. During rolling, the operating temperature of the tire increases due to energy dissipation. In consequence, the tire becomes softer and the vertical displacement increases again. In the long-term, additional thermal damage leads to a slowly increasing vertical displacement of point R if thermal damage is considered in the analysis.

3 Setup of a Cyber-Physical System

As outlined at the beginning, the submodel for thermal damage can be seen as one component of a larger approach to formulate a digital representation of the automobile tire during its different phases of its life cycle (Fig. 1).

Submodels might comprise the deformable road as an interaction partner [35], the vehicle or the close environment with material flows, e.g. stemming from TWP emission during driving [36, 37]. In order to name just one example of these complex interactions, the deformed surface of the pavement will cause different contact stresses in the contact patch and, finally, different wear characteristics of the tire tread. Geometrical changes as a function of the lifetime of both tire and road surface will consequently influence wear in the tire-road interface and, therefore, TWP emission as a function of the service life. On the structural scale, the choice of operational conditions in a certain range of application has also to be correctly assessed and assumed during lifetime predictions. In this context, concepts of accelerated material and structural tests [38] have to be developed. Their results including uncertainty regarding future (unknown) trends of boundary conditions (e.g., globally rising annual average temperatures due to climate change) might be represented by uncertain data modeling in the context of the digital representation of the rubber product. It becomes clear that the reliable optimization [39, 40] as well as the prediction based on several submodels and available information artifacts will require the use of advanced technologies, e.g. such as machine learning [41], in the future.

As future trend, the use of digital twins as digital representation of the object under investigation helps to set up such a complex ansatz as a combination of physical submodels, data-based models, and meta models. Digital twins are already discussed in the field of manufacturing [42], of product development [43], smart electric vehicles [44], fatigue life prediction of steel bridges [45], or the infrastructure of our future road-bound mobility [46]. In combination with their real object, they form a CPS with permanent links for information exchange (sensing and control).

In this context, a CPS (see Fig. 12) enables the setup of automated interfaces and optimal control by taking into account real-time information from sensors as well as additional input from material and structural tests, which are carried out in combination with predictive simulations of future states. Here, the digital representation of the object under investigation allows the use of different perspectives (views). As an example, numerical simulations aim at predicting the resistance against degradation from experimental investigations on the material and structural scale, whereas measurement data on the structural scale might help to calibrate and synchronize the state of the real object and its digital representation over its lifetime.

The CPS provides insights at the material and the structural level by focusing on different instants of time (real-time evaluation or prediction). This enables to already quantify impacts on mechanical rubber properties and the choice of ingredients during the manufacturing stage of new tires, e.g. to minimize pollution stemming from TWP emission in the environment [47].

Nowadays, there are still some challenges in setting up such a CPS and new questions arise in the process:

- cloud-based storage of data over time including data ownership, data accessibility, and access restrictions,
- setup of interactions between interoperable submodels [48–50] and information entities via ontologies [51],
- model setup and maintenance (software versioning) over time,
- physical coupling with real objects in terms of senor and control units,
- reliability of the digital infrastructure and downtime,
- synchronicity of digital and real objects of the CPS,
- test plan derivation and execution, range of testing, etc.

However, CPS will offer to formulate a reliable closed-chain description of complex systems and to overcome manual data treatment during the digital representation of the life cycle (Fig. 13).

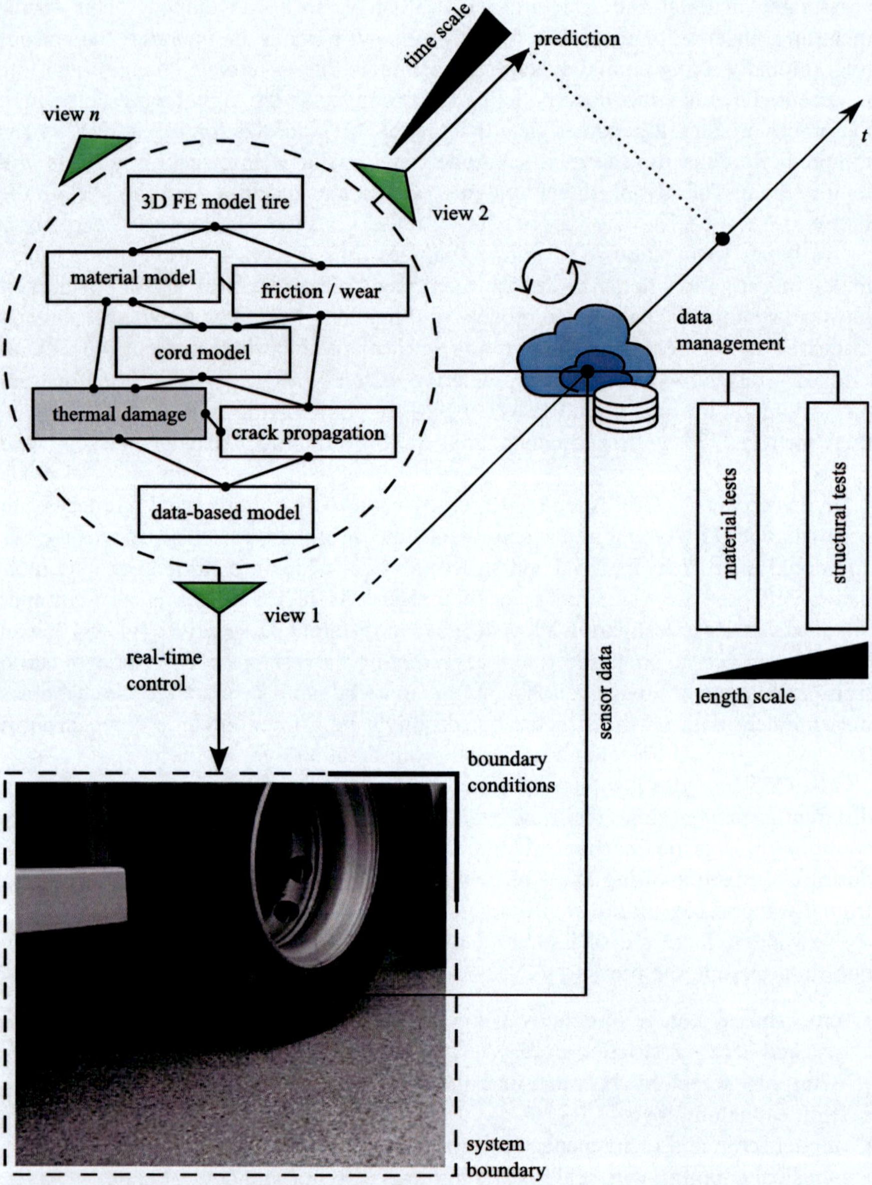

Fig. 12 Concept of a cyber-physical system (CPS) for a truck tire

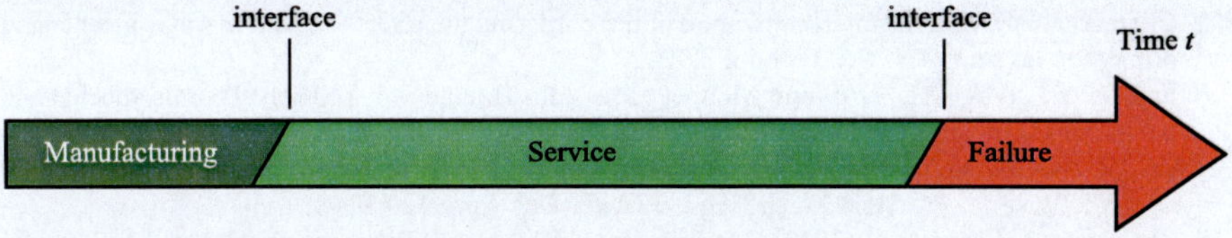

Fig. 13 Automated interfaces of the life cycle, compare to Fig. 1 [3]

4 Conclusion

In this chapter, thermal damage represented by a continuum mechanical model in the context of continuum damage mechanics (CDM) has been discussed as a potential submodel of a larger simulation environment for rubber products. With the help of the FEM, the thermal damage approach has been incorporated at the structural scale and stiffness loss over the lifetime of the automobile tire can be computed in combination with special solution techniques such as the ALE approach (rolling of the tire).

The combination with other interoperable submodels via automated interfaces and an underlying ontology enables to construct a CPS consisting of a digital representation of the tire and the real counterpart. The CPS can be used to gain new insights, plan, and carry out new test campaigns and characterization tests for an optimal design and service of, e.g., rubber components. The concept points out critical aspects and challenges of future developments.

Acknowledgements The authors gratefully acknowledge the financial support by the Deutsche Forschungsgemeinschaft (DFG) under grants KA 1163/30 and KA 1163/42.

References

1. Woo CS, Park HS (2011) Useful lifetime prediction of rubber component. Eng Fail Anal 18: 1645–1651
2. Wortberg J, Mistler M (2017) Lifetime prediction with nonlinear damage accumulation based on material stressing. Part I: introduction of a general procedure. Kautschuk Gummi Kunststoffe 4:25–29
3. Behnke R (2021) Computationally efficient modeling of dynamic multifield pavement interaction problems with different time scales. Habilitationsschrift, Technische Universität Dresden
4. Adam Q, Behnke R, Kaliske M (2020) A thermo-mechanical finite element material model for the rubber forming and vulcanization process: from unvulcanized to vulcanized rubber. Int J Solids Struct 185–186:365–379
5. Zreid I, Behnke R, Kaliske M (2021) ALE formulation for thermomechanical inelastic material models applied to tire forming and curing simulations. Comput Mech 67:1543–1557
6. Behnke R, Kaliske M (2018) Finite element based analysis of reinforcing cords in rolling tires: influence of mechanical and thermal cord properties on tire response. Tire Sci Technol 46:294–327

7. Chen B (2004) Material characterization of tire cords and the effects of cord thermal-mechanical properties on tires. Tire Sci Technol 32:2–22
8. Seghar S, Asaro L, Rolland-Monnet M, Aït Hocine N (2019) Thermo-mechanical devulcanization and recycling of rubber industry waste. Resour Conserv Recycl 144:180–186
9. Luo R, Mortel W, Wu X (2015) Investigation on rubber failure due to heat generation under dynamic loading. Proc Inst Mech Eng L J Mater Des Appl 229:77–87
10. Stevenson A, Campion R (2012) Durabiliy. In: Gent AN (ed) Engineering with rubber – how to design rubber components. Hanser Publishers, Munich, pp 139–176
11. Lion A, Johlitz M (2012) On the representation of chemical ageing of rubber in continuum mechanics. Int J Solids Struct 49:1227–1240
12. Nasdala L, Wei Y, Rothert H, Kaliske M (2008) Lifetime prediction of tires with regard to oxidative aging. Tire Sci Technol 36:63–79
13. Wang X, Yang K, Zhang P (2022) Evaluation of the aging coefficient and the aging lifetime of carbon black-filled styrene-isoprene-butadiene rubber after thermal-oxidative aging. Compos Sci Technol 220:109258
14. Wineman A, Shaw J (2007) Combined deformation- and temperature-induced scission in a rubber cylinder in torsion. Int J Non-Linear Mech 42(2007):330–335
15. Robertson CG, Stoček R, Kipscholl C, Mars WV (2019) Characterizing the intrinsic strength (fatigue threshold) of natural rubber/butadiene rubber blends. Tire Sci Technol 47:292–307
16. Johlitz M (2015) Zum Alterungsverhalten von Polymeren: Experimentell gestützte, thermo-chemomechanische Modellbildung und numerische Simulation. Habilitationsschrift, Universität der Bundeswehr München
17. Behnke R, Kaliske M (2018) Numerical modeling of thermal aging in steady state rolling tires. Int J Non-Linear Mech 103:145–153
18. Chagnon G, Verron E, Gornet L, Marckmann G, Charrier P (2004) On the relevance of continuum damage mechanics as applied to the Mullins effect in elastomers. J Mech Phys Solids 52:1627–1650
19. Miehe C (1995) Discontinuous and continuous damage evolution in Ogden-type large-strain elastic materials. Eur J Mech A Solids 14:697–720
20. Simo JC, Ju JW (1987) Strain- and stress-based continuum damage models – I. Formulation. Int J Solids Struct 23:821–840
21. Stoček R, Mars WV, Kratina O, Machů A, Drobilík M, Kotula O, Cmarová A (2017) Characterization of ageing effect on the intrinsic strength of NR, BR and NR/BR blends. In: Lion A, Johlitz M (eds) Constitutive models for rubber X. CRC Press, Taylor & Francis Group, London, pp 365–370
22. Nackenhorst U (2004) The ALE-formulation of bodies in rolling contact: theoretical foundations and finite element approach. Comput Methods Appl Mech Eng 193:4299–4322
23. Li F, Liu J, Yang H, Lu Y, Zhang L (2016) Numerical simulation and experimental verification of heat build-up for rubber compounds. Polymer 101:199–207
24. Liu Y, Wan Z, Tian Z, Du X, Jiang J, Yao M (1999) Fatigue of unidirectional cord-rubber composites. Tire Sci Technol 27:48–57
25. Behnke R, Kaliske M (2015) Thermo-mechanically coupled investigation of steady state rolling tires by numerical simulation and experiment. Int J Non-Linear Mech 68:101–131
26. Ebbott TG, Hohman RL, Jeusette J-P, Kerchman V (1999) Tire temperature and rolling resistance prediction with finite element analysis. Tire Sci Technol 27:2–21
27. Park HC, Youn S-K, Song TS, Kim N-J (1997) Analysis of temperature distribution in a rolling tire due to strain energy dissipation. Tire Sci Technol 25:214–228
28. Suwannachit A, Nackenhorst U (2013) A novel approach for thermomechanical analysis of stationary rolling tires within an ALE-kinematic framework. Tire Sci Technol 41:174–195
29. Li Z, Li Z, Wang Y (2019) An integrated approach for friction and wear simulation of tire tread rubber. Part I: friction test, characterization, and modeling. Tire Sci Technol 48:123–145
30. Li Z, Li Z, Wang Y (2019) An integrated approach for friction and wear simulation of tire tread rubber. Part II: wear test, characterization, and modeling. Tire Sci Technol 48:146–165

31. Araújo JPC, Palha CAO, Martins FF, Silva HMRD, Oliveira JRM (2019) Estimation of energy consumption on the tire-pavement interaction for asphalt mixtures with different surface properties using data mining techniques. Transp Res D 67:421–432
32. Friederichs J, Wegener D, Eckstein L, Hartung F, Kaliske M, Götz T, Ressel W (2020) Using a new 3D-printing method to investigate rubber friction laws on different scales. Tire Sci Technol 48:250–286
33. Wriggers P, Reinelt J (2009) Multi-scale approach for frictional contact of elastomers on rough rigid surfaces. Comput Methods Appl Mech Eng 198:1996–2008
34. Lévêque D, Schieffer A, Mavel A, Maire J-F (2005) Analysis of how thermal aging affects the long-term mechanical behavior and strength of polymer-matrix composites. Compos Sci Technol 65:395–401
35. Behnke R, Wollny I, Hartung F, Kaliske M (2019) Thermo-mechanical finite element prediction of the structural long-term response of asphalt pavements subjected to periodic traffic load: tire-pavement interaction and rutting. Comput Struct 218:9–31
36. Prenner S, Allesch A, Staudner M, Rexeis M, Schwingshackl M, Huber-Humer M, Part F (2021) Static modelling of the material flows of micro- and nanoplastic particles caused by the use of vehicle tyres. Environ Pollut 290:118102
37. Rausch J, Jaramillo-Vogel D, Perseguers S, Schnidrig N, Grobéty B, Yajan P (2022) Automated identification and quantification of tire wear particles (TWP) in airborne dust: SEM/EDX single particle analysis coupled to a machine learning classifier. Sci Total Environ 803:149832
38. Huang D, LaCount BJ, Castro JM, Ignatz-Hoover F (2001) Development of a service-simulating, accelerated aging test method for exterior tire rubber compounds: I. Cyclic aging. Polym Degrad Stab 74:353–362
39. Kaliske M, Serafinska A, Zopf C (2013) Optimized and robust design of tires based on numerical simulation. Tire Sci Technol 41:21–39
40. Serafinska A, Kaliske M, Zopf C, Graf W (2013) A multi-objective optimization approach with consideration of fuzzy variables applied to structural tire design. Comput Struct 116:7–19
41. Zhu J, Han K, Wang S (2021) Automobile tire life prediction based on image processing and machine learning technology. Adv Mech Eng 13:1–13
42. Cimino C, Negri E, Fumagalli L (2019) Review of digital twin applications in manufacturing. Comput Ind 113:103130
43. Fedorko G, Molnár V, Vasil M, Salai R (2021) Proposal of digital twin for testing and measuring of transport belts for pipe conveyors within the concept industry 4.0. Measurement 174:108978
44. Bhatti G, Mohan H, Raja Singh R (2021) Towards the future of smart electric vehicles: digital twin technology. Renew Sust Energ Rev 141:110801
45. Jiang F, Ding Y, Song Y, Geng F, Wang Z (2021) Digital twin-driven framework for fatigue life prediction of steel bridges using a probabilistic multiscale model: application to segmental orthotropic steel deck specimen. Eng Struct 241:112461
46. Kaliske M, Behnke R, Wollny I (2021) Vision on a digital twin of the road-tire-vehicle system for future mobility. Tire Sci Technol 49:2–18
47. Wagner S, Klöckner P, Reemtsma T (2022) Aging of tire and road wear particles in terrestrial and freshwater environments – a review on processes, testing, analysis and impact. Chemosphere 288:132467
48. Sattler L, Lamouri S, Pellerin R, Maigne T (2019) Interoperability aims in building information modeling exchanges: a literature review. IFAC PapersOnLine 52(13):271–276
49. Shehzad HMF, Ibrahim RB, Yusof AF, Khaidzir KAM, Iqbal M, Razzaq S (2021) The role of interoperability dimensions in building information modelling. Comput Ind 129:103444
50. Zhang F, Chen M, Kettner AJ, Ames DP, Harpham Q, Yue S, Wen Y, Lü G (2021) Interoperability engine design for model sharing and reuse among OpenMI, BMI and OpenGMS-IS model standards. Environ Model Softw 144:105164
51. Borsato M (2014) Bridging the gap between product lifecycle management and sustainability in manufacturing through ontology building. Comput Ind 65:258–269

Adv Polym Sci (2023) 289: 121–142
https://doi.org/10.1007/12_2022_128

Published online: 25 July 2022

Experimental and Numerical Description of the Heat Build-Up in Rubber Under Cyclic Loading

O. Peter, R. Stoček, and O. Kratina

Contents

Abstract This paper describes a new numerical method – within an ANSYS Workbench finite element method environment – for simulating the kinetics of the Heat Build-up process in rubber under cyclic multiaxial loading. A novel measuring setup controls a multiaxial load which forces the heat build-up by combining the

---

O. Peter
Centre of Polymer Systems, University Institute, Tomas Bata University in Zlín, Zlín, Czech Republic

Bonatrans Group a.s., Bohumín, Czech Republic

R. Stoček (✉)
Centre of Polymer Systems, University Institute, Tomas Bata University in Zlín, Zlín, Czech Republic

PRL Polymer Research Lab., s r.o., Zlín, Czech Republic
e-mail: stocek@utb.cz

O. Kratina
Centre of Polymer Systems, University Institute, Tomas Bata University in Zlín, Zlín, Czech Republic

dynamic alteration torque and multi-planar bending. In this method, the experimentally determined Heat Build-up is calculated numerically from the strain energy density with the determined hyperelastic material parameters at room temperature. A new quantity, the so-called dissipation constant, is introduced which makes it possible to determine the dissipation energy of cyclically loaded specimen. The experiments and calculations were carried out for a broad range of loading conditions, varying in both bending angle and rotational speed. Finally, the agreement between the experimentally determined and the calculated data is discussed.

Keywords Ansys · Dissipation constant · FEM · Finite element analyses · HBU · Heat build-up · Rubber

1 Introduction

In technical practice, a large number of rubber parts are subjected to cyclic dynamic loading, which places high demands on both the optimal product design and the quality of the material itself. The number of industrial and everyday applications of this class of polymeric materials has steadily increasing over last decades. Typical examples are synthetic and natural covalently cross-linked rubbers used in automotive tires or other applications such as drive belts, seismic isolators, fuel system hoses, gaskets, turbocharger hoses, cooling systems, engine bushings, spring pads, and more. In such applications, rubber materials are often subjected to a very complex load profile resulting from a variety of external actions, resulting in permanent small and large cyclic multiaxial deformations over a wide range of frequencies and amplitudes. In addition, rubbers exhibit high heat generation under cyclic loading due to dissipative mechanisms resulting from the viscoelastic nature of rubbers and the filler–filler and polymer–filler interactions [1]. However, the dissipated heat cannot be released to the environment fast enough due to the low thermal conductivity of rubber, resulting in a temperature rise in the rubber components, which is referred to as **H**eat **B**uild-**u**p (HBU) in the literature [2]. As rubber heats up under cyclic loading and heat becomes concentrated in the rubber due to its low thermal conductivity, it is necessary for the proper functioning of such rubber components to maintain their operating temperature below the critical temperature at which irreversible degradation processes occur. Exceeding this critical temperature leads to a significant degradation of the material and thus a shortening of its service life.

Figure 1 shows an example of the thermal degradation of a rubber segment of a tram wheel caused by cyclic mechanical loading. As the wheel rotation speed increases, the dissipative energy and thus the operating temperature of the rubber segment increases and the degradation of the rubber matrix progresses.

To avoid thermal degradation, it is necessary to predict the temperature development in the rubber part over time at a given load. This should be done, above all,

Fig. 1 Unused (left) and thermally degraded (right) rubber segment (Bonatrans Group a.s., Czech Republic)

simply, quickly, and as accurately as possible. Several publications have been written regarding the computational simulation of the development of heating in rubber under cyclic loading. Some of them dealt with cyclic compressive loading [3–5], others with tensile loading [6]. It is characteristic of all these works that in order to predict the heating development of rubber, it is necessary to know the hyperelastic and viscoelastic parameters of the material behavior, which places high demands on the number and accuracy of the tests performed.

From an experimental point of view, there are only a few techniques available to characterize the HBU of rubber. The most commonly used devices for HBU characterization, mainly for scientific purposes, are the Rubber Process Analyzer (RPA) and the Dynamic Mechanical Analyzer (DMA) and, for industrial applications with respect to heating under cyclic fatigue loading, the flexometer, also known as the Goodrich Flexometer (GF). While the RPA is used for the analysis of HBU in shear, the DMA can be used for the analysis of HBU under different types of loads such as compression, tension, and shear. The principle of the flexometer is based only on analysis in compression. As far as temperature measurement is concerned, only the surface temperature of the sample can be analyzed continuously with the RPA [7]. The DMA allows the continuous measurement of the surface temperature of the sample through a temperature sensor built into the press plate. In addition, the thermal needle probe is activated to penetrate inside the sample at the end of the test to obtain a temperature measurement [8]. The sample temperature at GF can be measured at the surface or inside the material [7]. Furthermore, due to the nature of the load in GF, it is not possible to continuously determine the temperature in the material because the thermal sensor can only be placed on the axis in the direction of the load, resulting in additional friction at the interface between the sensor and the rubber material. This additional friction could directly affect the real temperature value. In addition, customized devices have been developed which simulate different loads, mostly in compression, and where the surface temperature is sensed during loading by infrared temperature sensors, e.g. [9, 10]. Finally, for most rubber products, simple uniaxial loading does not correspond to actual multiaxial loading conditions in the field, which are defined by a combination of tension, compression, and shear.

Therefore, there is a very strong scientific and industrial interest in developing methods and laboratory test equipment for HBU analysis in simulating actual

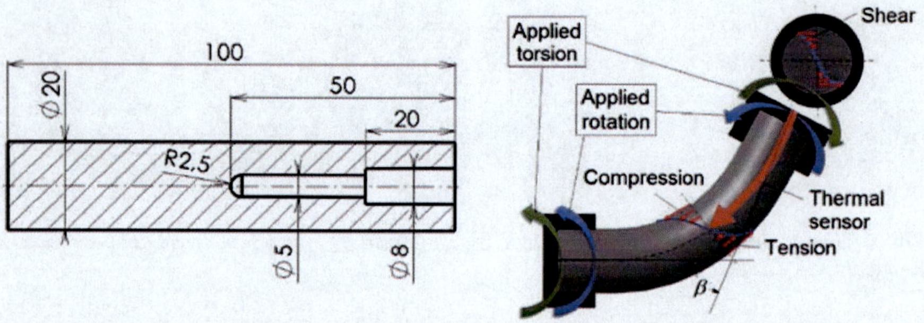

Fig. 2 Sample geometry (left) and scheme of the measuring principle (right) [11]

loading conditions applied to a rubber product in service, where the complex temperature at the surface and inside the rubber product is determined very accurately in situ during the analysis. Thus, a novel testing principle has been involved for the enhancement of data accuracy and in situ measurement of temperature development inside and outside of cylindrical rubber sample during cyclic fatigue loading, whereas the geometry of the rubber sample is presented in Fig. 2, left. The measurement principle is shown schematically in Fig. 2, right, which firstly has been presented by Stocek et al. [12]. In the first step, the cylindrical sample is fixed at both ends with a free distance of 60 mm. A flexible temperature sensor, which is an integral part of the testing device, is inserted into the inner bore of the specimen and contact is made between the sensor and the rubber in the middle part of the sample. In a second step, the sample is bent by an angle, β, and the angular position is fixed. The bending of the specimen causes deformation on both sides of the neutral axis, with the specimen in tension on one side and in compression on the other. It should be noted that as the angle increases, the upcoming deformation of the specimen in both tension and compression also increases and vice versa. Additionally, torque can be applied to the sample and therefore a shear stress could be generated in the specimen. Finally, the specimen is spun up and thus the dynamic alteration torque and multi-planar bending as a combined loading under fatigue condition are induced in the specimen. From the schematic representation of the deformed specimen, it can be seen that the maximum deformation, and hence the maximum stresses, occur at the surface of the specimen, while toward the center of the specimen they decrease until they reach zero at the neutral axis. Therefore, the temperature is expected to increase first at the surface of the specimen in the time and then concentrate further inside the sample due to the low thermal conductivity of the rubber. Due to the cooling of the surface by rotating the sample, the temperature inside will exceed the values occurring at the surface. By using an additional IR sensor, the temperature development at the surface can be measured non-contactually during the analysis to obtain information about the heat transfer from the surface to the center of the sample.

In the numerical simulation of Heat Build-up (HBU) based on the testing principle presented above, it is necessary to determine the hyperelastic properties of the cylindrical rubber sample under 3-point bending at room temperature and finally perform the actual HBU test to determine the temperature development both inside and on the rubber sample surface. A quantity called the dissipation constant is introduced to replace the knowledge of the viscoelastic behavior of the material. The dissipation constant serves as a converter between the strain energy density and the dissipation energy density, from which the dissipation power density is determined. The amount of dissipated energy is assumed to be constant and independent of temperature. By introducing a dissipation constant, it is possible to perform Finite Element Analysis (FEM) simulation in the time domain and obtain results in a reasonable period of time, thereby achieving a significant reduction in computational capacity requirements. Thermal material parameters such as thermal conductivity, specific heat, convections, and dissipation constant can be obtained from the HBU test for one load case using response surface optimization method [13] and the material parameters determined in this way can be further used to simulate other load cases. All work was performed in the FEM commercial software Ansys.

The actual experimental work then focused on the analysis of two rubber materials, namely butadiene rubber (BR) and styrene-butadiene rubber (SBR), which were chosen for their significant difference in values during heat development, with BR achieving significantly lower values than SBR [12, 14, 15].

2 Experimental

2.1 Material

Elastomers used in this study were neodymium-catalyzed butadiene rubber (High-Cis Nd-BR, Trinseo, Germany) and emulsion polymerized styrene-butadiene rubber (E-SBR 1500, Trinseo, Germany). These polymers were filled with 50 phr of N330 type carbon black (CB) to give a filler volume fraction (ϕ) of 0.19. Table 1 lists the complete formulations for the rubber compounds.

The cylindrical specimens (see Fig. 2, left) were cured in a heated press (LaBEcon 300 from Fontijne Presses, The Netherlands) at 150°C according to the optimum curing time *t90* + 1 min/1 mm thickness, where *t90* was determined from curing rheometer curves evaluated at 150°C with a moving die rheometer (MDR 3000 Basic from MonTech, Germany) according to ASTM 6204.

Table 1 Table of material formulas

Compound	E-SBR1500	Cis Nd-BR	CB N330	ZnO	Stearic acid	TBBS	6PPD	Sulfur
BR [phr]	–	100	50	2	1	2	1	2
SBR [phr]	100	–	50	2	1	2	1	2

2.2 *Three-Point Bending*

To determine the stiffness of cylindrical samples of both rubber materials, a 3-point bending test was carried out on a M350-5CT universal testing machine (Testometric, United Kingdom), with the loading scheme shown in Fig. 3. The bottom supports are fixed and their inner faces are spaced 60 mm apart. All supports have a radius of 5 mm at the point of contact with the sample. The sample was placed symmetrically in the device. The displacement of the upper jaw was 18 mm at a speed of 360 mm/min. The test was carried out at room temperature. The data output of the test is the force in depending on the displacement.

2.3 *Heat Build-Up (HBU) Experiment*

The HBU experiments were performed using the originally developed **H**eat **B**uild-**u**p **A**nalyzer (**HBuA**) by Coesfeld GmbH & Co. KG, Germany. The photo of the test chamber is shown in Fig. 4, left, while Fig. 4, right, shows the schematic arrangement of the individual components in the chamber for performing the measurement.

The test was carried out entirely according to the test procedure described above (see Fig. 2, right). A total of 6 varied loading conditions have been used for each compound, using two varied rotational speeds of 500 and 1,000 rpm and three bending angles, $\beta = 30$, 50, and 60°. Moreover, three replicates have been analyzed for each condition.

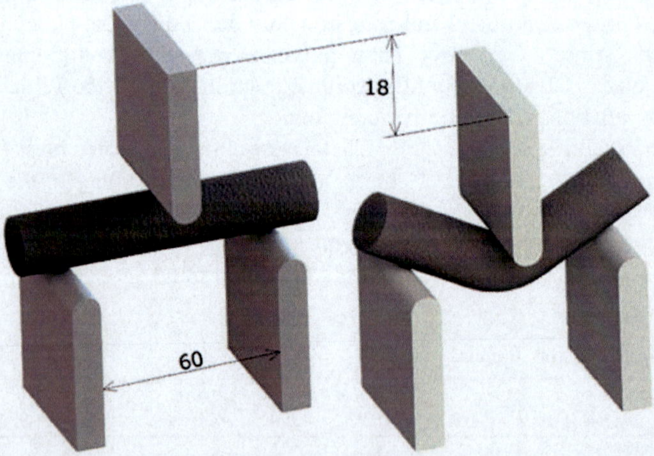

Fig. 3 Scheme of three-point bending

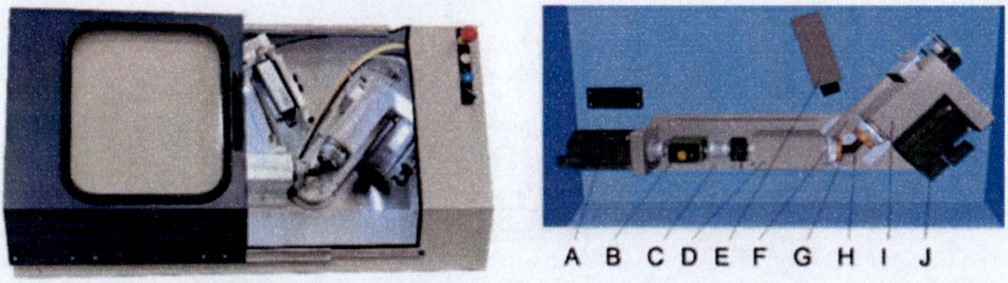

Fig. 4 Photograph of the HBuA (left) with the scheme of the individual components (right), where A – main engine, B – torque sensor, C – fixed support, D – main spindle, E – IR external thermal sensor, F – rotary shaft, G – rubber sample, H – internal thermal sensor, I – secondary spindle, J – secondary engine [11]

Fig. 5 Scheme for finding hyperelastic material parameters

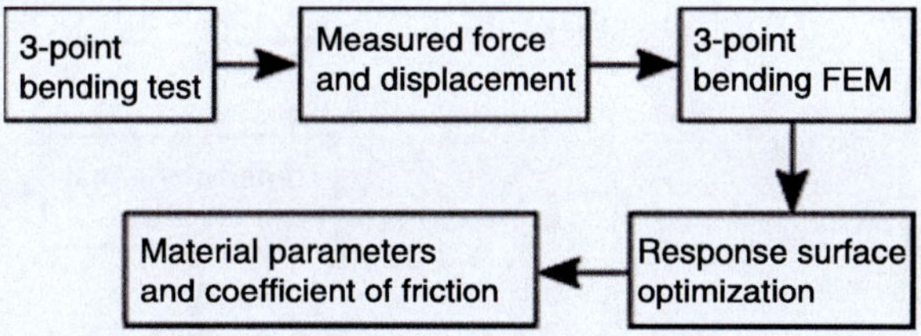

3 Heat Build-Up (HBU) Numerical Calculation

To be able to simulate HBU using the FEM, we first need to find the hyperelastic material parameters corresponding to the material stiffness of each rubber. For this purpose, we need to perform several steps shown in the scheme (Fig. 5).

Since we know the design of the cylindrical sample and the setting of the testing machine for the three-point bending test (see Fig. 3), we can simulate this test in the FEA environment. The objective of this simulation is to determine the hyperelastic material parameters at which the simulation results (force at a given displacement) will match the measured data. For this purpose, the response surface optimization [13] implemented in Ansys is used.

The computational model for three-point bending consists of a cylindrical specimen and a bottom and an upper support. This test was modeled as a contact problem with an unknown coefficient of friction between rubber sample and steel supports. Therefore, the friction coefficient was also varied and searched for along with the hyperelastic material parameters. A finite element mesh was created using linear hexahedral 8-node SOLID185 elements with CONTA174 and TARGE170 contact elements. The supports were considered as absolutely rigid (Fig. 6).

After determining the parameters of the hyperelastic material, the HBU simulation is performed according to the following scheme (Fig. 7).

First, HBU tests were performed, from which surface and internal temperatures were recorded. Subsequently, it was possible to proceed with finite element simulations of the HBU. In a first step, a structural analysis was performed in which the

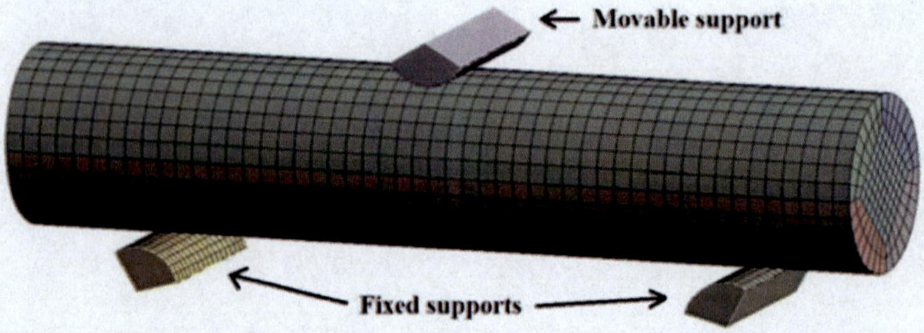

Fig. 6 Computational model of 3-point bending test

Fig. 7 Scheme for HBU simulation

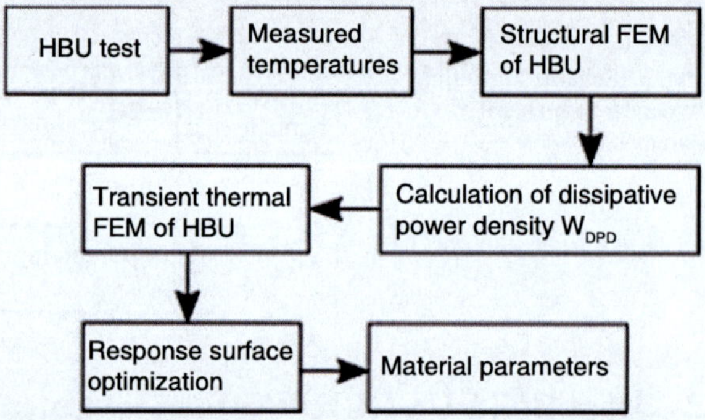

cylindrical sample was bent to the desired angle using joints on both ends of sample with only one end fixed and then spun for one revolution around its neutral axis in a time corresponding to one revolution of the sample during the test. A finite element mesh was created using linear hexahedral 8-node SOLID185. The desired output of this simulation is the strain energy density, W_{SED}, of each element and the time, t, during the spun.

The reason for these outputs is that the whole calculation methodology is based on the assumption that the dissipation energy is not the same at each point of the body, but depends on the magnitude of the local strain energy density. In the case of HBU testing of a rotating bent cylindrical sample, this means that most of the strain energy is converted to heat at the surface of the sample and, due to the low thermal conductivity of the rubber, is transferred toward the center of the sample during cyclic loading.

If we look at the strain energy density, W_{SED}, on one element per rotation, we find that it varies in maxima. This effect is well described by the theory of strongly curved beams [16]. If the slope is small enough, the axial stress course will be linear and corresponds to the theory of weakly curved beams [16]. As the slope of the cylinder ends increases, the axial stress will increasingly conform to a hyperbolic course along the cross section, and thus the difference of the strain energy density maxima in tension will increasingly diverge from that in compression (Figs. 8 and 9).

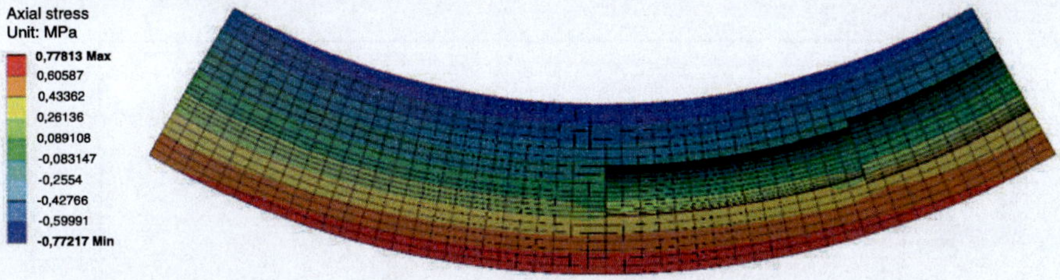

Fig. 8 Axial (bending) stress

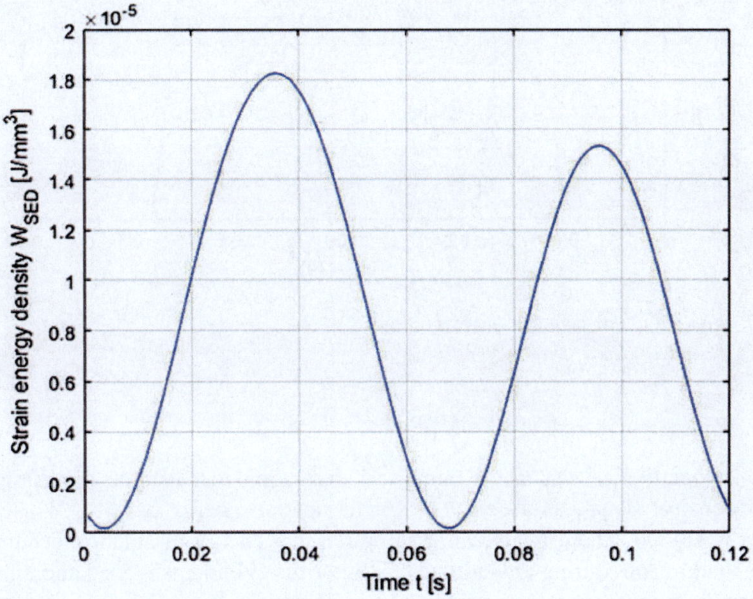

Fig. 9 W_{SED} of ith element per one revolution

In order to generate the inputs required for the transient thermal calculation, it is necessary to know both the strain energy density, W_{SED}, of each element and the time, t, obtained from the structural analysis, as well as the value of the dissipation constant, D. This dissipation constant, D, is a phenomenological constant here and serves as a converter between strain energy density, W_{SED}, and dissipation energy density, W_{DED}. The dissipation constant, D, is thus defined as:

$$D = \frac{W_{\mathrm{DED}}}{W_{\mathrm{SED}}} \tag{1}$$

From formula (1), it can be derived that the dissipation energy density, W_{DED}, is calculated as

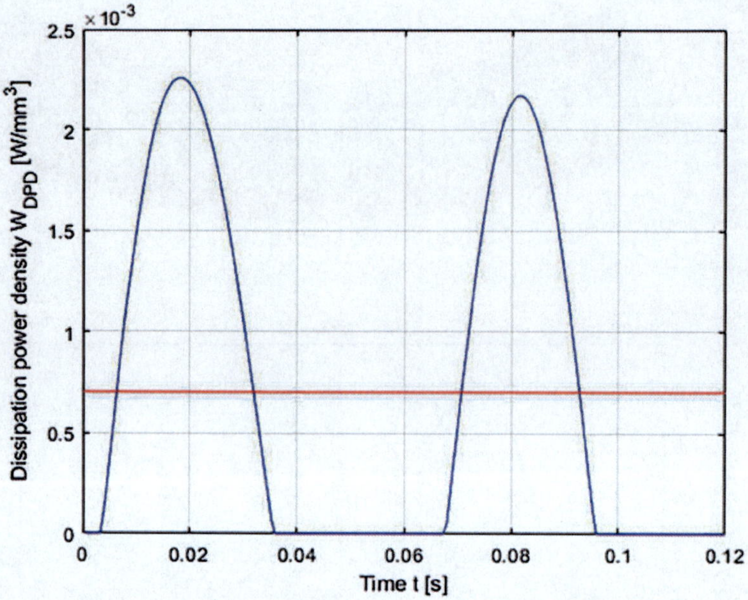

Fig. 10 W_{DPD} of ith element per one revolution

$$W_{DED} = D \cdot W_{SED} \tag{2}$$

Subsequently, we can use a numerical derivation to calculate the dissipation power density, W_{DPD}, as a derivative of dissipation energy density, W_{DED}, with respect to time, t. There is also an assumption that energy dissipation occurs only during loading, not during unloading [17]. Since the cylinder is loaded and unloaded twice per revolution, once in tension and once in compression, W_{DPD} has two peaks. W_{DPD} is set to zero at the time of unloading. In order not to have too small time step in the thermal transient calculation, the W_{DPD} value is averaged over one revolution (Fig. 10). This process is performed for all elements of the FEM mesh.

$$W_{DPD} = \frac{dW_{DED}}{dt} \tag{3}$$

Once the dissipation power density, W_{DPD}, has been determined for each element of the finite element mesh, it is possible to proceed to transient thermal FEM analysis. It is necessary to use the same finite element mesh as in the structural calculation. The thermal load is put into the calculation as internal heat generation [18] and corresponds to the average W_{DPD} value for each element. Regarding the heat dissipation to the surroundings, a total of three different convections are used. The first convection, C_1, is on the surfaces behind which the sample is fixed in the rotors and is marked in blue in Fig. 11. The second convection C_2 corresponds to the remaining outer surface of the sample and is shown in green. The last third

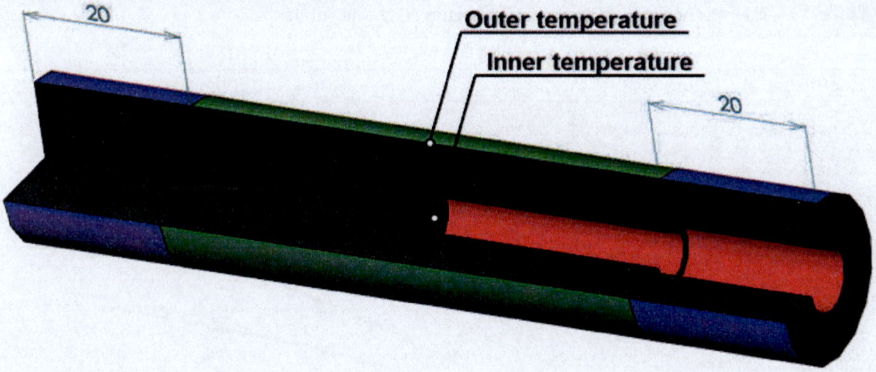

Fig. 11 Scheme of used convections

Table 2 Table of extended tube material model parameters

		SBR	BR
Cross-linked network modulus, *Gc*	[MPa]	1.6133	2.2317
Constraint network modulus, *Ge*	[MPa]	0.4333	0.3594
Empirical parameter, β	[–]	0.0800	0.0800
Extensibility parameter, δ	[–]	0.0314	0.0830
Incompressibility parameter, d_1	[MPa]	0	0

convection C_3 is marked in red and corresponds to the inner surfaces of the cylinder. The initial temperature in the calculations corresponds to the test temperatures. The temperatures from the finite element method calculations are taken from the nodes at the center of the outer and inner surfaces (see Fig. 11).

If the thermal parameters of the material are not yet known, including the dissipation constant, *D*, the response surface method was again used to determine them in relation to the actual HBU test results.

4 Results and Discussion

The Extended tube model [19] implemented in Ansys was chosen as the material model for the calculations. The parameters of hyperelastic material model have been obtained from the 3-point bending test together with the coefficient of friction between the steel and rubber components using the response surface optimization method [13] implemented in Ansys. The hyperelastic parameters are considered as fit parameters of our problem without going into discussions of the physical background of the model. All the resulting parameters are listed in Tables 2 and 3.

Figure 12 shows a comparison of measured and calculated force in relation to displacement. The correlations clearly show a significant agreement between the

Table 3 Table of coefficients of friction between steel and rubber

		SBR	BR
Coefficient of friction, *f*	[−]	1.89	0.52

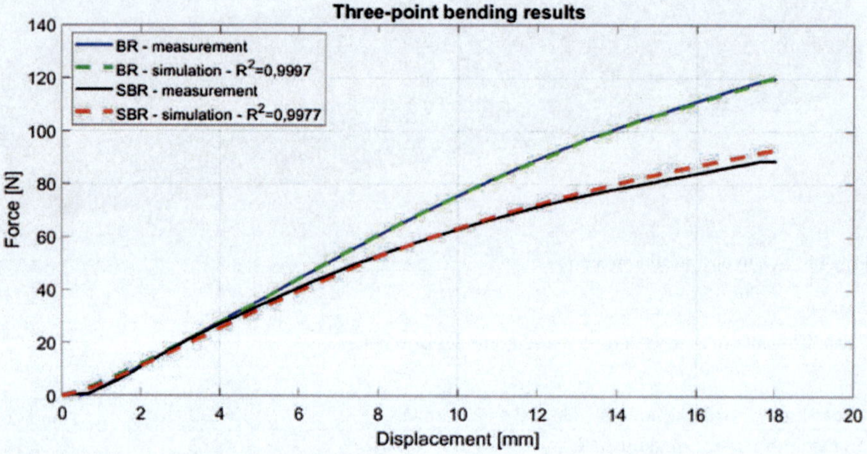

Fig. 12 Three-point bending results

Table 4 Table of found thermal material parameters

Compound	Thermal conductivity k [W/(m K)]	Specific heat C_P [J/(kg K)]	Dissipation constant D [−]
BR	0.15	1,600	1.25
SBR	0.30	1.700	2.550

experimental and calculated values over the entire range of applied displacement. The regression reaches the value $R^2 = 0.9997$ for BR material and $R^2 = 0.9977$ for SBR material.

The thermal material parameters for both BR and SBR were determined using the response surface optimization method, where variables such as thermal conductivity, k, specific heat, C_p, dissipation constant, D, outer surface convection C_2, and inner surface convection C_3 were varied. Convection C_1 was set to 100 W/(m^2 K). The material parameters found were determined so that the calculation results for the test with the smallest bending angle $\beta = 30°$ at speeds of 500 and 1,000 rpm achieved a high agreement. For the thermal material parameters and convections thus determined, all measured variations were subsequently calculated and are listed in Tables 4 and 5.

Before a broader explanation of the experimental HBU data, Fig. 13 shows the general character of HBU development both on the surface and inside the sample as a function of loading cycles. The curves in this figure represents the HBU data obtained for the bending angle $\beta = 50°$ and rotation speed 500 rpm. The solid line

Table 5 Table of found convections

Rotation speed [rpm]	C_1 [W/(m^2 K)]	C_2 [W/(m^2 K)]	C_3 [W/(m^2 K)]
500	100	30	20
1,000	100	41.5	26

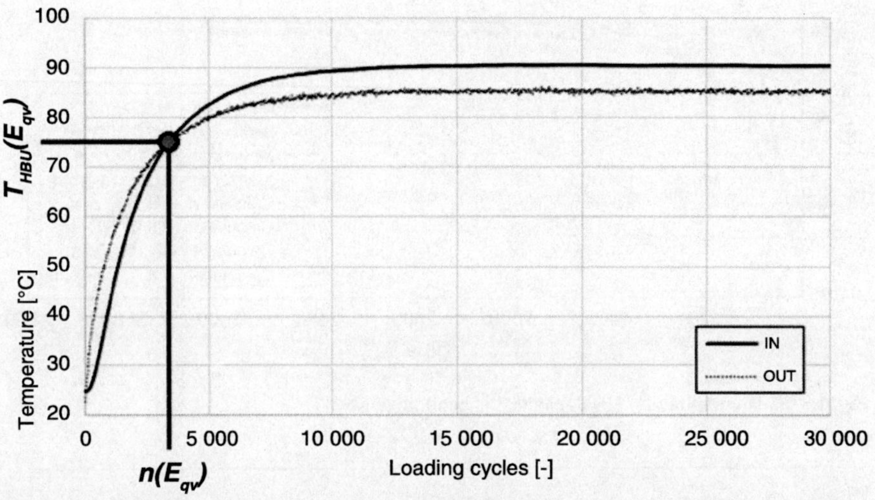

Fig. 13 HBU, T_{HBU} related to loading cycles for the SBR compound at bending angle, $\beta = 50°$ and rotation speed 500 rpm

shows the temperature inside, $T_{HBU(IN)}$ (labeled "IN" in the graphs) and the dotted line outside $T_{HBU(OUT)}$ (labeled "OUT" in the graphs) of the sample. It is clear from the nature of the curves that the temperature starts to rise at the surface during the first load cycles and the temperature rise inside the sample is delayed and slower. The temperature rise on the outside and inside of the sample gradually slows down, with the outside temperature reaching a slower acceleration compared to the inside temperature. Therefore, at a certain number of cycles, the absolute values of the inside temperature exceed the outside temperature. This point, labeled as "E_{qv}," is called the equilibrium or transient temperature point between the surface and internal temperature. The corresponding temperature, $T_{HBU}(E_{qv})$, can be linked to the number of cycles, $n(E_{qv})$, or the time, $t(E_{qv})$, at which a constant temperature occurs throughout the radial cross section of the sample. Beyond this transition point, the temperature value in the interior space is already dominant, with both curves increasing until constant values are reached. This finding fully confirms the above assumption of a higher temperature rise at the surface due to its maximum deformation and a temperature concentration inside the sample due to the low thermal conductivity of the rubber-like materials.

Since the material parameters were found such that the results of the calculations corresponded to the measured course for the bending angle $\beta = 30°$ of the sample, it

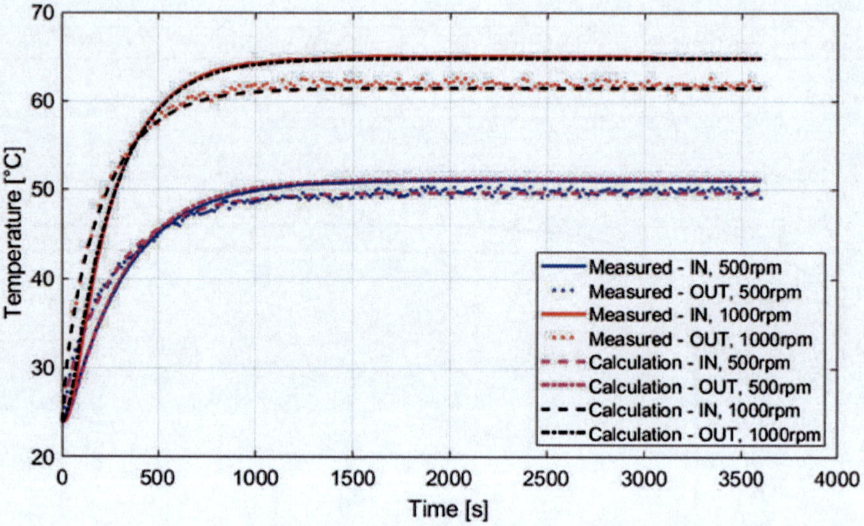

Fig. 14 SBR compound – HBU results 30° bending angle

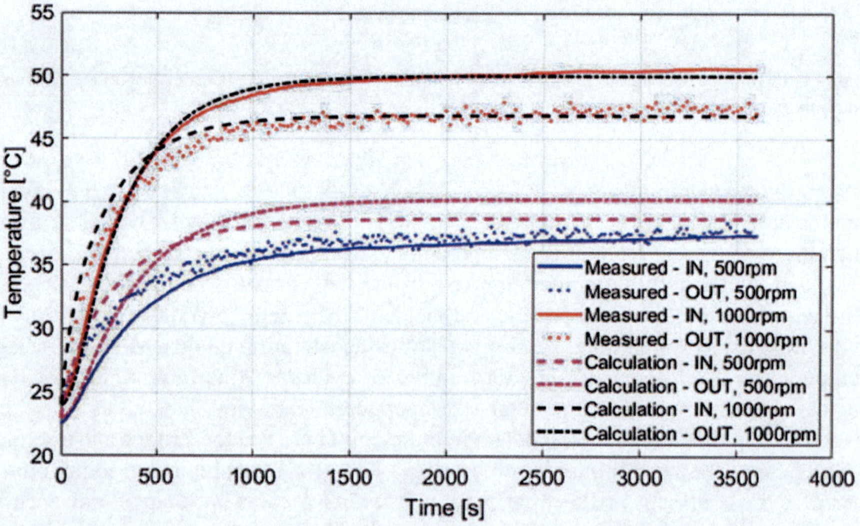

Fig. 15 BR compound – HBU results 30° bending angle

can be seen in the following graphs (Figs. 14 and 15) that the calculated HBU course largely corresponds to the measured course.

For higher bending angles, the differences between measured and calculated HBU values are larger compared to the bending angle $\beta = 30°$ (see Figs. 16, 17, 18, and 19). The calculated HBU results for the SBR compound are always higher

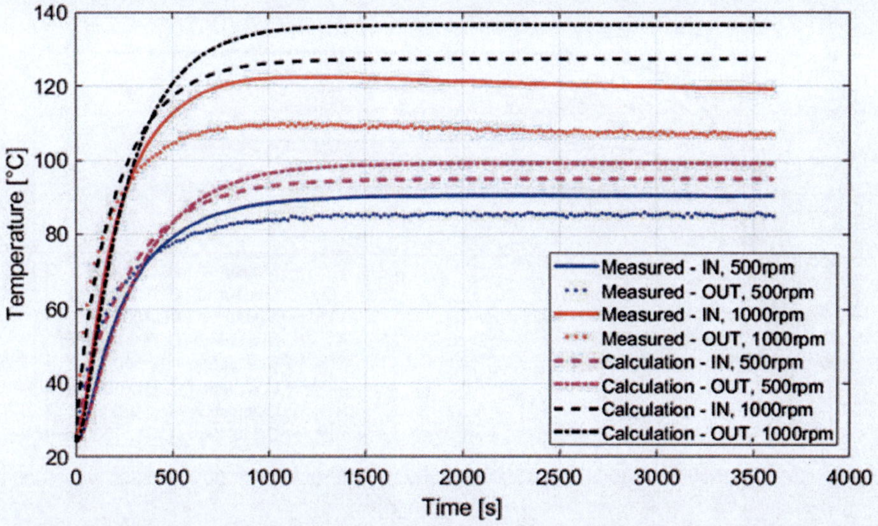

Fig. 16 SBR compound – HBU results 50° bending angle

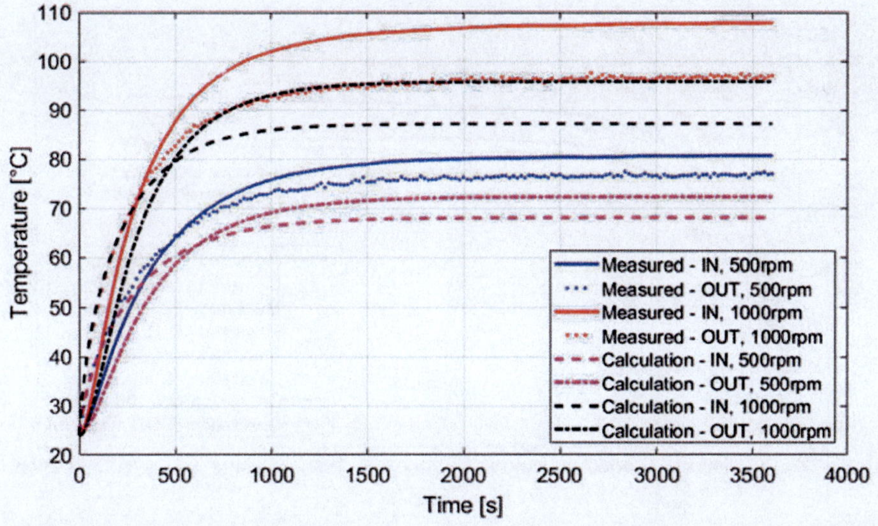

Fig. 17 BR compound – HBU results 50° bending angle

than the experimentally measured HBU, which can be explained by the fact that the dissipation energy of this compound decreases with increasing temperature. Conversely, this is the case for the BR compound, where the calculated HBU results are always lower than the measured HBU, so that the dissipation energy in the material increases with increasing temperature.

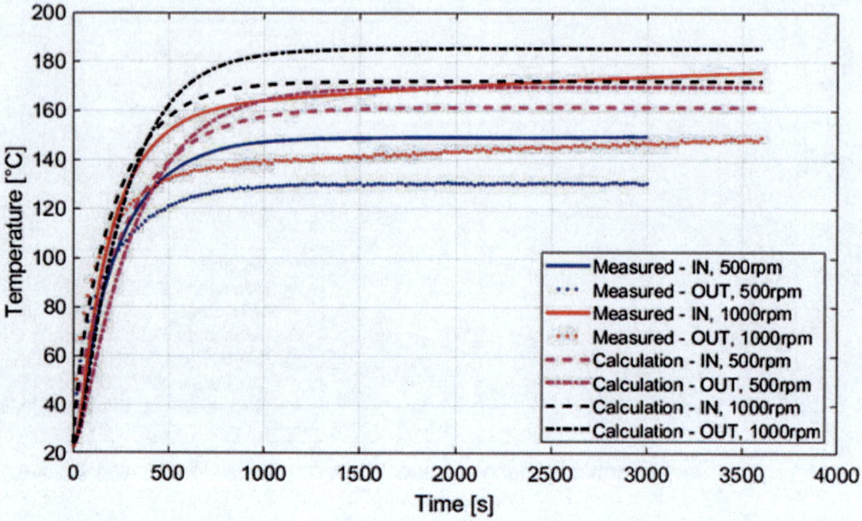

Fig. 18 SBR compound – HBU results 60°(1,000 rpm) and 70°(500 rpm) bending angle

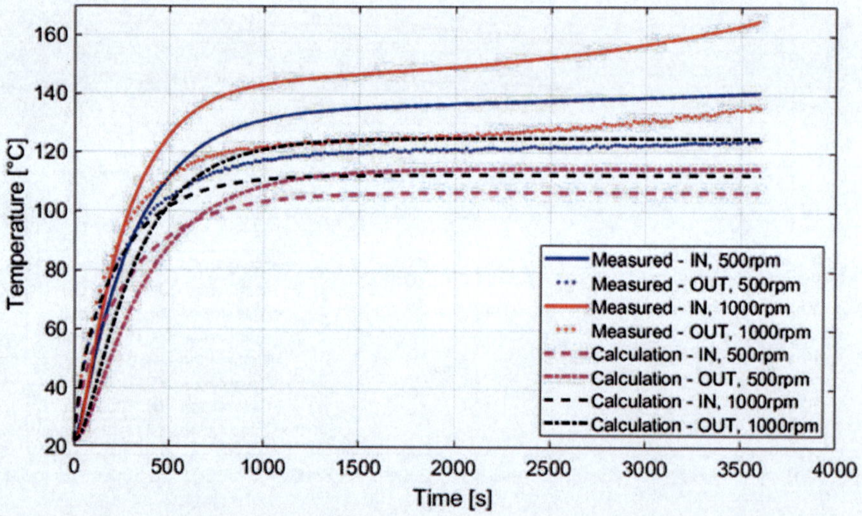

Fig. 19 BR compound – HBU results 60°(1,000 rpm) and 70°(500 rpm) bending angle

The comparison between measured and calculated temperatures at the end of the test for both compounds is summarized in Tables 6 and 7. The difference was calculated as

Table 6 SBR compound, comparison between measured and calculated temperature at the end of the test

Rotation speed	[rpm]	500			1,000		
Bending angle	[°]	30	50	70	30	50	60
T, measured – IN	[°C]	51.0	90.3	149.5	64.8	119.0	175.5
T, calculated – IN	[°C]	51.2	99.0	169.6	64.8	136.5	185.3
Difference	**[%]**	**0.4**	**9.6**	**13.4**	**0.0**	**14.7**	**5.6**
T, measured – OUT	[°C]	49.5	85.0	130.0	61.5	106.5	148.0
T, calculated – OUT	[°C]	49.6	94.7	161.3	61.4	127.2	172.0
Difference	**[%]**	**0.2**	**11.4**	**24.1**	**−0.2**	**19.4**	**16.2**

Table 7 BR compound, comparison between measured and calculated temperature at the end of the test

Rotation speed	[rpm]	500			1,000		
Bending angle	[°]	30	50	70	30	50	60
T,measured – IN	[°C]	37.4	80.7	140.4	50.6	107.7	165.0
T,calculated – IN	[°C]	40.3	72.2	114.6	50.0	95.8	124.9
Difference	**[%]**	**7.8**	**−10.5**	**−18.4**	**−1.2**	**−11.0**	**−24.3**
T,measured – OUT	[°C]	37.5	77.0	124.0	47.1	96.9	136.0
T,calculated – OUT	[°C]	38.8	68.0	106.5	46.9	87.1	112.5
Difference	**[%]**	**3.5**	**−11.7**	**−14.1**	**−0.4**	**−10.1**	**−17.3**

Table 8 SBR compound, comparison between measured and calculated temperature and time of equilibrium point

Rotation speed	[rpm]	500			1,000		
Bending angle	[°]	30	50	70	30	50	60
$T\ (E_{qv})$, measured	[°C]	45.3	76.6	100.0	55.6	93.8	110.5
$T\ (E_{qv})$, calculated	[°C]	45.8	84.1	140.6	55.3	110.5	148.1
Difference	**[%]**	**1.1**	**9.8**	**40.6**	**−0.5**	**17.8**	**34.0**
$t\ (E_{qv})$, measured	[s]	515	442	216	390	278	201
$t\ (E_{qv})$, calculated	[s]	520	520	520	387	388	388
Difference	**[%]**	**1.0**	**17.6**	**140.7**	**−0.8**	**39.6**	**93.0**

$$\text{Difference} = \frac{T_{\text{calculated}} - T_{\text{measured}}}{T_{\text{measured}}} .100\ [\%] \tag{4}$$

Regarding the equilibrium point, large differences between the measured and calculated temperatures $T(E_{qv})$ and times $t(E_{qv})$ are observed, as can be seen in Tables 8 and 9. The measured times $t(Eqv)$ decrease with increasing bending angle β (see Fig. 20), but the calculated times $t(Eqv)$ remain constant. It seems that the assumptions for transferring the convection values from the bending angle $\beta = 30°$ to other load cases may be incorrect, so that the convection values may differ at different bending angles.

Table 9 BR compound, comparison between measured and calculated temperature and time of equilibrium point

Rotation speed	[rpm]	500			1,000		
Bending angle	[°]	30	50	70	30	50	60
T (E_{qv}), measured	[°C]	–	64.3	92.7	41.0	73.5	92.0
T (E_{qv}), calculated	[°C]	36.8	62.6	96.0	43.9	78.9	100.7
Difference	**[%]**	**–**	**−2.6**	**3.6**	**7.1**	**7.3**	**9.5**
t (E_{qv}), measured	[s]	–	510	320	370	320	250
t (E_{qv}), calculated	[s]	622	623	624	480	480	481
Difference	**[%]**	**–**	**22.2**	**95.0**	**29.7**	**50.0**	**92.4**

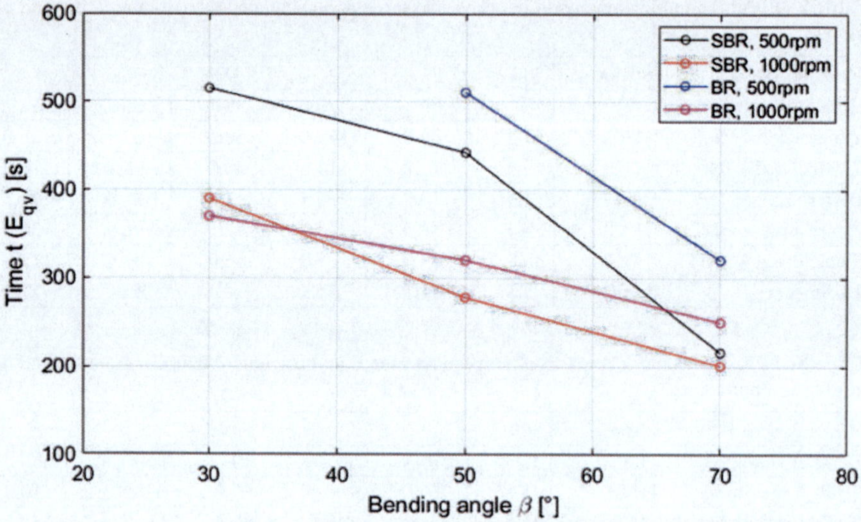

Fig. 20 Measured times, *t (Eqv)* according to the bending angle, *β*

The development of the temperature over time in the computational model is presented in Fig. 21. It can be seen that at the beginning of the calculation ($t = 5$ s) the temperature on the outer surface increases. The temperature maximum gradually moves to the center of the body over time until the temperature stabilizes.

The maximum error of the calculation results in relation to the measurements is approximately 25%. To keep this error as small as possible, it will be necessary to investigate the temperature dependence of the dissipation energy. In addition, a sensitivity analysis should be carried out to see how the individual material parameters affect the temperature development and equilibrium point.

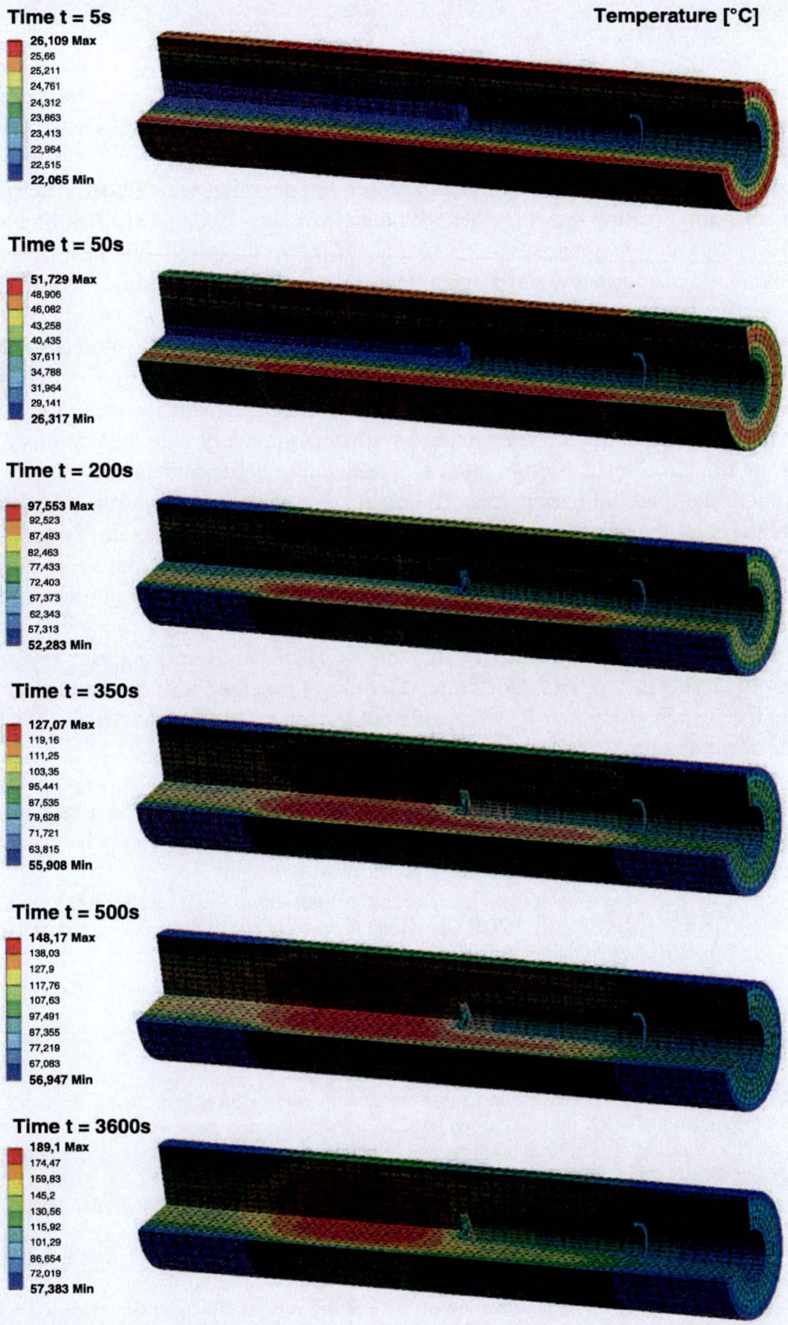

Fig. 21 Temperature redistribution in model over time

5 Conclusion

The fundamentals of a new approach for the calculation of HBU in the time domain were developed and implemented in the commercial FEM software Ansys. Results were obtained knowing only engineering hyperelastic materials parameters according to the Extended tube model at room temperature, which were determined by the 3-point bending test, together with data from the HBU test at a bending angle of 30° and rotational speeds of 500 and 1,000 rpm. Based on these tests, all mechanical and thermal material parameters were determined using the numerical response surface method in FEM software.

The results of HBU calculations compared with measurements indicate that for the dissipation energy of SBR compounds decreases with increasing temperature, while the dissipation energy of BR compounds increases with increasing temperature. It also appears that the convection parameters may vary with the bending angle β during the HBU tests. Further analysis is needed to better understand the effect of the individual material parameters. The main advantage of this approach is that the calculation is done in the time domain. This is possible thanks to the constant introduced, the so-called dissipation constant, D, which replaces the knowledge of the viscoelastic properties of the material. The use of this constant allows different loads to be simulated sequentially at different frequencies.

Furthermore, the unique test facility of the Heat Build-up Analyzer (HBuA – Coesfeld GmbH & Co. KG, Dortmund, Germany) together with the numerical tool presented in this study can be very efficiently used to predict the HBU of rubber under cyclic loading.

Acknowledgement The authors would like to thank Bonatrans Group a.s., Czech Republic, for information on the behavior of temperature development of rubber parts in operation.

Funding This work was supported by the Ministry of Education, Youth and Sports of the Czech Republic – DKRVO (RP/CPS/2022/006) and within the project OP RDE Junior Grants of TBU in Zlín, Reg. No. CZ.02.2.69/0.0/0.0/19_073/0016941.

References

1. Medalia AI (1991) Heat generation in elastomer compounds: causes and effects. Rubber Chem Technol 64(3):481–492. https://doi.org/10.5254/1.3538565
2. Gent AN, Hindi M (1988) Heat build-up and blow-out of rubber blocks. Rubber Chem Technol 61(5):892–905. https://doi.org/10.5254/1.3536225
3. Luo W, Yin B (2018) Modeling of the heat build-up of carbon black filled rubber. Polym Test 69:116–124
4. Li F, Liu J (2016) Numerical simulation and experimental verification of heat build-up for rubber compounds. Polymer 101:199–207
5. Riahi E, Allou F (2017) Quantification of self-heating and its effects under cyclic tests on a bituminous binder. Int J Fatigue 104:334–341
6. Tong X, Chen X (2018) The heat build-up of a polymer matrix composite under cyclic loading: experimental assessment and numerical simulation. Int J Fatigue 116:323–333

7. Soltani S, Sourki FA (2005) Effect of carbon black type on viscous heating, heat build-up, and relaxation behaviour of SBR compounds. Iran Polym J 14(8):745–751
8. Esmaeeli R, Nazari A, Aliniagerdroudbari H, Hashemi SR, Alhadri M, Zakri W, Farhad S (2018) Heat built up during dynamic mechanical analysis (DMA) testing of rubber specimens. Proceedings of the ASME 2018 international mechanical engineering congress and exposition. Volume 9: mechanics of solids, structures, and fluids. Pittsburgh, Pennsylvania, USA. November 9–15, 2018. V009T12A059. ASME https://doi.org/10.1115/IMECE2018-88627
9. Schieppati J, Schrittesser B, Wondracek A, Robin S, Holzner A, Pinter G (2018) Impact of temperature on the fatigue and crack growth behavior of rubbers. Procedia Struct Integr 13:642–647. https://doi.org/10.1016/j.prostr.2018.12.106
10. Le Saux V, Marco Y, Calloch S, Doudard C, Charrier P (2010) Fast evaluation of the fatigue lifetime of rubber-like materials based on a heat build-up protocol and micro-tomography measurements. Int J Fatigue 32:1582–1590. https://doi.org/10.1016/j.ijfatigue.2010.02.014
11. Stoček R, Kratina O, Voldánová J, Kipscholl R, Heinrich G (2022) Realistische Charakterisierung der Wärmeentwicklung von Gummi unter hochdynamischen, multiaxialen zyklischen Belastungen. GAK Gummi-Fasern-Kunststoffe 4:14–21
12. Stoček R, Stěnička M, Kipscholl R (2019) Heat build-up characterization under realistic load. In: Constitutive models for rubber XI – proceedings of the 11th European conference on constitutive models for rubber. CRC Press/Balkema, Nantes, pp 157–162. https://doi.org/10.1201/9780429324710-28
13. Raymond H. Myers, Douglas C. Montgomery, Christine M. Anderson-Cook, Response surface methodology: process and product optimization using designed experiments. 4th edn. Wiley, 2016, 856. 978-1-118-91601-8
14. Stocek R, Heinrich G, Kipscholl R, Kratina O (2021) Cut & chip wear of rubbers in a range from low up to high severity conditions. Appl Surf Sci Adv 6:100152. https://doi.org/10.1016/j.apsadv.2021.100152
15. Rifdi Rizuan MI, Abdul Wahab MA, Romli AZ (2015) Effect of carbon black structures towards heat build-up measurements and its dynamic properties. Adv Mat Res 1134:131–137. ISSN: 1662-8985
16. Timoshenko S (1953) History of strength of materials. McGraw-Hill, New York
17. Akutagawa K, Hamatani S (2015) The new interpretation for the heat build-up phenomena of rubbery materials during deformation. Polymer 66:201–209
18. Ansys® Mechanical Enterprise, Release 2020R2, Help System, Internal Heat Generation, ANSYS, Inc.
19. Kaliske M, Heinrich G (1999) An extended tube-model for rubber elasticity: statistical-mechanical theory and finite element implementation. Rubber Chem Technol 72:602–632

Adv Polym Sci (2023) 289: 143–166
https://doi.org/10.1007/12_2022_122

Published online: 31 May 2022

The Effect of Thermal Ageing on the Fatigue Resistance of Hydrogenated Acrylonitrile Butadiene Rubber (HNBR) Compounds

Barnabas Shaw, Julien Ramier, and James J. C. Busfield

Contents

Abstract Compounds incorporating hydrogenated acrylonitrile butadiene rubber (HNBR) are widely adopted in applications that operate under extreme conditions such as at high temperatures, often whilst in contact with oil and other chemicals. They also work effectively when subjected to high strains and high strain rates. The detailed mechanisms that allow these materials to perform so well under these specific conditions are not yet fully understood. In this study a high acrylonitrile (ACN) content HNBR is incorporated into four different rubber compounds that are subjected to different thermal ageing conditions to investigate their impact on fatigue crack growth resistance of the cured compound. The resulting compounds are

B. Shaw and J. J. C. Busfield (✉)
School of Engineering and Materials Science, Queen Mary University of London, London, UK
e-mail: j.busfield@qmul.ac.uk

J. Ramier
Schlumberger Cambridge Research Ltd, Cambridge, UK

subjected to detailed DSC, TGA and FTIR analysis to help understand the changes that take place in the compounds during ageing for the first time. Counterintuitively, in this work ageing under specific set of conditions can produce an increase in the fatigue crack resistance in some of the elastomer compounds and when this happens it is observed that this occurs when the HNBR compounds develop knottier tearing profiles. Analysis of the DSC, FTIR and TGA results confirm that during ageing there is a steady evolution of the some of the ingredients in the compound such as the processing oils and the plasticiser which appears to have a significant effect on the fatigue properties of the HNBR materials.

Keywords Crack growth · Elastomer · Fatigue · Hydrogenated acrylonitrile butadiene rubber (HNBR) · Tearing energy · Thermal ageing

1 Introduction

Understanding of the fatigue properties of HNBR after ageing is of great importance as HNBR compounds are widely adopted elastomeric materials used in a wide range of extreme environments. However, there is little prior research investigating the ageing behaviour of typical HNBR compounds, especially when thermally aged in air. Researchers such as Shaw et al. [1] have investigated this thermal ageing in air. This paper is a continuation of that work using similar materials. Research by Beranger et al. [2] previously showed that NBR and HNBR compounds that are thermally aged and tested in fatigue have intriguing results. Their NBR compounds, for example, indicated a potential increase in fatigue properties after ageing when tested at the ageing temperature, however no detailed explanation was offered. HNBR materials that are used in sealing applications [3] particularly in the oil and gas or geothermal applications are often exposed to high temperatures in excess of 150°C whilst being in contact with organic solvents and other aggressive chemicals. These materials have to operate robustly at high strains and at high strain rates whilst subjected to these extreme conditions. Understanding the origin of these surprisingly good properties might help enable HNBR elastomers to be improved still further by exploiting their significant toughening mechanisms.

A limited number of research papers into HNBR products aged in oil have been published in the last decade including papers by Weitao et al. [4], Alcock and Jorgensen [5], and Alcock et al. [6] who have performed studies on how HNBR compounds respond to a combination of different chemicals at elevated temperatures. Ulu et al. [7] examined the fatigue of HNBR compounds under thermal ageing using a limited Wöhler curve analysis. Beyond this work there is very limited previous research investigating the effect of thermal ageing in air on commercial HNBR compounds.

Researchers like Gent et al. [8] and Hamed [9] have demonstrated how crystallisation at large strains in tear test piece made from polymers such as Natural

Table 1 The elastomer formulations. Numbers are parts per hundred rubber (PHR)

Material	Name	HNBR0	HNBR60	HNBR75	HNBR75NP
Elastomer	Therban 4364	100	100	100	100
Filler	N550		60	75	75
Antidegradants	X	1.5	1.5	1.5	1.5
Plasticisers	Y	10	10	10	
Crosslink	Peroxide	8	8	8	8
Co-agent 1 & 2	Z1 + Z2	10	10	10	10

Rubber, that exhibit strain induced crystallisation, can produce crack tip blunting. This crystallisation in advance of the crack tip can cause directional changes in crack growth, which has the effect of dramatically increasing the radius of the crack tip and which toughens the materials very effectively. The ability of HNBR to crystallise either at low temperatures or under strain is an on-going debate in the scientific literature. The extent to which HNBR compounds can exhibit crystallinity is dependent upon the ACN content. Hayashi [36] showed that at low ACN contents HNBR can consist of repeated units of tetramethylene spaced between the ACN units. Kobatake et al. [10] developed this further and demonstrated that polymer segments with five or more repeating units of tetramethylene exhibited the ability to strain crystallise, in a manner that was similarly to NR. By increasing ACN content, these units were spaced further from each which ultimately inhibited the development of crystallinity. The ability for high ACN, HNBR materials to crystallise has also been investigated by Severe and White [11]. Using X-Ray Diffraction (XRD) and differential scanning calorimetry (DSC), they proposed that high ACN content HNBR materials also formed crystals due to sequential tetramethylene and ACN units, forming ordered structures. The detailed nature of these structures has not yet been widely investigated.

The base polymer used in this study is an HNBR with 43% ACN content (approximately 5.5% ± 1 residual double bonds) cured using a peroxide curing system. The carbon black filler N550 has an STSA surface area of 40 m^2/g and an average particle size of 39–55 nm. The rest of the formulation used widely adopted plasticisers and antidegradants that are typically used for O&G high temperature sealing applications. The four compound formulations used are shown in Table 1.

The four formulations are broadly similar except for the filler and plasticiser content. The number in each case relates to the compound's filler content in parts per hundred rubber (phr). Further details regarding formulations cannot be disclosed due to commercial considerations.

These products were mixed using an internal mixer and the compounds were finished off using a two-roll mill. Before samples were cured, a sample from each compound was tested to identify the optimum cure characteristics using a standard oscillating disk rubber cure rheometer. All four materials exhibited a plateau type of cure. Vulcanisation and moulding were performed using a hot press at 170°C. The preparation of the testing specimen involved moulding 2 mm thick sheets and then using standard cutters to cut out the test samples. The materials were tested in both their unaged condition and then after ageing. Two different ageing conditions were used in this study as follows:

- Aged in an air oven at 100°C for 168 h (1 week).
- Aged in an air oven at 150°C for 168 h.

Samples were placed in an unstrained state during the ageing.

2 Experimental

All the cyclic fatigue crack growth tests were carried out on an Instron 8801 Servo-Hydraulic Test Machine. The mechanical properties of each of the compounds were all tested using an Instron 5967 Testing Machine equipped with a video-extensometer.

2.1 *Tensile Tests*

Tensile test pieces were cut out using an ASTM D412 type C dumbbell die geometry and testing was undertaken to measure each material's initial unstrained performance in uniaxial tension. The dumbbell specimen features wider gripping ends with a larger cross-section area and a working section at the centre of approximately 30 mm length, 6 mm width and approximately 2 mm thickness. Two contrasting ink dots were placed on the surface of the working sections which were used to measure the strain of the test pieces in the gauge length accurately using a video-extensometer. For the initial stress vs. strain characterisation, the testing parameters were:

- Maximum Strain: 50%
- Strain rate: 0.01/s (slow quasi-static conditions)

2.2 *Cyclic Plane Strain Fatigue*

Plane strain fatigue crack growth (FCG) test pieces were used to characterise cyclic FCG behaviour as has previously been described by Stocek et al. [12]. The material model developed related the rate of crack growth per fatigue cycle in the elastomer to the strain energy release rate. The research is based on work by Griffith [13] who suggested that a crack in glass would grow if the elastic energy released by the growth was greater than the surface free energy of the surfaces created. Rivlin and Thomas [14] hypothesised that the energy release rate required to propagate a crack at a specific rate was a material characteristic which was therefore independent of the test piece geometry. If this was the case, then geometric changes should not influence

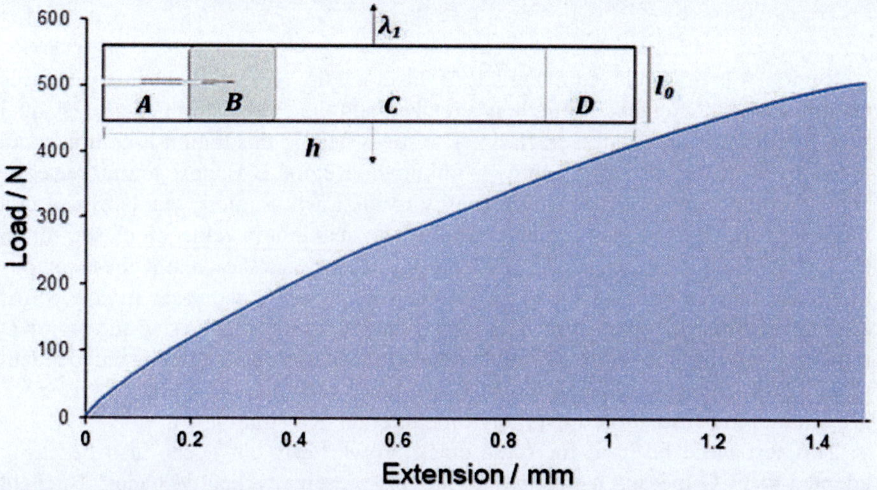

Fig. 1 Typical load extension loading curve of sample HNBR60, used to find the elastic strain energy, *U*, in the test piece and a diagram of the tearing regions in a fatigue crack growth test piece

the tear rate. Experiments were carried out by Greensmith and Thomas [15], Thomas [16] and Lake [17] to investigate this with various testing geometries to establish relationships for strain energy release rate. Tsunoda et al. [18] showed that the behaviour was also true for non-strain crystallising materials. The plane strain FCG test specimen adopted in this work is one of the most commonly adopted geometries used to characterise the crack growth behaviour.

A test specimen is prepared with a horizontal length that is significantly larger than the vertical height between the long sample grips and the thickness is significantly smaller than the other two dimensions. The specimen is gripped along its two longest sides and it is strained perpendicularly to them at an extension ratio λ_1. Due to the constraints from the grips, during deformation the other extension ratios are constrained and $\lambda_2 = 1$ and due to the incompressibility constraint $\lambda_3 = 1/\lambda_1$. If a test piece features a horizontal length that is 10 times longer than its vertical height length, as is the case with a pure shear test specimen, then a pure shear region can be found anywhere away from the vicinity of the free edge. This approximation has been adopted here to calculate the tearing energy.

This test piece was adopted, to measure the crack growth behaviour of rubber by introducing a crack or cut in the mid plane part of the way through the sample, by Rivlin and Thomas [14]. This test piece geometry shown in Fig. 1 allows the strain energy release rate, also often widely known as the tearing energy, to be simply derived from the elastic strain energy, *W*, measured in the region in the middle of the uncut test piece multiplied by the unstrained height of the sample, l_0.

$$T = Wl_0 \tag{1}$$

Region A is fully relaxed to the left of the crack tip. Here the rubber experiences no strain when the cracked sample is stretched and the elastic strain energy is equal to 0. Region B is around the crack tip. The stress field in this region is complex and the value of the elastic strain energy is unknown. Region D is close to the free edge and again the value of elastic strain energy is unknown. Region C is deformed in a more uniform homogeneous plane strain. The maximum value of elastic strain energy is W in this region. As the crack propagates through the sample, these regions move and change size. As the crack progresses, region C decreases in size whilst region A increases. The position of B moves but it does not change size and region D remains constant. It is worth noting that the strain energy release rate is independent of the crack length when using this FCG test piece geometry. This enables the crack growth rate to be monitored over any time interval and strain level.

This test piece be used for static crack growth tests but it can also be easily adopted to FCG measurements, where the test piece is cyclically loaded. Busfield et al. [19] suggested an alternative way to calculate the strain energy release rate using the direct measurement of the force and displacement data from an individual loading cycle measured during a cyclic FCG test. This allows for the effects such as cyclic stress softening to be accounted for throughout the test. To calculate the total elastic strain energy, *U*, it is simply a case of integrating the loading curve of the load vs. deflection graph at a particular point during the test. The strain energy density, *W* can be approximated by the integral of the curve shown in Fig. 1 by an appropriate volume.

$$W = \frac{U}{V} \tag{2}$$

The equation for volume, *V*, here can be expressed ignoring edge effect as:

$$V = l_0 t(h - c) \tag{3}$$

where *t* is the thickness of the sample, *h* is the longitudinal length of the sample and *c* is the length of the crack. But this calculation ignores edge effects. To account for these edge effects, De [20] proposed that an additional length, *x*, could be added to the uncracked sample width when calculating the strained sample volume to compensate for the edge effects. Measuring this experimentally he found that this length was approximately equal to 28% of the unstrained sample height l_0 and was independent of crack length. This then adapts the volume of material remaining in the plane strain region to be:

$$V = l_0 t(h - c + x) \tag{4}$$

This can then be incorporated as the effective volume into Eq. (1) to derive a strain energy release rate expression measured as the cyclic crack growth test is progressing:

$$T = Wl_0 = \frac{U}{t(h - c + x)} \tag{5}$$

This FCG experiment was used to measure the relationship between the strain energy release rate and the crack growth rate per cycle to help characterise the FCG resistance of a material. The crack growth rate is measured by capturing time lapse images of the sample during the test. The photos monitor the length of the crack at specified time intervals that can then be processed using computer software, such as 'Image J'. The software scales pixels on a photograph to physical lengths that were calibrated using a suitable length scale captured in the image. The time for each photo is known and by knowing the test frequency allows the cycle number to be calculated which then allows the average crack growth rate per cycle d*C*/d*N* to be determined.

All the values for the crack propagation rates and strain energy release rates were made after the test piece had been through an initial 1,000 conditioning cycles [21]. In fixed deformation dynamic tests, cycle 1,000 is often used as an approximation of when cyclic stress softening effects are no longer significant. Therefore, to reduce complexities during the characterisation of the compounds, the first 1,000 cycles of FCG were always neglected [22].

Plane strain FCG test specimens used in this body of work were rectangular in shape measuring 175 × 45 × 2 mm. The crack of 30 mm was inserted via a blade and then propagated a small amount by hand to roughen the crack tip to a more typical steady state crack tip Papadopoulos et al. [22]. The top and bottom grips both gripped the sample along the full longitudinal length leaving a sample height between the grips of approximately 15 mm. Therefore, the working section of the FCG test piece was reduced from the cut shape dimension of 45 mm to just 15 mm between the grips. This ensured a working section length to height ratio that was approximately 10 times longer, which was the amount required to ensure that the central region of the test piece remained in plane strain. The fatigue test parameters ranged accordingly:

- Maximum strain values: 10–50%
- Minimum strain value: 0% (under fully relaxing conditions – indicating that the minimum load in any given cycle was always 0 N)
- Strain rate values: 20–100%/s
- Testing frequency: 1 Hz sinusoidal wave
- Typical number of cycles at a specific strain amplitude: 1,000–5,000

To enable samples to be compared, crack growth rate was maintained within a certain range for all the tests reported in this work. Very rapid crack growth behaviour was hard to measure as at least 1,000 cycles were adopted to stabilise the cyclic strain softening effects at each strain cycle. The unfilled elastomer was

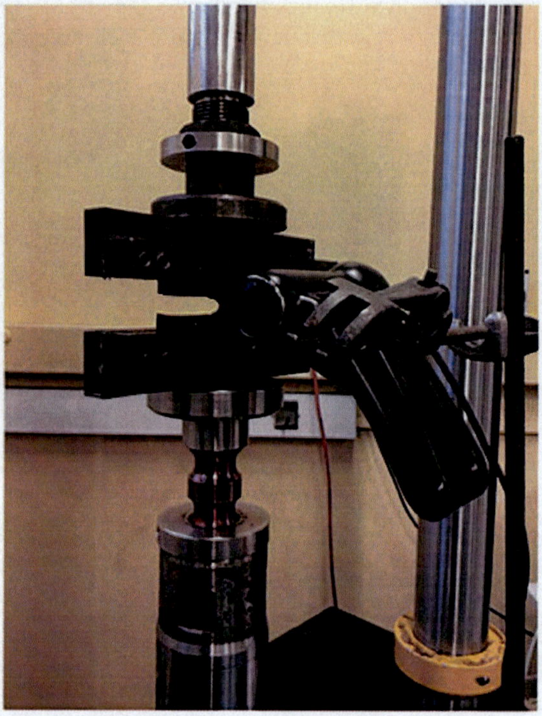

Fig. 2 Cyclic FCG test setup using the hydraulic Instron testing machine

much less stiff than the filled elastomer compounds and so different strain ranges were needed to compare the crack growth rates over broadly comparable FCG rates.

The test procedure consisted of accurately measuring the specimen's dimensions, including an average thickness over the sample. The sample was placed in the grips and tightened. As the specimen was gripped tighter it would start to buckle as volume was forced out between the grips. The grips were then moved to adjust for this so that the test would start at zero load. A line was drawn onto the sample along the grip edge. This helped to identify if slippage had arisen in the grips. The camera was set up to take pictures of the crack as it propagated every minute. As the testing frequency was set at 1 Hz, a picture was taken every 60th cycle. A length scale was caught in the frame for each image. In the first cycle of every test, the camera was initially triggered to begin taking pictures at the maximum displacement of every FCG cycle. After this, data was collected and processed to determine the strain energy release rate and crack growth rate for each specific crack growth rate after the 1000th cycle. A photo of the test facilities is shown in Fig. 2.

All four compounds were initially characterised in a virgin, unaged state at room temperature to establish a baseline characteristic for their cyclic fatigue behaviour. Samples were then tested under a similar range of conditions after ageing at either 100°C or 150°C for a week.

For cyclic fatigue loaded parts, the approach introduced by Gent et al. [23] showed that a characteristic crack growth rate per cycle can be measured. They showed that the extent of crack growth during each loading cycle was essentially determined by the maximum strain energy release rate encountered during the cycle. A range of researchers such as Lake [24], Busfield et al. [25], Asare and Busfield [26] have characterised the relationship between the strain energy release rate and crack propagation rate over the engineering FCG regime using a power law dependency:

$$\frac{\mathrm{d}c}{\mathrm{d}N} = B\left(\frac{T}{T_u}\right)^{\beta} \tag{6}$$

where $T_u = 1$ kJ/m^2 and is included only to make T/T_u dimensionless, B and β are material constants for the specific compound. β can be characterised from the slope of a graph of dC/dN vs. T/T_u when they are plotted on logarithmic scales. The value of β is known to reflect the elastomer properties such as the amount of hysteresis or the onset of strain induced crystallisation [22]. Stocek [27] has proposed an alternative equation to fit the stable FCG data more reliably, but the data analysis of the work presented here was completed in advance of that paper being published.

In this investigation each elastomer compound was cyclically fatigue tested pre and post ageing with the crack growth per cycle propagation rates measured with respect to a range of different strain energy release rates. The oven temperature was measured continuously with a thermocouple to ensure the correct temperature was maintained. After ageing, the samples were left to cool and equilibrate to room temperature conditions before the initial cut was inserted into the sample. These data are all plotted on logarithmic scales to allow the crack growth behaviour to be characterised over the range of strains encountered in dynamic seals.

2.3 Thermogravimetric Analysis (TGA)

Thermogravimetric (TGA) tests were undertaken to help identify the composition of each sample before and after ageing. This involves measuring the mass loss as a sample was heated to a high temperature in an inert nitrogen atmosphere. The thermal decomposition of the material was measured as the percentage of mass loss. Samples were cut using a scalpel into approximately 10 mg pieces, which were placed in a crucible for testing. Tests began at room temperature and the temperature was raised by 10°C/min up to 800°C. Changes in mass could be highlighted by plotting the differential of the mass loss. This highlighted the regions where the rate of mass lost changed so that specific regions where the mass loss rate was greatest could be identified and examined more carefully.

2.4 Differential Scanning Calorimetry (DSC)

Differential Scanning Calorimetry (DSC) testing was performed on all samples in order to measure glass transition temperature (Tg). Sample temperatures were lowered to −90°C and left to equilibrate for 10 min. A standard temperature ramp was then performed from −90 to 100°C at a rate of 10°C/min. Samples were left to equilibrate for 10 min before returning back to room temperature.

2.5 Fourier Transform Infrared Spectroscopy (FTIR)

Both unaged and aged samples were characterised using Fourier Transform Infrared Spectroscopy (FTIR). Samples were tested before and after ageing, on the surface and within the cross-section of the sample. This provided a wide analysis with which spectrum could be analysed before and after ageing. The use of FTIR on elastomer samples is common and has been used for comparing elastomer formulations before and after ageing, for example, by Lou et al. [28] and Fernandez-Berridi et al. [29]. FTIR spectrum of all individual ingredients from all samples were provided and used to match the peaks generated by the formulations used in this work. This enabled a direct correlation from individual ingredients to full formulations. By comparing the different trace before and after ageing allows the changes to the chemistry and hence any potential loss of ingredients during ageing to be identified.

Each sample, that was either unaged or aged, was tested at least three times so as to look for any potential changes within the samples and the results with the strongest signals were used. All samples were both tested on the exterior surface and in the cross-section of the specimen that had been cut using sharp scissors. This enabled a comparison of chemistry at both the sample surface and within the cross-section where external materials such as dirt on surface can be eliminated. The results shown here are all from the freshly cut cross-sections of the samples.

3 Results

Figure 3 shows the first cycle tensile behaviour for all four cured compounds up to 50% strain at room temperature.

As anticipated the unfilled sample, HNBR0, had the lowest modulus, as it has no reinforcing filler present. The HNBR60 and HNBR75 both had a higher measured modulus which increased as reinforcing filler volume fraction was increased. HNBR7 NP, the compound without plasticiser, had the highest modulus of all the four cured compounds. This contract between HNBR75 and HNBR75NP highlights the significant role the plasticiser has on reducing the compound modulus.

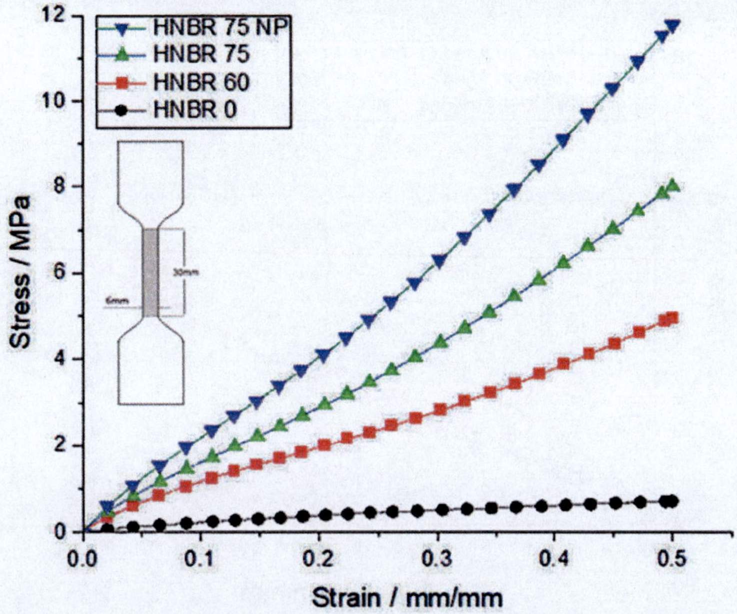

Fig. 3 Tensile tests of HNBR0, HNBR60, HNBR75, and HNBR75NP to 50% and a typical dumbbell specimen with length and width dimensions

Samples were also tested in tension at room temperature after ageing using the same conditions as the initial modulus test. Figure 4 shows the data for HNBR60, which was similar to the behaviour seen for all four compounds.

It is evident that ageing introduced a significant change to the stress vs. strain behaviour. When aged in air, the HNBR60 samples showed an increase in modulus, with a larger increase after ageing at 150°C. This increase in stiffness of the compound might result from oxidation effects that might be activated more readily at higher temperatures. In addition to potential oxidation during ageing, potential changes in the elastomer composition might also be taking place. Equilibrium swelling tests and the plateau rheology curves suggested that additional crosslinks were not being formed. Therefore, the likely reason for the increase in modulus is due to either ageing mechanisms, loss of material such as plasticiser and low molecular weight material or potential filler flocculation [30]. Previous research by O'Rourke [31] has shown that ester plasticisers are lost from rubber compounds when aged at temperatures above 150°C.

Figure 5 shows the fatigue crack growth tests plots for the four unaged materials, and Table 2 summarises the calculated B and β parameters.

When plotted in this way, the tougher materials are the ones with the slower crack growth rate per cycle at a given strain energy release rate. In this case the toughest materials are the ones that are furthest to the bottom right on the graph.

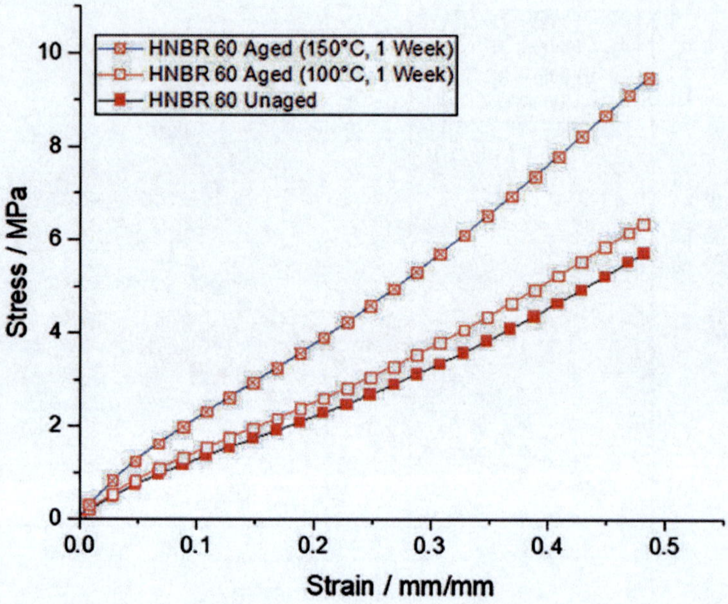

Fig. 4 Room temperature tensile tests of HNBR60 unaged, aged at 100°C in air and aged at 150°C in air

Understandably, as carbon black content was increased between HNBR60 and HNBR75, an increase in toughness was seen alongside a decrease in material constant β. Both of these properties correspond to an increase in hysteresis, mirrored in the tensile test results. It is worth noting that testing beyond these strain energy release rates results in a catastrophic tearing regime. The independently measured tear strength of the material measured using a trouser tear test piece is shown by the dotted vertical line. Therefore, it appears the unfilled material performs better in fatigue than the filled materials, however this was only over a very narrow strain energy release rate range and the benefit of the carbon black filler becomes more evident at the higher strain energy release rates.

Surprisingly, HNBR0 was tougher than the filled samples for strain energy release rates between 1.5 and 3.0 kJ/m^2. Despite this outcome not being anticipated, it can be attributed to a number of factors. The β value of the unfilled sample is much lower than the filled samples. This typically indicates a higher level of hysteresis, particularly at the crack tip. For some elastomers like natural rubber that are known to exhibit strain induced crystallisation, the β values are often seen to be in the range of 2–2.5 [32].

Strain induced crystallisation creates regions of knotty tearing behaviour, where the crack growth profile is not always smooth, but the crack growth path becomes more convoluted. This usually makes the measured FCG data much more scattered. When it arises, the knotty tearing slows down the crack growth behaviour. This data scatter can be a result from change in the rate of testing [33] or the temperature of the

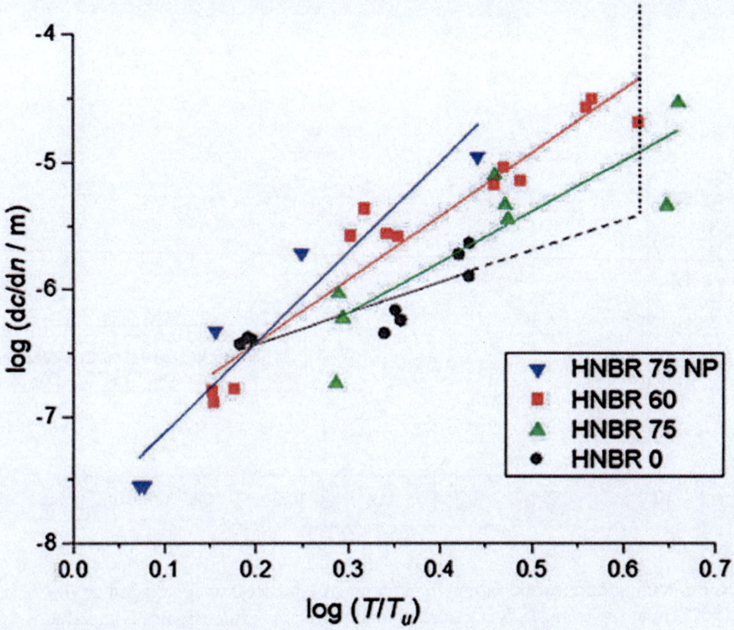

Fig. 5 Room temperature fatigue crack growth behaviour of all four compounds unaged at 1 Hz. T_u is included to make the axis dimensionless. Its value is 1 kJ/m^2

Table 2 Material constants found for unaged compounds tested in cyclic fatigue

	B/m	β
HNBR0	9.83×10^{-4}	2.5
HNBR60	5.97×10^{-4}	5.0
HNBR75	6.19×10^{-4}	4.0
HNBR75NP	3.96×10^{-4}	7.1

testing [34] both of which alter the ability of the elastomers such as Natural Rubber to exhibit strain induced crystallisation during tear or fatigue testing.

However, it can be assumed that the knotty tearing behaviour, whilst being unpredictable, results in a material with a tougher fatigue response. It is also apparent that the scatter range increases as strain energy release rate increases, showing that this toughness response does not simply translate the strain energy release rate vs. crack propagation rate relationship higher or lower, it affects the gradient of this relationship and more specifically the β material parameter. Another noticeable difference was in the crack tip shape between regions of knotty tearing and smooth tearing in all the samples. In regions of knotty tearing, the crack tips became much blunter and rounder whereas in smooth tearing regions the advancing crack tip was much sharper.

The sample with no plasticiser, HNBR75 NP had a significantly lower fatigue resistance than its counterpart HNBR75 that incorporated a plasticiser. In fact,

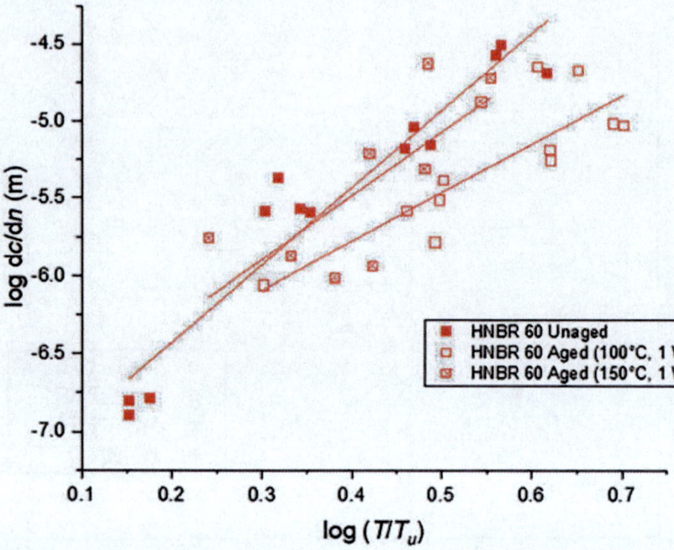

Fig. 6 Room temperature cyclic fatigue behaviour of HNBR60 unaged, aged at 100°C in air and aged at 150°C in air at 1 Hz. T_u is included to make the axis dimensionless. Its value is 1 kJ/m^2

results for this sample could only be tested between the range of 6–8% strain before catastrophic tearing took place. A very clear change in β value was observed for this compound. The plasticiser has a significant effect on chain mobility. The effect here on hysteresis is to lower the glass transition temperature T_g, of the compound. By removing some of the plasticiser during ageing, the T_g will increase somewhat, with a knock-on effect on increasing the hysteresis. The tensile tests showed this to be the case and was confirmed by the DSC testing reported later on in this report. It therefore gives evidence that the plasticiser's influence of reducing chain mobility in the polymer matrix and potentially reducing the interactions between the polymer matrix and the filler network [35] both have a significant effect on the FCG rate. HNBR0, HNBR60, HNBR75 and HNBR75NP were then tested under similar conditions after ageing. These tests were all done at room temperature using the same testing parameters as the unaged samples. The results are shown in Figs. 6 and 7, and B and β parameters are summarised in Tables 3 and 4.

HNBR60 aged at 100°C in air not only shows a surprising and unexpected increase in toughness but also features a lower β value. There is also a reasonable level of scatter in the results, especially at higher strain energy release rates. This suggests that an increase in the knotty behaviour produced an increase in the scatter at higher strain energy release rates, resulting in a lower value of β. It can, therefore, be argued that ageing in air at 100°C increases the incidence of knotty behaviour which introduces an additional, unidentified, toughening mechanism. Another question this raises is whether the toughening behaviour is evident over the entire range of strain energy release rate that has been tested.

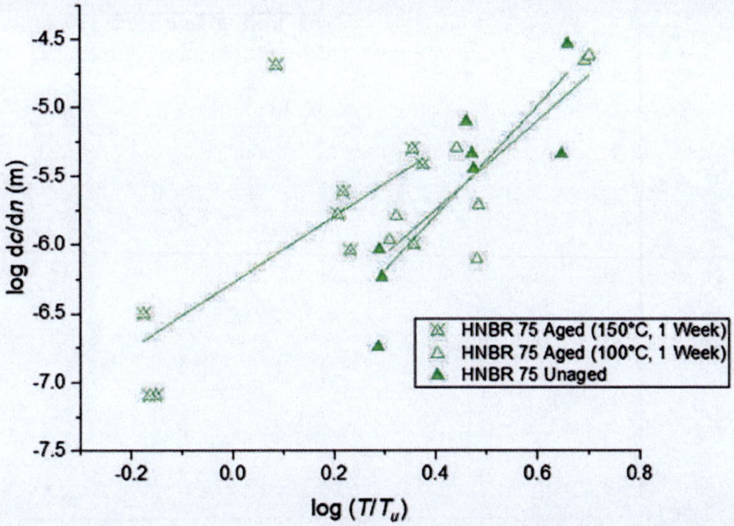

Fig. 7 Room temperature cyclic fatigue behaviour of HNBR75 unaged, aged at 100°C in air and aged at 150°C in air at 1 Hz. T_u is included to make the axis dimensionless. Its value is 1 kJ/m^2

Table 3 Material constants found for HNBR60 unaged and aged compounds in cyclic fatigue

	B/m	β
HNBR60 unaged	5.97×10^{-4}	5.0
HNBR60 aged (100°C, air, 1 week)	8.84×10^{-4}	3.1
HNBR60 aged (150°C, air, 1 week)	8.00×10^{-4}	4.1

Table 4 Material constants found for HNBR75 unaged and aged compounds in cyclic fatigue

	B/m	β
HNBR75 unaged	6.19×10^{-4}	4.0
HNBR75 aged (100°C, air, 1 week)	8.67×10^{-4}	3.3
HNBR75 aged (150°C, air, 1 week)	1.88×10^{-3}	2.4

HNBR60 aged at 150°C in air as anticipated showed a reduction in fatigue resistance compared to materials that were aged at 100°C in air. However, the gradient of the plot was similar to the unaged material. There was however an increase in the measured experimental scatter resulting from an increase in knotty tearing. This scatter was present both above and below the unaged results, indicating that the general behaviour of the elastomer after ageing at 150°C in air was weaker than the unaged results, but that the frequency at which knotty tearing occurs increased.

HNBR75 aged at 100°C in air featured FCG results similar to that of the unaged material, except with a broader scatter in the measured behaviour. Any lowering of β value here should be understood with that in mind, similarly to the aged results seen in HNBR60.

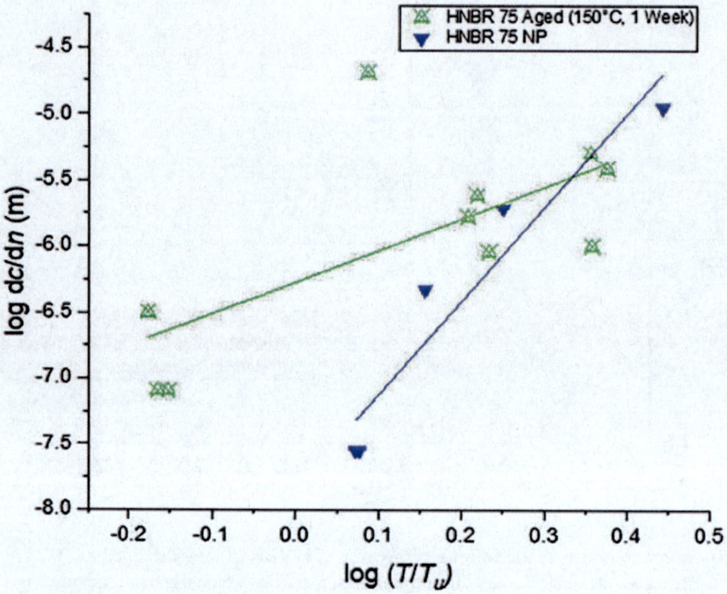

Fig. 8 Room temperature cyclic fatigue behaviour of HNBR75 aged at 150°C in air and HNBR75NP unaged. T_u is included to make the axis dimensionless. Its value is 1 kJ/m^2

HNBR75 aged at 150°C in air showed a significant reduction of the fatigue resistance, together with a significant increase in the experimental scatter. It might be expected that once a certain threshold temperature had been exceeded during ageing that the fatigue properties would inevitably decrease. In this case, 150°C was clearly above this threshold when compared to the tests aged at 100°C. In this case, oxidation might have become a significant contributor to the weakening of the elastomer.

It was thought possible that the ageing process might have also removed some of the plasticiser, potentially via evaporation. Therefore, Fig. 8 shows the HNBR75NP unaged sample plotted with the HNBR75 aged at 150°C in air in order to compare their behaviour.

The HNBR75NP samples were so weak after either ageing process that it was impossible to measure the crack growth resistance of the materials after ageing.

The results at higher strain energy release rates are broadly comparable. However due to a dramatic change in β value, likely due to oxidation, the lower strain energy release rates are not. This clearly shows that whilst the plasticiser might be lost, there are obviously additional effects taking place such as oxidation and loss of other low molecular weight material. TGA and FTIR experiments were undertaken to examine this in greater detail.

DSC tests were performed on all four formulations including the HNBR60 formulation aged in air at 100°C to measure the T_g differences. All formulations

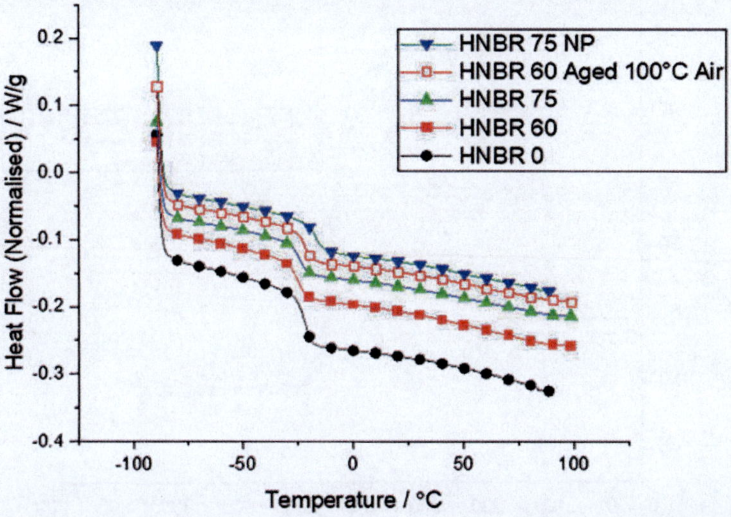

Fig. 9 DSC tests performed on unstrained samples of unaged HNBR0, unaged HNBR60, unaged HNBR75, unaged HNBR75NP and HNBR60 aged at 100°C in air

Table 5 Resulting T_g measurements from DSC tests

Material	T_g (°C)
HNBR0	−22.6
HNBR60	−24.4
HNBR60 aged 100°C in air	−21.9
HNBR75	−24.3
HNBR75NP	−17.9

were tested five times for an average reading of T_g in an unstrained state. The median results of these DSC for each cured compound can be seen below (Fig. 9, Table 5).

T_g was measured as the midpoint of the dramatic heat flow drop found at approximately −22°C for the unfilled material. As can be seen from the results, the addition of carbon black reinforcement to the samples lowered the T_g slightly by approximately 1.7–1.8°C. The removal of plasticiser in HNBR75 NP has a greater effect on the T_g as it had an average increase of 6.4°C. Changes in T_g can cause changes in hysteresis during the room temperature testing of the crack growth behaviour. The ageing process of HNBR60 also produced a T_g increase of approximately 2.5°C. This could be a result of net loss of some of the plasticiser.

TGA tests were undertaken to attempt to understand the detailed composition of each sample before and after ageing. Results were plotted as the percentage mass loss with temperature. Changes in mass could be highlighted by plotting the differential of the mass loss. This highlighted the gradient changes observed in the results so that specific regions where the mass loss rate was greatest could be identified and analysed individually.

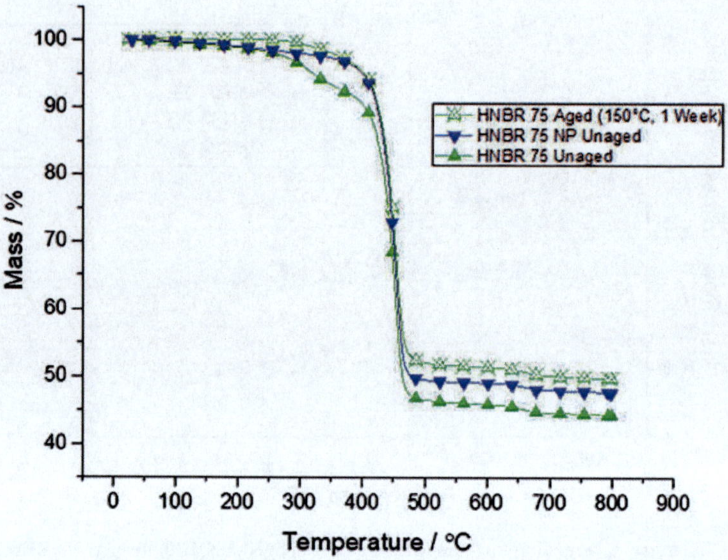

Fig. 10 TGA results for HNBR75 aged at 150°C in air, HNBR7NP unaged and HNBR75 unaged

It is important to note that the TGA results were normalised so that the initial test mass is 100%. It was clear from the earlier experiments that the mass of the test piece may have been altered by the ageing process. So after ageing, different samples even from the same initial compound may have different inorganic residual percentage masses depending on the detailed composition of the compound prior to TGA testing (Fig. 10).

The general behaviour is similar for all the samples: there is a small mass loss of up to ~10% below 360°C, followed by the main mass loss from 360 to 480°C due to the evolution of the polymer. In the unfilled sample, this is 70–78% of the initial mass. There is then a smaller additional mass loss up to 800°C.

The largest difference between the TGA results arose between 25 and 360°C. In this temperature range the losses are a result of losing low molecular weight materials. The percentage of the overall mass loss within the approximate region of 25 and 360°C for these two samples is approximately:

- HNBR75 Unaged: 7.7%
- HNBR75 Aged: 2.9%
- HNBR75 NP Unaged: 2.9%

The HNBR75NP TGA tests can also help identify any differences in plasticiser content before and after ageing for other samples. Firstly, for HNBR75NP samples aged in air, very small differences were seen in the region of 25 and 360°C. These were approximately:

- HNBR75NP Unaged: 2.9%
- HNBR75NP Aged at 100°C in air: 1.4%
- HNBR75NP Aged at 150°C in air: 0.8%

This percentage mass loss difference can be almost entirely be attributed to plasticiser content, the evolution of which begins at approximately 225°C.

The HNBR75NP TGA tests can also help identify any differences in plasticiser content before and after ageing for other samples. Assuming the approximate moisture content is the same, these differences must arise from small losses of extra crosslinking ingredients not used during vulcanisation and ingredients with a low initial mass fraction such as the antidegradants.

By comparing the unaged sample test of HNBR75NP with the sample test of HNBR75 aged at 150°C in air, we can identify the degree to which plasticiser was lost in the ageing process. The similarity between these two tests is striking. Small differences can be seen between 25 and 360°C with an increased mass loss for the unaged HNBR75NP sample.

To investigate this further, FTIR was used to characterise all the unaged and aged samples. This was done to attempt to identify any significant changes in composition before and after ageing to confirm the results from the fatigue experiments and the TGA. The results shown here are all from the freshly cut cross-sections of the samples. All the materials were characterised, however results from filled materials HNBR60, HNBR75 and HNBR75NP, which showed a rising baseline with derivative-like peaks, were much more difficult to analyse. So only the unfilled materials are shown here (Fig. 11).

The FTIR spectra were assigned using the literature and by comparison of the spectra of components with the elastomer compounds. The detailed identification of the components that contributed to IR peaks will not be presented here. Not all components could be detected, especially those that are present in small mass fractions such as antidegradants.

The FTIR peaks associated with the most significant changes were between 1,765–1,660, 1,330–1,220, 1,150–1,050, and 820–750 cm^{-1}. These peaks all reflect the plasticiser. The FTIR results primarily identify that the peaks most affected by ageing are associated with changes to the plasticiser content. This confirms the results from the TGA tests that ageing in air removed a significant amount of the plasticiser, especially at 150°C for a week.

Another way of analysing these results is through ratios of the areas under specific peaks. This allows semi-quantitative analysis that can be used to understand the mechanisms that are occurring during ageing. In this analysis, the peak at wavenumbers 2,260–2,215 cm^{-1}, indicating $C \equiv N$ bond from the HNBR base elastomer, is used as a reference, and is assumed not to be affected by ageing under these conditions. This shows the ratio of the area under the peak from 1,765 to 1,660 cm^{-1} indicating the $C = O$ bond for the plasticiser, compared to the area under wavenumbers 2,260–2,215 cm^{-1} indicating the $C \equiv N$ bond for the HNBR base polymer. These areas in the peaks are shaded on the HNBR0 unaged result. The ratios for each test are:

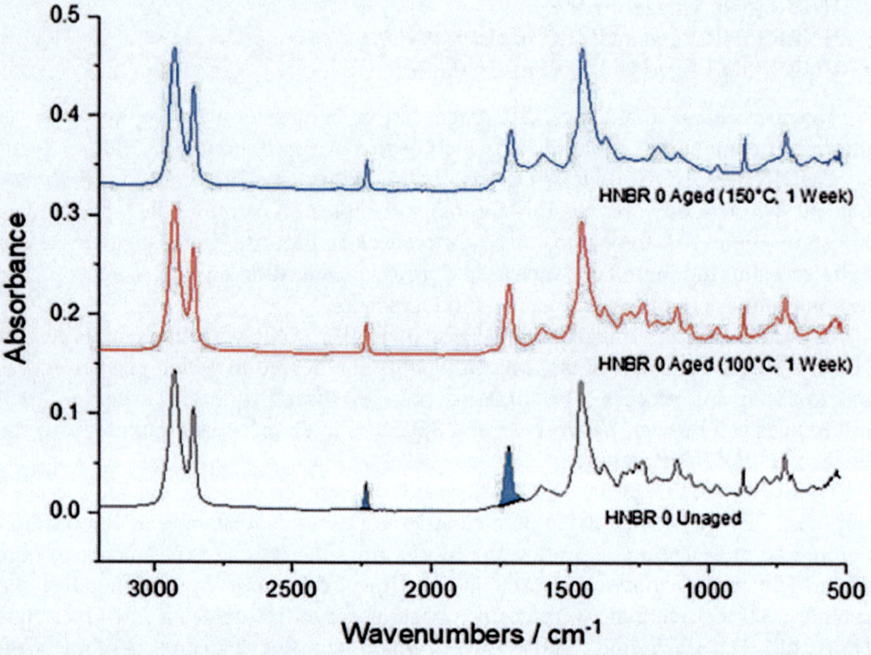

Fig. 11 FTIR comparison of HNBR0 after different ageing conditions, spectra offset for clarity. Peaks used for quantitative analysis highlighted in HNBR0 unaged spectrum

- HNBR0 unaged: 5.6
- HNBR0 aged 100°C in air: 5.4
- HNBR0 aged 150°C in air: 3.8

A small loss of plasticiser is seen in the samples aged in air at 100°C as expected. Here it can be confirmed that the ageing effects have the greatest effects on the samples aged at 150°C, showing that the increase in temperature is a very influential factor in removing the plasticiser.

4 Conclusions

This work looked at a selection of mechanical and chemical experiments investigating how the fatigue properties of typical HNBR compounds are altered when aged at high temperatures in air. This work shows that:

- All the compounds featured an increase in the modulus after ageing in air.
- The materials had a fatigue response that varied from a smooth tearing to a knotty tearing behaviour depending upon the detailed strain energy release rate conditions or ageing conditions.

- Aged samples can become tougher or weaker depending on the detailed ageing conditions. In general, the samples aged at 150°C in air had a reduced fatigue resistance and samples aged at 100°C in air unexpectedly became slightly tougher.
- DSC results showed that the sample without plasticiser had, as anticipated, higher T_g than the compounds incorporating a plasticiser.
- DSC showed that ageing at 150°C caused an increase in T_g for each material. TGA results showed that samples aged in air featured a loss of the plasticiser, especially when aged for a week at 150°C.
- Ageing increases the measured β values in the crack growth tests and the loss of plasticiser may in part help explain why some samples cyclic fatigue tests after ageing show an increased toughening behaviour. This mechanism is proposed here for the first time to explain this unexpected behaviour.
- FTIR results confirmed the significant loss of plasticiser after ageing in air and showed that the loss of plasticiser was primarily a temperature driven process.

The detailed mechanism that results in the toughening of some of the HNBR compounds when aged in air particularly at 100°C is still not clear from this investigation and will be the main enquiry for future work. To examine this further, it is proposed that simple trouser tear testing would help identify if ageing induces a knotty tearing and if so detailed X-ray analysis might help identify if the ageing process alters the rate of strain induced crystallisation in these materials. Both of which might result in an increase in the toughening of the material observed in the FCG data presented.

Acknowledgements The authors would like to thank EPSRC for funding this research as an Industrial CASE studentship EP/N509322/1 in collaboration with Schlumberger.

References

1. Shaw B, Busfield JJC, Jerabek J, Ramier JL (2017) Characterising the cyclic fatigue performance of HNBR after aging in high temperatures and organic solvents for dynamic rubber seals. Conference: the 10th European conference on constitutive models for rubber, ECCMR X
2. Beranger AS, Qin J, Heuillet P, Baurier H (2018) Fatigue crack growth behavior of NBR, HNBR, HNBR ZSC compounds. Procedia Engineering 213:145–152
3. Windslow RJ, Busfield JJC (2019) Viscoelastic modelling of extrusion damage in elastomer seals. Soft Mater 17(3):228–240
4. Weitao L, Weifang Z, Xuerong L, Wei D, Dan X (2017) Degradation of hydrogenated nitrile rubber (HNBR) O-rings exposed to simulated servo systems conditions. IOP conference series materials science and engineering: polymer degradation and stability, vol 265, p 012003
5. Alcock B, Jorgensen JJ (2015) The mechanical properties of a model hydrogenated nitrile butadiene rubber (HNBR) following simulated sweet oil exposure at elevated temperature and pressure. Polym Test 46:50
6. Alcock B, Peters TA, Gaarder RH, Jorgensen JK (2015) The effect of hydrocarbon ageing on the mechanical properties, apparent crosslink density and CO2 diffusion of hydrogenated nitrile butadiene rubber (HNBR). Polym Test 47:22

7. Ulu KN, Huneau B, Verron E, Heuillet P, Béranger A-S (2017) Fatigue of HNBR blends and the effects of thermal ageing. Procedia Eng 213:153–160
8. Gent AN, Razzaghi-Kashani M, Hamed GR (2003) Why do cracks turn sideways? Rubber Chem Technol 76:122
9. Hamed GR (2005) Tearing of vulcanized rubber. Rubber Chem Technol 78(3):548
10. Kobatake T, Kodama K, Hayashi S (1997) Improvement of low-temperature flexibility of hydrogenated nitrile-butadiene rubber. Rubber Chem Technol 70:839
11. Severe G, White JL (2000) Physical properties and blend miscibility of hydrogenated acrylonitrile-butadiene rubber. J Appl Polym Sci 78:1521
12. Stocek R, Stenicka M, Maloch J (2020) Determining parametrical functions defining the deformations of a plane strain tensile rubber sample. Adv Polym Sci 286:19–38
13. Griffith AA (1921) The phenomena of rupture and flow in solids. Philos Trans R Soc A 221:582
14. Rivlin RS, Thomas AG (1953) Rupture of rubber. Part 1. Characteristic energy for tearing. J Polym Sci. https://doi.org/10.1002/pol.1953.120100303
15. Greensmith HW, Thomas AG (1955) Rupture of rubber. Part III. Determination of tear properties. J Polym Sci 18:189
16. Thomas AG (1960) Rupture of rubber. Part VI. Further experiments on the tear criterion. J Appl Polym Sci 3:168
17. Lake GJ (1969) Fracture mechanics of rubber. Proceedings 2nd international conference on fracture
18. Tsunoda K, Busfield JJC, Davies CKL, Thomas AG (2000) Effect of materials variables on the tear behaviour of a non-crystallising elastomer. J Mater Sci 35:5187–5198
19. Busfield JJC, Ratsimba CHH, Thomas AG (1997) Crack growth and strain induced anisotropy in carbon black filled natural rubber. J Nat Rubber Res 12:131
20. De DK (1994) The effect of particulate fillers on the strain energy function and the crack growth in rubbers. PhD thesis. Queen Mary University of London
21. Asare S, Thomas AG, Busfield JJC (2009) Cyclic stress relaxation (CSR) of filled rubber and rubber components. Rubber Chem Technol 82:104–112
22. Papadopoulos IC, Thomas AG, Busfield JJC (2008) Rate transitions in fatigue crack growth of elastomers. J Appl Polym Sci 109:1900–1910
23. Gent AN, Lindley PB, Thomas AG (1964) Cut growth and fatigue of rubbers. I. The relationship between cut growth and fatigue. J Appl Polym Sci 8:455
24. Lake GJ (1983) Aspects of fatigue and fracture of rubber. Prog Rubber Technol 45:89
25. Busfield JJC, Jha V, Liang H, Papadopoulos IC, Thomas AG (2005) Prediction of fatigue crack growth using finite element analysis techniques applied to three dimensional elastomeric components. Plast Rubber Compos 34:349–356
26. Asare S, Busfield JJC (2011) Fatigue life prediction of bonded rubber components at elevated temperature. Plast Rubber Compos 40:192–198
27. Stocek R (2020) Some revisions of fatigue crack growth characteristics of rubber. Adv Polym Sci 286:1–18
28. Lou W, Zhang W, Liu X, Dai W, Xu D (2017) Degradation of hydrogenated nitrile rubber (HNBR) O-rings exposed to simulated servo system conditions. Polym Degrad Stab 144:464
29. Fernandez-Berridi M, Gonzalez N, Mugica A, Bernicot C (2006) Pyrolysis-FTIR and TGA techniques as tools in the characterization of blends of natural rubber and SBR. Thermochim Acta 444:65
30. Tunnicliffe LB, Kadlcak J, Morris MD, Shi Y, Thomas AG, Busfield JJC (2014) Flocculation and viscoelastic behaviour in carbon black-filled natural rubber. Macromol Mater Eng 299: 1474–1483
31. O'Rourke (2008) Reactive ester plasticizer for elastomers. Rubber World 239:16
32. Gent AN (2001) Engineering with rubber: how to design rubber components.3rd edn. Carl Hanser Verlag GmbH & Co. KG

33. Sakulkaew K, Thomas AG, Busfield JJC (2011) The effect of the rate of strain on tearing in rubber. Polym Test 30:163–172
34. Sakulkaew K, Thomas AG, Busfield JJC (2013) The effect of temperature on the tearing of rubber. Polym Test 32:86–93
35. Busfield JJC, Deeprasertkul C, Thomas AG (2000) The effect of liquids on the dynamic properties of carbon black filled natural rubber as a function of pre-strain. Polymer 41:9219–9225
36. Hayashi S, Sakakida H, Oyama M, Nakagawa T (1991) Low-temperature properties of hydrogenated nitrile rubber (HNBR). Rubber Chem Technol 64(4):534–544. https://doi.org/10.5254/1.3538571

Adv Polym Sci (2023) 289: 167–184
https://doi.org/10.1007/12_2022_117

Published online: 25 March 2022

Effect of Antioxidants on the Aging Behavior of NR and S-SBR Materials

Katja Oßwald and Katrin Reincke

Contents

Abstract The increase in raw material prices and customer requirements for product reliability and service life lead to a high interest in increasing the service life of elastomer products. For oxidative aging processes and their modeling, the material-dependent oxygen uptake capacity and the oxygen permeability play a major role. Therefore, the aging resistance in the case of the thermo-oxidative attack is dependent on the oxygen diffusion rate in the material volume. The influence of time-dependent thermo-oxidative aging of natural rubber (NR) and solution styrene-butadiene raw elastomer (S-SBR) on the final properties of the vulcanizates produced from them was investigated. It was observed that the storage of the raw elastomers has an influence on the tensile strength and hardness properties. Nuclear magnetic resonance spectroscopy (NMR spectroscopy) was used to investigate the aging processes and the developing network structure of raw elastomer after thermo-oxidative aging. The polymer network formed during vulcanization, which is influenced by aging of the raw elastomer was observed. Differential scanning

K. Oßwald (✉) and K. Reincke
Polymer Service GmbH Merseburg, Merseburg, Germany
e-mail: katja.osswald@psm-merseburg.de

calorimetry (DSC) analysis showed that the oxidation behavior of NR and S-SBR is different from each other.

Keyword Antoxidant · Long-term aging · NR · OIT time · SBR · Short-term aging

1 Introduction

Elastomer materials, when used in the molded parts or components made from them (tires, dampers, seals, conveyor belts, etc.), are exposed not only to mechanical stresses but also to a wide variety of environmental influences such as humidity, high or low temperatures, radiation or the influence of chemical substances, with combinations of these factors frequently occurring. This can lead to a change in properties, which can be summarized under the term "aging." According to [1], aging is defined as "the totality of chemical and physical changes over time that lead to a change in mechanical properties such that the applicability of the products is reduced over time." These property changes can lead to the point where the component/molded part fails completely and can no longer fulfill its function [2]. Aging processes can occur at any time. Various influences can result in physical and irreversible chemical processes at the molecular level, such as polymer chain scission, crosslinking, or the breaking and reforming of covalent bonds (recombination) [3, 4]. While chain scission leads to a decrease in viscosity, crosslinking stiffens the material [5–8]. A combination of both mechanisms leads to [9] to the formation of microcracks. Materials in which chain scission occurs tend to form surface cracks [10].

Direct and indirect methods are used to evaluate the aging resistance or the effects of any kind of (artificial) aging. Indirect evidence of aging-related material changes is usually provided by mechanical tests. Frequently, the literature [11, 12] describes the performance of tensile tests and/or the application of hardness tests in order to quantitatively detect aging defects and thus indirectly demonstrate aging. Another method of investigation described in the literature is the evaluation of aging effects by means of dynamic-mechanical analysis (DMA) [13]. With the direct methods such as nuclear magnetic resonance spectroscopy (NMR spectroscopy), the differential scanning calorimetry (DSC) [8, 12, 14], and the chemiluminescence (CL) [15], aging-related changes, like (re)crosslinking, chain scission, formation and degradation of functional groups, are detected directly. CL and DSC can be used to determine oxygen induction time (OIT), although the method of determination is different. The static DSC method is an isothermal process in which the time period from the first contact with oxygen to the start of oxidation is referred to as the oxidation induction time (OIT) can be measured.

In this paper, the effects of time variation in thermal oxidative aging on the mechanical properties of natural rubber (NR) and styrene-butadiene rubber (S-SBR) materials are investigated. Samples were stored in an oven at 100°C for a maximum of 168 h (rapid short-term aging) and at 80°C for up to 2,000 h (long-term aging). DSC was used as a supporting method to show the aging of the materials.

2 Aging of NR and SBR Materials and the Function of Antioxidants

In general, it is assumed that an aging process is a multi-stage radical mechanism. In the literature [16] this is described as a three-stage process, with initiation, radical formation, and termination. In principle, all organic polymers are subject to degradation by light and oxidation. The degree and speed of oxidation of rubbers depend, among other things, on the double-bond content in the polymer chain. It was described in [17] that the thermal stability of a polymer depends on the resistance of the C–C bonds, with the degradation of the polymer being influenced by the substituents in the backbone. A higher number of substituents, such as oxygen atoms and branching, reduce thermal stability, while aromatic groups increase it. Polymers such as natural rubber (NR) or styrene-butadiene rubber (SBR) are particularly susceptible to oxidative processes. Their polymer networks are easily altered by reaction with oxygen, resulting in chain scission and crosslinking. The extent and ratio of these two processes depend not only on the elastomer type and compound, but also on the temperature and oxygen concentration. Aging studies [11] on carbon-black (CB) reinforced SBR and NR vulcanizates showed that at the early stage of the aging process, the formation and destruction of the crosslink chains are similar, so the crosslink densities do not change. Investigations on NR-based vulcanizates [18] also showed that chain breakage and chain formation take place during thermo-oxidative aging. At the beginning of the aging process, chain breakage predominates, and carbonyl by-products are formed, which relatively quickly reach a stationary state. During aging, network chains can form due to the presence of residual crosslinking additives, the rearrangement of cross-linked sulfur, or as a result of oxidative coupling [11]. The aging of elastomeric materials is usually accompanied by a change in mechanical properties [11–13], which is directly related to the change in crosslink density [15] and structure [19, 20], whereby polysulfidic sulfur bridges are re-built to mono- and di-sulfide bonds by dissociation [21–23].

In order to prevent premature aging of the materials, they are stabilized during the compounding process and antioxidants are added to the rubber compound. An antioxidant generally serves to chemically bind the attacking medium or to render it ineffective, whereby its molecular structure and chemical reactivity influence its effectiveness against thermo-oxidative aging [8]. As a result of exposure to mechanical stress or temperature loading, the oxidation process is initiated by radicals. It is assumed that initially peroxy radicals are formed, which by abstraction of an H-atom form hydroperoxides and polymer radicals. The decomposition of the hydroperoxides proceeds very slowly, and their formation can already take place during rubber manufacturing [24]. Secondary antioxidants react directly with the hydroperoxides in a redox reaction, forming oxidation products, while primary antioxidants act as radical scavengers. The aging process is interrupted, with the peroxide radicals being saturated by abstraction of hydrogen [8].

Fillers have a further influence on the aging resistance of elastomer materials. For carbon-black reinforced vulcanizates, for example, it is known that resistance to thermo-oxidative aging decreases with the carbon-black content, since the oxygen absorption capacity of sulfur-crosslinked vulcanizates is improved by the carbon black. This means that the more oxygen available, the greater the degradation of the polymer [25].

3 Materials and Methods

Raw elastomers natural rubber (NR) of the type SMR 10 (Standard Malaysian Rubber) and solution styrene-butadiene rubber (S-SBR) of the type SPRINTAN™ SLR 4602-SCHKOPAU (Synthos S. A.) were investigated. CB-reinforced NR and S-SBR vulcanizates with different antioxidants were also analyzed. The compounds were prepared in a Plasticorder PL 2000 laboratory internal mixer (Brabender GmbH & Co. KG, Duisburg, Germany) with a chamber volume of 75 cm^3. The rubber compounds were prepared according to the formulation in Table 1 with an initial kneader temperature of 50°C, a rotor speed of 50 min^{-1}, and a chamber filling level of 75%.

In one part of the study, the effects of storage of the raw rubber on the final properties of the CB-reinforced vulcanizates produced from it were investigated. For this purpose, the raw elastomers were stored at 8°C, 23°C, and 35°C up to a maximum time period of 1,344 h. In the second part of the study, various antioxidants were mixed into the compounds to investigate their influence on the aging behavior of the vulcanizates. For this purpose, the vulcanizates were subjected to a thermo-oxidative rapid aging process and a long-term aging process. The samples were stored in a heating oven at 100°C (rapid short-term aging) and 80°C (long-term

Table 1 Formulation

Material	Content (phr)	Addition time (min)
NR or S-SBR	100	0
Carbon black (N234)	0 I 20 I 60	1
IPPD/6PPD/TMQ[a]	1.50	1
ZnO	3	1
Stearic acid	1	1
Sulfur	1.7	5
CBS[b]	2.5	5 stop: 10

[a] N-Isopropyl-N′-phenyl-p-phenylendiamine (IPPD); N-(1,3-Dimethylbutyl)-N′-phenyl-p- phenylenediamine (6PPD); 2,2,4-Tri-methyl-1,2-dihydrochinolin (TMQ)
[b] N-Cyclohexyl-2-benzothiazyl-sulfenamide (CBS)

aging). They were taken at different times up to a maximum of 2,000 h. After the aging, the mechanical and physical properties were investigated.

The evaluation of aging-related material changes was carried out by determining the characteristic values in the quasi-static tensile test according to ISO 37 [26] and the determination of the Shore A hardness according to DIN ISO 48-4 [27]. Dynamic-mechanical analysis (DMA) tests on virgin and aged vulcanizates were carried out at a temperature of 60°C, in an amplitude range from 0.01% to a maximum of 100% and a frequency of 1 Hz.

Using NMR spectroscopy, the relaxation times T_1 and T_2 can be determined via the exponential, relaxation-time-dependent decrease in magnetization. T_1 is the longitudinal or spin-lattice relaxation time, which is usually much larger than T_2. A strong relaxation process is the spin–spin interaction, which is described by the relaxation time T_2 (also called transverse relaxation time). The relaxation time T_2 depends, among other things, directly on the molecular mobility of the sample. A higher mobility decreases this interaction, thus the relaxation time T_2 increases [1, 2]. The decay of transverse magnetization, measured as a function of time, contains information about the mobility of fractions of chains inside the sample. The T_2 relaxation time can be determined from this function using mathematical models (see Eq. 1) [28].

$$M(t) = A \times \exp\left(-t/T2\mathrm{A}\right) + B \times \exp\left(-t/T2B\right) + C \tag{1}$$

M(*t*) is the magnetization, where *A* and *B* are the rate constants. *C* is a constant which, according to [7], is considered only for oil-stretched polymers. The mobility of the polymer chains has an influence on the magnetization decrease. This means that the more mobile the polymer chains are, the weaker the spin–spin interaction and the slower the magnetization decrease. $T_{2\mathrm{A}}$ and $T_{2\mathrm{B}}$ determine the rate of magnetization decrease. Thus, the smaller $T_{2\mathrm{A}}$ and $T_{2\mathrm{B}}$ are, the faster the decrease occurs [28].

Furthermore, the oxygen induction time (OIT) as a measure of the resistance to thermo-oxidative aging or degree of oxidative pre-damage was determined by using differential scanning calorimetry measurements (DSC). During the measurement, the sample was heated under a nitrogen atmosphere up to the desired temperature. After reaching this temperature, the measurement was proceeded as an isothermal test under oxygen atmosphere so that thermo-oxidative processes could take place. The oxidation induction time was determined according to ASTM D3895 [29]. A heating rate of 20 K/min, a gas flow of 50 ml/min, and a temperature of 130–150°C were used.

4 Characterization of Aging Process and Its Influence on the Mechanical and Physical Properties of Elastomeric Materials

4.1 Analysis of Aging Processes in NR and SBR Raw Elastomer

The degradation of polymers and polymer networks by aging takes place with energy supplied in the form of heat, light, oxygen, and ozone. The oxidation processes, which are initiated by free radicals, are terminated by recombination or by inactivation of the radicals using stabilizers [30]. In principle, it is assumed that the aging of polymer networks is associated with or can be explained by the change in network structure and it is therefore important in determining the technically relevant properties of elastomers [2, 31].

In Fig. 1a, b, the tensile strength and Shore A hardness of the investigated CB-reinforced NR vulcanizates as a function of exposure time are shown. The raw polymer was stored under defined temperature conditions to simulate different

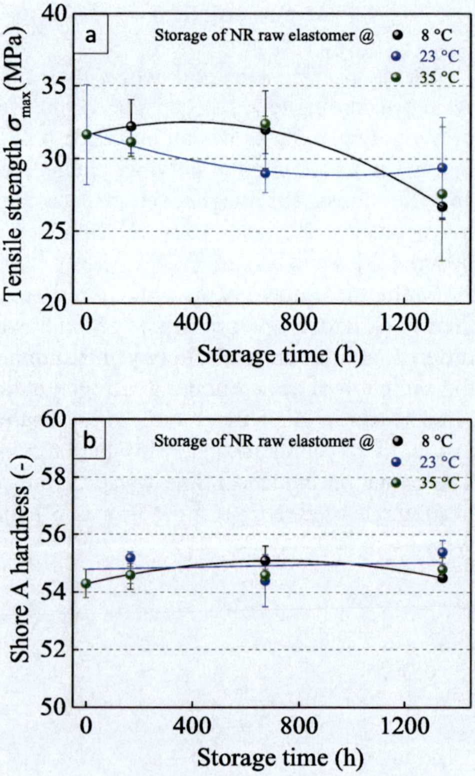

Fig. 1 Tensile strength (**a**) and Shore A hardness (**b**) characteristics as a function of storage time and temperature at 8°C, 23°C, and 35°C of the raw elastomer for NR vulcanizates with 20 phr N234 and IPPD as antioxidant

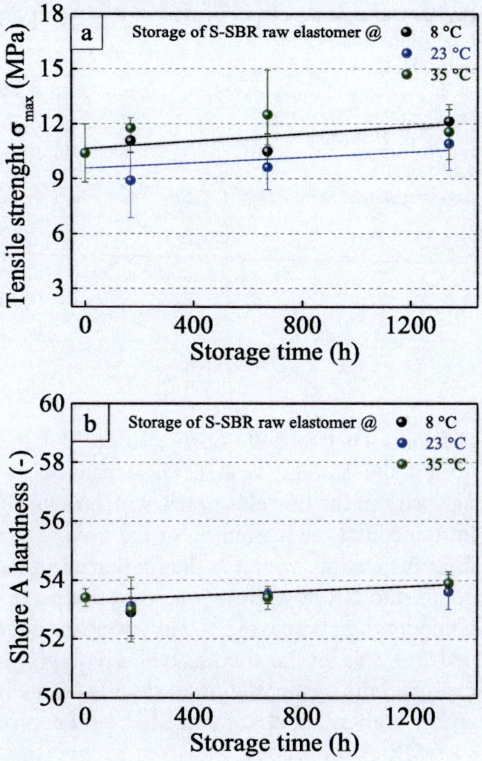

Fig. 2 Tensile strength (**a**) and Shore A hardness (**b**) characteristics as a function of aging time and temperature at 8°C, 23°C, and 35°C of the raw elastomer for S-SBR vulcanizates with 20 phr N234 and IPPD as antioxidant

conditions in industrial stocks or during transport. If the raw elastomer is stored at 8°C or 35°C (compare Fig. 1a), this initially doesn't lead to a change in the tensile strength σ_{max} until 672 h. From 672 h onwards, a slight reduction of σ_{max} can be observed. If, on the other hand, the rubber is stored at 23°C, the tensile strength reaches a minimum after 672 h and does not change significantly thereafter. After 1,344 h, σ_{max} is comparable for all vulcanizates, irrespective of the previous storage temperature of the raw elastomer. The Shore A hardness (compare Fig. 1b) values are not significantly affected by the storage time and temperature of the raw elastomer and remain at the same level. Possibly there could be an influence of further storage conditions such as permanent ventilation.

Compared to NR, however, SBR shows a different behavior. Figure 2 shows the relationship between the tensile strength and the Shore A hardness values as a function of the storage time and temperature of the SBR raw elastomer for the related carbon-black reinforced SBR vulcanizate. If the polymer is previously stored at 8°C or 23°C, a linear dependency with a slight increase in strength and hardness over the whole range can be observed. In contrast, storage at 35°C causes a slight increase in tensile strength up to 672 h and a reduction thereafter.

During storage of raw natural rubber, it is known that the carbonyl groups incorporated in the chains react with the amino acids of the stabilizing latex shell

Fig. 3 Dependence of the relaxation time T_{2A} of NR vulcanizates on storage time of the NR raw elastomer at storage temperatures of 8 and 35°C for NR vulcanizates

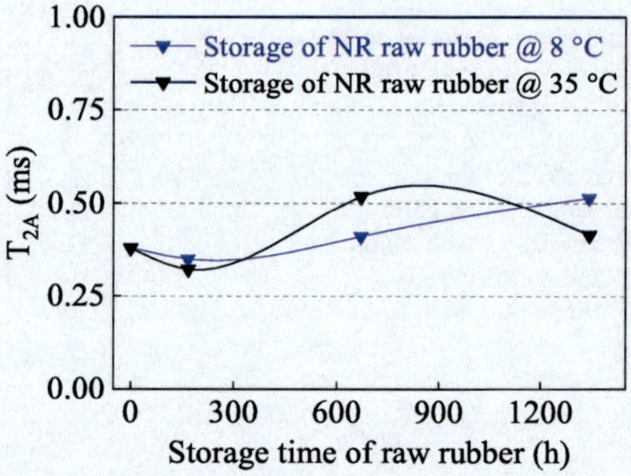

and form a network. It is also assumed that the average molar mass increases slightly during the storage time [24]. It can be assumed that both factors increase the viscosity of the raw elastomer, which in turn leads to the change of the shear energy introduced at the beginning of the mixing process. This could eventually affect the filler dispersion, which is then reflected in the final mechanical properties. SBR, in contrast to NR, is already protected against premature aging by stabilizers during the compounding process [32]. However, the concentration of double bonds also plays a decisive role in the thermo-oxidative aging resistance of the raw elastomers. For example, the concentration of double bonds in NR with 13.90 mmol/cm^3 is considerably higher than that in an SBR with approx. 2.91 mmol/cm^3 [16].

NMR spectroscopy was used to investigate how thermo-oxidative aging of the raw elastomer affects the developing network structure. In Fig. 3 the relaxation time T_{2A} as a characterizing parameter of the rigid network is shown as a function of the aging time of the NR raw elastomer for two storage temperatures. As already mentioned, the relaxation times that can be determined by NMR depend directly on the molecular mobility of the chains [33]. In an elastomer, however, the polymer chains do not have the same mobility everywhere. Chain segments that are far away from crosslinking points and free chain ends, for example, are more mobile than chain segments near crosslinking points [28]. The results in Fig. 3 indicate that the polymer network formed during vulcanization is influenced by aging of the raw elastomer. As the raw elastomer ages, there is a slight increase in the molecular mobility of the chains incorporated in the network, which could indicate, for example, a change in the spacing of the network nodes. The temperature of storage affects the relaxation times somewhat differently. A clear correlation between the relaxation time T_{2A} and the tensile strength (see Fig. 1), especially under consideration of the storage temperature, however, cannot be proven. From the results, it can be concluded that longer storage times may lead to changes in the raw elastomer that have a direct effect on the polymer network of the vulcanizates.

The polymer chains of SBR and NR degrade at different rates under the influence of oxygen and temperature. This is shown by investigations by using DSC to determine the oxidation induction time (OIT). In Fig. 4a, b, results of these OIT

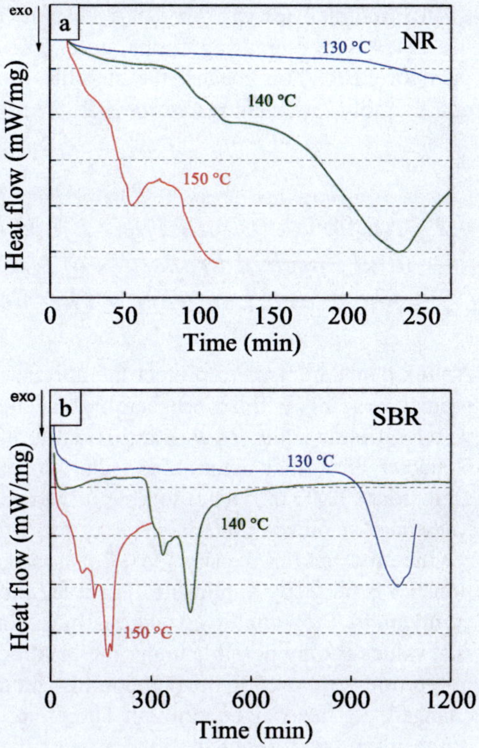

Fig. 4 DSC thermograms for NR (**a**) and SBR (**b**) raw elastomers to determine the oxidation induction time (OIT) at different temperatures

Table 2 OIT values determined in DSC for NR and S-SBR raw elastomer at different temperatures

Rubber	Temperature (°C)	OIT (min)
NR	130	199
	140	76
	150	7
S-SBR	130	917
	140	276
	150	40

measurements at different temperatures are shown. Natural rubber degrades relatively quickly compared to SBR. The trend also correlates with the concentration of double bonds in the main chain [16]. The higher this value is, the lower the OIT values are (compare Table 2). In principle, however, it can be stated for both elastomers that the resistance decreases with increasing temperature. The decrease is more pronounced for NR than for SBR. This means, the stability of NR with respect to heat influence is considerably lower compared to SBR. Reasons for that were discussed above. With increasing temperature, the thermal stability of both elastomers decreases. The courses of the thermograms for SBR are striking (Fig. 4b) starting at 140°C because they show various peaks. This indicates that the stabilizer in the polymer is starting to degrade in stages.

In principle, however, it can be stated for both elastomers that the resistance decreases with increasing temperature. The decrease is more pronounced for NR than for SBR. This means, the stability of NR with respect to heat influence is considerably lower compared to SBR. Reasons for that were discussed above.

4.2 Antioxidants and Their Effect on the Mechanical and Physical Properties of NR and SBR Blends After Short- and Long-Term Thermo-Oxidative Aging

Antioxidants are used to protect the polymer as well as the entire vulcanizate from premature aging. In this work, carbon-black reinforced compounds were produced in which different additives with antioxidative effects were mixed in. Table 3 shows the characteristic values determined from the crosslinking kinetics. The addition of the antioxidant TMQ thus leads to a slightly lower value of the curing time t_{90} compared to the non-stabilized material. The torque difference ΔM is often used as a measure for the crosslinking density [34]. For non-reinforced rubber compounds, this relationship is perfectly applicable. There are, however, limitations for filler-reinforced compounds. Previous investigations have shown that a relative comparison of the ΔM values is only possible under the condition that the same carbon-black type and proportion were used in the compounds and no other components of the mixture are changed, e.g. plasticizer addition. Therefore, ΔM should be applicable here to get a rough impression about the influence of the antioxidant type on the crosslinking behavior. As can be seen in Table 3, the torque difference ΔM is lower when antioxidant is used. Thus, the crosslink density is influenced generally by the antioxidant addition and the lowest level is reached with TMQ.

Figure 5 shows the storage modulus *G*' as a function of amplitude for carbon-black reinforced NR vulcanizates with and without antioxidant (6PPD) before and after thermo-oxidative aging at 100°C for 168 h. It can be observed that in the initial condition, the modulus values are comparable for the investigated vulcanizates, independent of the use of the antioxidant. Generally, the oxidative influence process leads to aging that is connected to a stiffening of the material, as expected. Significant differences can be observed for the thermo-oxidative loaded samples. Interestingly, the addition of 6PPD leads to higher modulus values. This means, the processes at the molecular scale leading to a stiffening of the material such as post-crosslinking or re-building of sulfur bridges seem to be more pronounced in the material with an antioxidant.

Table 3 Characteristic values of crosslinking kinetics determined at 160°C

Materials	t_{90} (min)	ΔM
NR60	4.84	2.91
NR60/6PPD	4.71	2.81
NR60/IPPD	4.67	2.78
NR60/TMQ	4.34	2.76

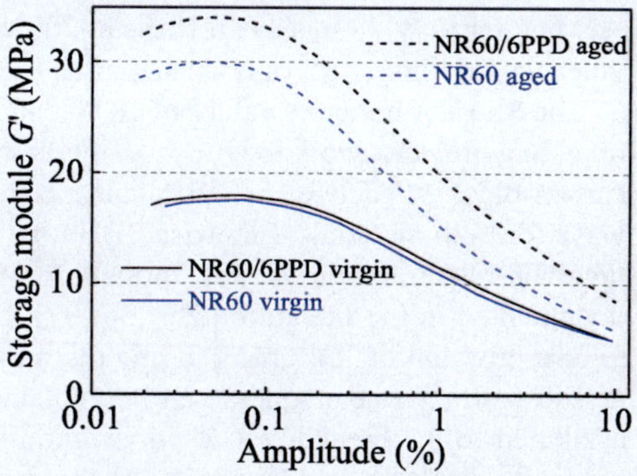

Fig. 5 Storage module *G*' after short-term aging for NR with 60 phr CB N234 and with and without antioxidant in the initial state and after 168 h thermo-oxidative aging at 100°C (RPA test with frequency of 1 Hz and at a temperature of 60°C)

Table 4 Summary of characteristic values from the quasi-static tensile test, Shore A hardness and OIT values for carbon-black reinforced NR vulcanizates without and with antioxidant before and after rapid thermo-oxidative aging at 100°C

Materials	Aging time (h)	σ_{max} (MPa)	ε_R (%)	σ_{100} (MPa)	σ_{200} (MPa)	Shore A (–)	OIT (min)
NR60	0	24.7 ± 0.7	293 ± 7.3	7.1 ± 0.3	16.7 ± 0.5	76 ± 0.7	78
	24	21.8 ± 1.3	219 ± 13	8.7 ± 0.5	19.8 ± 0.8	78 ± 0.9	–
	48	16.4 ± 2.3	160 ± 19	9.4 ± 0.2	–	78 ± 1.0	–
	72	12.3 ± 2.3	122 ± 17	9.8 ± 0.5	–	80 ± 0.9	–
	168	9.8 ± 1.0	109 ± 10	8.9 ± 0.0	–	79 ± 0.9	54
NR60/ 6PPD	0	29.6 ± 1.3	379 ± 15	5.6 ± 0.1	13.7 ± 0.1	72 ± 0.4	64
	24	25.4 ± 3.3	252 ± 29	9.3 ± 0.2	20.5 ± 0.7	80 ± 0.9	–
	48	19.0 ± 3.7	179 ± 35	10.2 ± 0.3	–	82 ± 0.1	–
	72	21.2 ± 2.3	203 ± 25	9.9 ± 0.2	20.9 ± 0.2	81 ± 1.2	–
	168	16.7 ± 2.7	148 ± 23	11.0 ± 0.1	–	82 ± 1.0	34
NR60/ IPPD	0	29.3 ± 1.5	369 ± 25	6.1 ± 0.2	14.5 ± 0.4	74 ± 0.6	78
	24	24.1 ± 1.8	250 ± 18	8.4 ± 0.2	19.1 ± 0.2	79 ± 0.6	–
	48	19.1 ± 4.3	170 ± 33	10.8 ± 0.5	–	82 ± 1.0	–
	72	19.4 ± 1.7	190 ± 1.6	9.5 ± 0.1	–	81 ± 0.6	–
	168	16.2 ± 2.9	135 ± 6.8	11.6 ± 0.2	–	84 ± 0.8	23
NR60/ TMQ	0	26.0 ± 1.0	323 ± 16	6.8 ± 0.1	15.8 ± 0.1	75 ± 0.3	86
	24	20.6 ± 2.3	214 ± 25	8.9 ± 0.2	19.4 ± 0.4	80 ± 0.4	–
	48	21.0 ± 0.9	200 ± 9.1	9.9 ± 0.1	20.9 ± 0.3	82 ± 0.5	–
	72	18.2 ± 1.6	188 ± 16	8.9 ± 0.1	–	80 ± 0.5	–
	168	16.5 ± 2.1	148 ± 14	10.7 ± 0.6	–	82 ± 0.8	35

Also other mechanical properties, e.g., for instance tensile strength, are influenced by the antioxidant addition, as it is shown in Table 4. It is noticeable that for the NR60 vulcanizate without antioxidant lower tensile strength σ_{max} and strain at break ε_R values were found as a function of the aging time. The stress values at 100% and 200% strain, on the other hand, are at a comparable level for all vulcanizates. Although it can be noted that for the NR60 vulcanizate, embrittleness apparently sets in after only 48 h, since a stress at 200% was no longer achieved. For all other vulcanizates, this effect only occurs after 168 h.

The Shore A hardness values of all vulcanizates show an enhancement with aging time. In a previous work [35], also an increase in Shore A hardness was observed for carbon-black filled NR and SBR vulcanizates, although this occurred in different ways for both materials. Likewise, IRHD-m hardness testing showed that a surface layer of higher hardness was formed for the NR vulcanizate. In some studies documented in the literature [5, 7, 36, 37], the increase in hardness was attributed to post-crosslinking occurring during thermo-oxidative aging. The increase in stress at 100% strain and hardness can be explained according to [25], but can also be attributed to the breakdown of polysulfidic bonds with simultaneous formation of mono- or disulfidic bonds. In [38], for sulfur-crosslinked elastomers, in addition to the thermo-oxidative cleavage of polymer chains, the anaerobe transformation of the network node structure represents one of the main causes of aging. This means that the network itself may still be intact, but the network node structure has already changed, which can ultimately lead to a change in physical properties. In addition, free sulfur can contribute to post-crosslinking, which can then be associated with an increase in the crosslink density and thus with a change in the molecular relaxation processes.

After the rapid thermo-oxidative aging, the OIT times were determined (also shown in Table 4). It is striking that the addition of the antioxidant 6PPD reduces the OIT time slightly. In contrast, the additives IPPD and TMQ do not change or extend the resistance times. After short-term aging, a reduction of the OIT times can be observed for all vulcanizates. This means, a pronounced oxidation takes place during aging.

In addition to rapid aging, long-term aging at 80°C, up to 2,000 h, was carried out on selected NR and SBR vulcanizates with 60 phr CB N234 and various antioxidants. In principle, it can be stated that NR and SBR vulcanizates age in different ways. In dynamic-mechanical tests by using an RPA (not shown here), it was seen for all vulcanizates that above 500 h aging for NR and 1,500 h for SBR a pronounced material stiffening took place leading to a reduction of the maximum possible amplitude of 100% [39]. Also in the tensile tests, a clear difference in the thermo-oxidative aging behavior of SBR and NR is reflected. However, the effect of the different antioxidants is comparable, as it can be illustrated with Fig. 6. For both polymers, the vulcanizate strength is decreased during this thermo-oxidative aging and the type of the antioxidant seems to have no influence under these aging conditions. As discussed above, NR is more sensitive to thermo-oxidative aging due to its molecular structure, and this explains the much stronger decrease of the mechanical properties up to 2,000 h aging time. NMR studies [40] have also shown that NR and SBR age differently. The reason for this is the higher degree of unsaturation in NR, which makes it less resistance to thermal and oxidative attack. In the first 200 h, it was observed that polysulfidic bonds are separated into mono- and dislufidic bonds. Furthermore, the investigations have shown that the defect fraction in the unaged state of SBR is less than 1% and does not increase during aging.

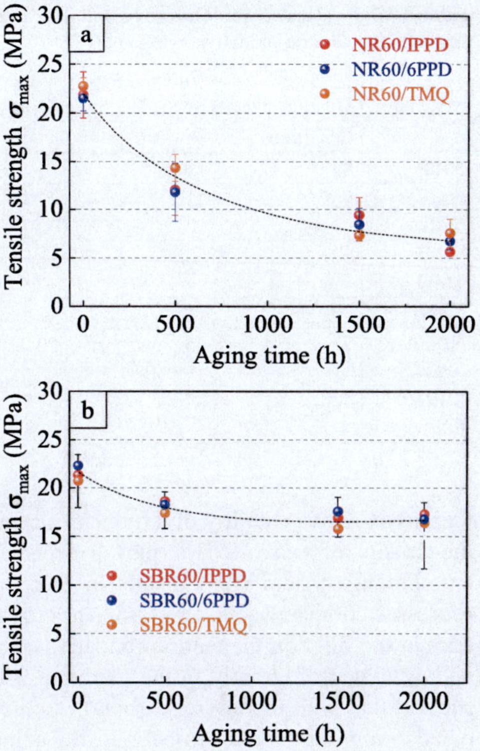

Fig. 6 Tensile strength depending on the aging time for carbon-black filled NR (**a**) and SBR (**b**) vulcanizates with different antioxidants

In contrast to the results reported here, in previous investigations [39], it has been shown that the aging behavior in terms of mechanical properties like tensile strength and elongation at break change can vary depending on the antioxidant used. In [41], a different reaction of antioxidants to thermo-oxidative aging and under the influence of light was reported. Further, a reason for the effective differences between the antioxidants can be, on the one hand, the diffusivity and, on the other hand, the molecular structure and the resulting reactivity compared to oxygen. The antioxidant 6PPD has longer alkyl groups compared to IPPD and so a slower diffusion to the surface is generated [16]. In Table 5 it can be seen that pre-aging of the material already occurs after compound production and the subsequent vulcanization process, which leads to a reduction in the OIT. In general, the addition of the antioxidants 6PPD and TMQ in a non-reinforced NR material leads to a significant increase in the OIT. For the carbon-black reinforced SBR material with the antioxidant 6PPD, a significantly higher OIT was determined as well as in the unvulcanized, vulcanized, and after thermo-oxidative aging than comparatively for an NR material. It can be further observed that the long-term aging process generally leads to a reduction of the OIT, with the reduction being the most pronounced for the NR materials. In [42] it was reported that stabilizers are consumed in the mixing process and during vulcanization, which influences the OIT accordingly. Regarding the

Table 5 OIT values (@150°C) for NR and SBR materials determined before and after vulcanization and after thermo-oxidative aging at 80°C

Material	Unvulcanized	Vulcanized/ unaged	Vulcanized/aged 500 h @ 80°C	Vulcanized/aged 2,000 h @ 80°C
	OIT (min)			
NR0	83	104	61	39
NR0/ 6PPD	239	110	126	103
NR0/ TMQ	321	–	–	–
NR60	90	78	–	–
NR60/ 6PPD	75	45	23	10
SBR60/ 6PPD	420	478	375	290

thermo-oxidative stability of a rubber vulcanizate, the crosslinking system [42] and the density of network [15] play a decisive role. It was found that an efficient crosslinking system (EV) results in more stable materials concerning oxidative processes compared to a semi-efficient crosslinking system (SEV) [42]. The influence of the filler on the thermo-oxidative stability of an elastomer material is found and evaluated differently in the literature. Chemiluminescence measurements [42] showed that with increasing carbon black content, the OIT values decrease. Compared to non-reinforced compounds, the aging of carbon-black reinforced materials starts and ends later. In [43] it was discussed by using results of X-ray absorption spectroscopy that the presence of carbon black leads to the reduction of sulfur bridge degradation. From Table 5 it can be seen that the addition of carbon black prolongs the durability of the unvulcanized NR material (NR60), compared to the unvulcanized non-reinforced NR material (NR0).

5 Conclusion

The influence of thermo-oxidative aging on the mechanical and physical properties of carbon-black reinforced NR and SBR compounds was investigated. It became clear that the duration and temperature during the storage of raw polymers can already influence the mechanical and physical properties of the vulcanizates produced from them. It was observed that NR and SBR react differently to thermo-oxidative stress. Short-term aging tests on carbon-black reinforced NR vulcanizates showed a stiffening with aging time, associated with increase in hardness characteristics and decrease in stability times. It was found that already the vulcanization process leads to a reduction of the oxidation induction times. After subsequent long-term aging, a significant decrease in stability with time was determined. It was found that SBR is thermally more stable compared to NR. Reasons for that were discussed in detail.

The investigations carried out in this work show that rapid aging is also a possibility to obtain information about the aging behavior of the materials and its effect on the property level in a short time. The determination of the OIT time can give an indication of whether aging has taken place. It cannot give an indication of the change in properties. The combination of rapid and long-term aging and the additional determination of material properties finally offer the possibility to predict the aging behavior of materials at different temperatures and times. The OIT time can be used here as an indirect method to support the statements.

References

1. Ehrenstein GW, Pongratz S (2007) Beständigkeit von Kunststoffen. Carl Hanser Verlag, Munich
2. Azura AR, Ghazali S, Mariatti M (2008) Effects of the filler loading and aging time on the mechanical and electrical conductivity properties of carbon black filled natural rubber. J Appl Polym Sci 110:747–752
3. Ha Anh T, Vu-Khanh T (2005) Effects of thermal aging on fracture performance of polychloroprene. J Mater Sci 40:5243–5248
4. Chou H-W, Huang J-S, Lin S-T (2007) Effects of thermal aging on fatigue of carbon black-reinforced EPDM rubber. J Appl Polym Sci 103:1244–1251
5. Lucas P, Baba M, Lacoste J, Gardette J-L (2002) Crosslinking of dienic elastomers resulting from ageing: X-ray diffraction and refractometry measurements. Polym Degrad Stab 76:449–453
6. Gillen K, Celina M, Bernstein R (2003) Validation of improved methods for predicting long-term elastomeric seal lifetimes from compression stress-relaxation and oxygen consumption techniques. Polym Degrad Stab 82:25–35
7. Mostafa A, Abouel-Kasem A, Bayoumi MR, El-Sebaie MG (2009) The influence of CB loading on thermal aging resistance of SBR and NBR rubber compounds under different aging temperatures. Mater Des 30:791–795
8. Nakazono T, Ozaki A, Matsumoto A (2011) Phase separation and thermal aging behavior of styrene-butadiene rubber vulcanizates using liquid polymers as plasticizers studied by differential scanning calorimetry and dynamic mechanical spectroscopy. J Appl Polym Sci 120:434–440
9. Tuampoemsab S, Sakdapipanich J (2007) Role of naturally occurring lipids and proteins on thermal aging behaviour of purified natural rubber. Kautsch Gummi Kunstst 60:678–684
10. Tomer NS, Delor-Jestin F, Singh RP, Lacoste J (2007) Cross-linking assessment after accelerated ageing of ethylene prpopylene diene monomer rubber. J Polym Degrad Stab 92:457–463
11. Hamed GR, Zhao J (1999) Tensile behavior after oxidative aging of gum and black-filled vulcanizates of SBR and NR. Rubber Chem Technol 72(4):721–730
12. Spratte T, Plagge J, Wunde M, Klüppel M (2017) Investigation of strain-induced crystallization of carbon black and silica filled NR based on mechanicat and temperature measurements. Polymer 115:12–20
13. Wei YT, Nasdala L, Rothert H, Xie Z (2004) Experimental investigations on the dynamic mechanical properties of aged rubbers. Polym Test 23:447–453
14. Abdel-Aziz MM, Basfar AA (2000) Aging of ethylene-propylene diene rubber (EPDM) vulcanized by γ-radiation. Polym Test 19:591–602

15. Giese U, Homeier I, Navarro Torrejon Y, Kautz S (2013) Aging prozesses – mechanismen and quantitative characterization concerning polymer structure, antioxidants and crosslinking. Chem List 107:S3–S11
16. Santoso M, Navarro Torrejon Y, Giese U, Schuster RH (2008) Untersuchung thermischer und oxidativer Alterungsprozess von Elastomeren. Kautsch Gummi Kunstst 06:306–311
17. Ranimol S, Seno J, Kuruvilla J, Sabu T, Zachariah O (2006) Thermal stability and ageing properties of sulphur and gamma radiation vulcanized natural rubber (NR) and caroxylated styrene butadiene rubber (XSBR) latices and their blends. Polym Degrad Stab 91:1717–1725
18. Kumar A, Commerence S, Verney V (2004) Aging of elastomers: a molecula approach based on rheological characterization. Polym Degrad Stab:751–757
19. Fan R, Zhang Y, Huang C, Zhank T, Fan Y, Sun K (2001) Effect of crosslink structure on dynamic mechanical properties of natural rubber vulcanizates under different aging conditions. J Appl Polym Sci 81(3):710–718
20. Pimolsiriphol V, Soeomi P, Sirisinha C (2007) Relationship among thermla ageing degradation dynamic properties, cure systems and antioxidants in naturla rubber vulcanizates. Polym-Plast Technol Eng 46 2:113–121
21. Yahya YR, Azura AR, Ahmad Z (2001) Effect of curing systems on thermal degradation behaviour of natural rubber (SMR CV 60). J Physiol Sci 22(2):1–14
22. Khanlari S, Kokabi M (2011) Thermal stability, aging properties, and flame resistance of NR based nanocomposites. J Appl Polym Sci 119(2):855–862
23. Mosch A (2002) Untersuchung zur beschleunigten Schwefelvernetzung von Poly(isorpen) – Reaktionsprodukte der Vulkanisation mit Sulfinamiden. , Universität Hannover
24. Röthemeyer F, Sommer F (2006) Kautschuk Technologie Werkstoffe – Verarbeitung – Produkte. 2. Revised edition, Chap. 3, Carl Hanser Verlag, Munich, pp 339–340
25. Ismail H, Ishiaku US, Azhar AA, Mohd Ishak ZA (1997) A comparative study of the effect of thermo-oxidative aging on the physical properties of rice husk ash and commercial fillers in Epoxidized natural rubber compounds. J Elastomers Plast 29:270–289
26. DIN 53504 (2017-03) Testing of rubber – determination of tensile strength at break, tensile stress at yield, elongation at break and stress values in a tensile test
27. DIN ISO 48-4 (2021-02) Rubber, vulcanized or thermoplastic - determination of hardness – part 4: indentation hardness by durometer method (shore hardness) (DIN ISO 48-4:2018)
28. Prieß C, Katzenmaier V, Kreiselmaier R, Traber B, Beck K (2004) Analyse des oxidativen alterungsverhaltens elastomerer werkstoffe. Kautsch Gummi Kunstst 9:16–21
29. ASTM D3895:2019 Standard test method for oxidative-induction time of polyolefins by differential scanning calorimetry
30. Keller RW (1989) Oxidation and ozonation of rubber. Rubber Chem Technol 58(3):637–652
31. Chou HW, Hunag JS, Lin ST (2007) Effects on thermal aging on fatigue of carbon black reinforced EPDM rubber. J Appl Polym Sci 103:1244–1251
32. SPRINTAN™ SLR 4602-SCHKOPAU Technical data sheet, Trinseo Deutschland GmbH Schkopau, Germany, state 24 Jun 2021
33. Somers AE, Bastow TJ, Burgar MI, Forsyth M, Hill AJ (2000) Quantifying rubber degradation using NMR. Polym Degrad Stab 70:31–37
34. Müller M (2002) Einflussgrößen der Verstärkung elastomerer Werkstoffe durch polymere und klassische Füllstoffe, Dissertation, Universität Hannover
35. Reincke K, Langer B, Döhler S, Heuert U, Grellmann W (2014) Alterung und Beständigkeitsuntersuchungen von Elastomerwerkstoffen. Kautsch Gummi Kunstst 67(10): 60–67
36. Kumar A, Commereuc S, Verney V (2004) Ageing of elastomers: a molecular approach based on rheological characterization. Polym Degrad Stab 85:751–757
37. Chakraborty S, Kar S, Dasgupta S, Mukhopdhyay R (2010) Effect of ozone, thermo, and thermo-oxidative aging on the physical property of styrene butadiene rubber-Organoclay nanocomposites. J Elastomers Plast 42:443–452

38. Huneke B, Klüppel M (2006) Analyse der thermischen Alterung von Elastomeren. Teil 2: Mechanische Charakterisierung der Netzwerktopografie. Kautsch Gummi Kunstst 59:242–250
39. Osswald K, Reincke K, Schossig M, Sökmen S, Langer B (2019) Influence of different types of antioxidants on the aging behavior of carbon-black filled NR and SBR vulcanizates. Polym Test 79:106053
40. Karekar A, Oßwald K, Reincke K, Langer B, Saalwächter K (2020) NMR studies on the phase-seolved evolution of cross-link densities in thermo-oxidatively aged elastomer blends. Macromolecules 53:11166–11177
41. Ihlemann J, Giese U (2014) Experimentelle Analyse und virtuelle Nachbildung oxidativer Alterungsvorgänge in Elastomerbauteilen. Final report on the IGF projects 16313 BG
42. Kautz S (2019) Einfluss der Vernetzung und von Füllstoffen auf diffusionsbedingte Alterungsprozesse. Dissertation, University Hannover
43. Modrow H, Zimmer R, Visel F, Hormes J (2000) Monitoring thermal oxidation of sulfur crosslinks in SBR-elastomers using sulfur K-edge XANES: a feasibility study. Kautsch Gummi Kunstst 6:328–337

Adv Polym Sci (2023) 289: 185–208
https://doi.org/10.1007/12_2022_132

Published online: 22 August 2022

Degradation of Tires During Intended Usage

R. Kipscholl and R. Stoček

Contents

Abstract There are several factors that influence the wear behaviour and service life of tires. Three of them are closely interrelated. They are the tire design, the operating conditions and the rubber material from which the tire is made. One of these factors is of utmost importance and less tangible. It is the choice of the appropriate elastomer, fillers, additives and also vulcanising agents to meet the specific requirements for a particular tire. There are expectations for noise, grip, rolling resistance and comfort, stability as well as safety at the limit (braking, cornering). Alongside this, the biggest challenge in tire development is to extend the lifetime of the tire by increasing resistance to tire degradation, which leads to cracking and wear, as much as possible to reduce overall costs and pollution and to increase sustainability to save

R. Kipscholl
Coesfeld GmbH & Co. KG, Dortmund, Germany

R. Stoček (✉)
PRL Polymer Research Lab, s r.o., Zlín, Czech Republic

Centre of Polymer Systems, University Institute, Zlín, Czech Republic
e-mail: stocek@utb.cz

resources. All these challenges are the focus of scientists and engineers as long as rubber is used for transport and locomotion of cars, trucks, etc. Moreover, it is difficult to establish correlations between available laboratory tests and real tire tests in the field. Therefore, this chapter gives a brief introduction to the theory of fracture mechanics of rubber leading to wear and shows why analyses with long known laboratory equipment predict a different wear behaviour of rubber than that of a tire in service. Finally, a broad overview is given of novel, high-performance technical measuring equipment with which very reliable experimental results could be obtained that correspond very well to the theory of fracture mechanics and much better to the real tire wear behaviour in service. The recent experimental research impressively demonstrates what a huge step has been taken in the meantime to improve the prediction of wear under specific operating conditions and the design and production for tailor-made tires.

Keyword Degradation · Instrumented chip & cut analyser · Intrinsic strength analyser · Lifetime · Rubber · Tear and fatigue analyser · Tire · Wear

1 Introduction

Considering the lifetime of tires, the wear resistance is of undisputed importance for tire tread material, which is in dynamic contact with the rough surface of the road. But also, the shoulder of a tire is subject of wear forcing effects by permanent deflection. Real dynamic loading conditions of the tire treads and shoulders are based on the load in the contact region between the tire footprint and the driving ground. The derivation of wear driving theory presented here is based on the brush model [1, 2]. By the detailed view in Fig. 1, the mechanisms inside the tire tread while the tire rolls are shown. The deformation of a rolling tire in the contact patch is caused by shear force, which acts parallel to the driving direction between the tire tread and the surface of the driving ground. The shear forces are always generated by friction force resulting from rolling, accelerating, decelerating and curving and the applied friction coefficient. Especially if the road surface is uneven or the tire rolls off the road in rough terrain, the tire surface realises the contact with the driving ground over the edges of the asperities as it is indicated in Fig. 1. Therefore, we can say that the load on a tire is the integral of all local forces generated in the contact between asperities and tire tread. For further discussion, it is important to keep in mind that the textures of driving grounds even at smooth road surface are not uniform. The surfaces of driving grounds cover a wide range of qualities from very smooth to very harsh with asperities of different sizes. From this fact, besides the shear force there are also radial forces which are acting very locally between the asperities and the tire tread. They serve, on the one hand, for the necessary friction and, on the other hand, they penetrate the tire footprint orthogonal to the shear force and generate there local deformations of different sizes. From Fig. 1 it can be seen

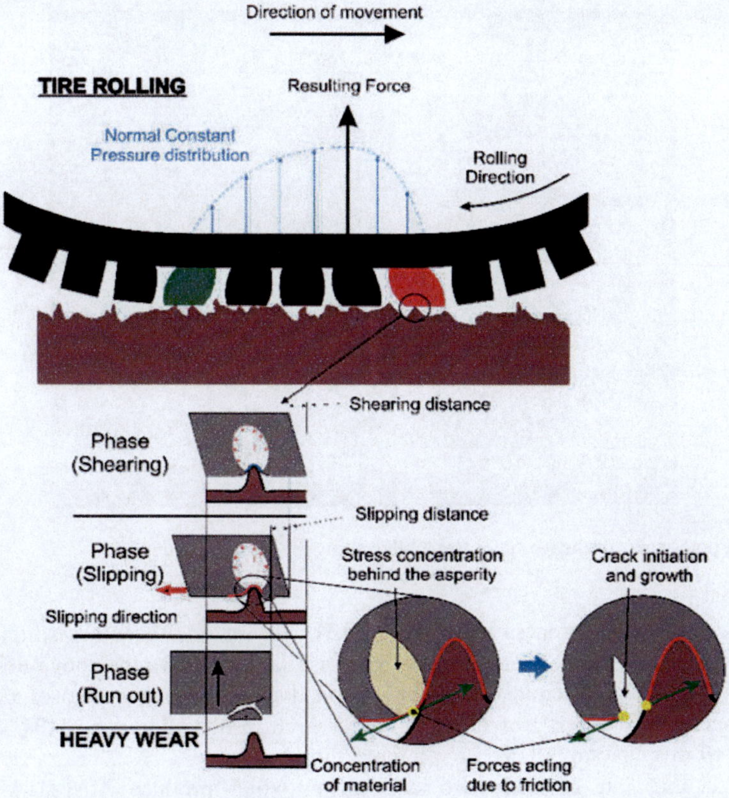

Fig. 1 Illustration of the deformation of the rubber layer between the tire carcass and the road according to the brush model with detail on local deforming impacts caused by asperities

how the rubber matrix is deflected by stress. Due to the permanent multiaxial stretching of chains parallel over the contact area and even locally rectangular by asperities, the intrinsic strength of the rubber matrix (T_0) will be exceeded and chains rupture consequently. A crack is initiated. Broken chains weaken the rubber in their vicinity. By permanent multiaxial stretching, the stress concentrates close to the initiated crack and further chains may break, preferably chains closest to the already broken ones. Thus, the fracture processes with micro-crack initiation in the tire tread have been started. As the tire goes through millions of fatigue cycles, these initiated micro-cracks continuously propagate and lead to the detachment of rubber particles ranging in size from very small (~1–100 μm) [3] to larger pieces, depending on how rough the ground is on which the tyre is mainly moved.

By the viscoelasticity of rubber, a more or less large portion of energy introduced by stretching and radial deformations will be dissipated. Dissipation occurs in the tread material and even in the shoulder of the tire and results in a phenomenon well known as Heat Build-up (HBU). The heating-up together and chemical degradation

Fig. 2 Photographs of the worn tire treads illustrating the wear mechanism

influences the wear process too. Due to HBU the material becomes softer, more energy will be dissipated and less energy is available to drive the above-described fractional process. Chemical degradation, on the other hand, decreases rubber's resistance against crack growth. Detailed research results addressing HBU are not subject of this chapter but can be referenced in [4, 5].

In fact there are basically two mechanisms which produce different kinds of mechanical wear. The first one is the friction generated one, which is motivated as above. The only reason for the different appearance and severity of wear damages is the roughness of the driving ground and the load on the tire. On smooth roads, such as public roads, very small rubber particles are created and lost through an even distribution of small asperities in height and size. They cause only small impacts on the tire surface. This wear phenomenon is called fatigue wear or abrasion. When driving on very rough terrain, e.g. mines, dirt roads or farmland, the so-called chip and cut (CC) wear prevails. In addition to the friction-initiated wear there are sharp asperities like edgy stones, graves or similar parts on the driving ground. They have a by far stronger disruptive impact on the rubber matrix, even to larger volumes of the tire tread. Those harmed regions are heavily weakened by the concentration of broken chains within the matrix and larger particles break out of the tread, more cuttings in the material and peel out of chips from the tread may also occur [6–10]. Last but not least, fatigue crack growth which is initiated by aging is provoked by chemical and thermal influences or exposure to sunlight, artificial light and UV light. Those effects reduce the crack growth resistance and appears even on the sidewall of tires under cyclic dynamic load [11].

The photographs shown in Fig. 2 illustrate the damage on tires produced by intended usage. In Fig. 2 left is a photo of a multi-purpose tread (MPT) shown, where the effects of fatigue wear from running on a dull surface causing smooth abrasion and from running over sharp edges that cause large amounts of rubber volume to be cut and chipped (CC) out of the tread are visible. Also cracks propagating from the corner of the profile to the centre of the tread are visible. Another example shown in Fig. 2, right, presents a photograph of a typically worn tire tread of moto-cross/ enduro (MX) tire, where a combination of generated fatigue wear during rolling over common roads and partial CC damages, caused by operating in hard terrain, can be seen.

Besides those mechanisms, smearing abrasion of rubber also is observed. Smearing generates a soft and sticky rubber layer on the contact surfaces. It is believed that smearing is fostered by thermal degradation and oxidation of the rubber material [6]. Moreover, the fatigue wear also increases with increasing polymer glass temperature, increasing ambient temperature and decreasing filler concentration [7, 12]. The most significant effect on rubber abrasion in terms of material composition can be attributed to reinforcing fillers with respect to carbon black (CB) in general [7, 8, 10] and type of rubber used [7, 8, 10, 13]. By focusing the most common rubbers applied in tires, which are natural rubber (NR), styrene-butadiene rubber (SBR) and butadiene rubber (BR) it is widely known in the tire industry that BR compounds have significantly greater fatigue wear resistance than NR compounds, whereas the resistance of SBR-based compounds is in between. To find an answer why BR has the highest fatigue wear resistance, the very low glass transition temperature has to be considered. BR has a *Tg* about −100°C compared to two other types of rubbers, which are SBR and NR with *Tg* about −10°C and −60°C, respectively. A low *Tg* leads to the fact that the main dispersion regime of the material is located at very high frequencies ($\sim 10^8 \ldots 10^{10}\ s^{-1}$), in this case the most of the viscoelastic energy will be dissipated and is therefore not available to produce defects like initiation of cracks and propagate already existing cracks. The typical local frequencies of tread polymers of an operating tire on nano-length scales are in the MHz range because of the locally very high strain rates ($\sim 10^7 \ldots 10^{10}\ s^{-1}$) of these processes. Due to the very low *Tg* of BR, the dispersion regime of BR falls into a high frequency regime; i.e. most of the introduced cyclic energy is dissipated, and crack initiation resistance is high as well as the fatigue threshold, T_0. That results in a very low crack growth speed and inhibited crack initiation. Therefore, most of the events acting on the rubber while the tire rolls result in a tearing energy that is below the fatigue threshold T_0 and will not contribute to accrual damage and related wear. Thus, the wear advantage of BR over SBR and NR can be understood in terms of a lower Tg and higher fatigue threshold T_0 for BR.

In the past, NR-based compounds have been experimentally attributed with higher wear compared to SBR and BR-based compounds, when the increase in tear energy is considered to be the cause of increasing abrasion or cutting wear. However, in the case of off-the-road tires, this contradicts the latest scientific findings reported in [14–17] that NR is the most resistant material to cutting wear due to its crystallisation at high strains. Thus, while for tires operating on smooth road surfaces, where the treads are loaded with low local deformations causing low local stresses in the tread material, the wear trends for the base polymers are quite

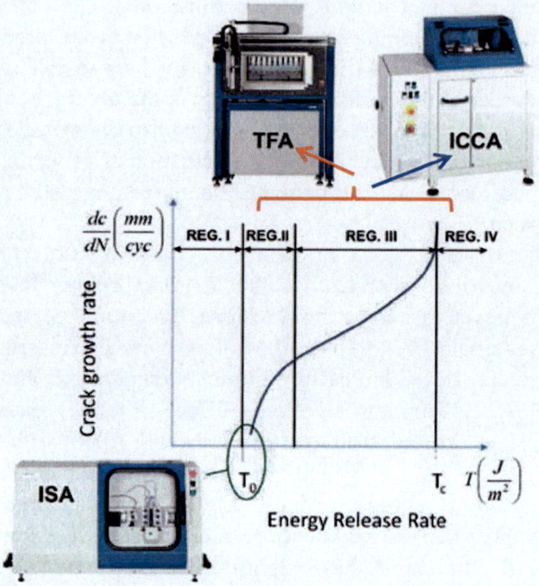

Fig. 3 FCG characteristic with the assignment of the testing devices to specific regions

clear, i.e. in terms of fatigue abrasion resistance BR > SBR > NR, the abrasion trends for these base polymers have been found to be exactly opposite NR > SBR > BR for tires operating in hard terrain [14, 18], which was also confirmed by the tire field test [19]. In [16], the total fatigue crack growth (visualised in Fig. 3 and described later in the chapter) for pure NR and binary NR/SBR mixture from fatigue threshold to ultimate strength was first correlated with CC behaviour. It is known that wear damage is related to fracture mechanics processes in the range from minimum crack energy near the fatigue threshold to critical crack energy near the ultimate strength. First of all, it could be shown that the fatigue threshold is in full agreement with the already mentioned higher fatigue wear resistance of SBR material compared to NR, i.e. the non-crystallising SBR rubber has a higher threshold compared to the crystallising NR. In agreement with the CC wear, the CC resistance of NR is lower than that of NR/SBR blends at lower force levels. On the other hand, the trend reverses at higher ranges of cracking energy [14–17]. The CC behaviour at low and also at high cracking energy is consistent with the ranking of FCG resistance of these rubber compounds [16]. These experimental results show that it is possible to interpret the abrasion mechanism based on the complex FCG properties in a predictive laboratory analysis of the rubber matrix on simple test samples.

FCG characteristics, represented due to double logarithmic plot in the Fig. 3, experimentally describes fatigue fracture behaviour with a general function of the fatigue crack growth (FCG) rate, $r = da/dn$ (in which da is the increment of crack growth and dn is the difference of the relevant cycle counts between which the crack increases by da). It depends on the energy release rate or tearing energy, T, correspondent to the rubber materials, which are loaded cyclically. This concept for the description of FCG behaviour is based on the preliminary work from Irwin (1957) in [20] by introducing the stress intensity factor. Paris in [21] applied this approach to the phenomenon of fatigue crack propagation. The FCG characteristic can be split into four different regions. In the first region, there is a minimum energy requirement for the fracture process. Due to independence of time, temperature and the degree of swelling is often called fatigue threshold or intrinsic strength, T_0 [22]. The mechanically caused and therefore relevant FCG proceeds in a broad range of the tearing energies between the intrinsic energy, T_0, and the critical tearing energy or ultimate strength, T_C. In the region II the dependence of the FCG rate, r, on the tearing energy, T, is linear and in the 3rd region changes into the formulation of the crack propagation equation in the form of a power-law [23–26]. The equation, today called the Paris equation reads:

$$\frac{da}{dn} = \beta T^m, \tag{1}$$

where a is the crack length, n represents the actual cycle count, β and m are the material and environment parameters. The exponent m is called the Paris exponent and represents the value of the exponent of the power-law. The law can be written also as follows:

$$\log\left(\frac{da}{dn}\right) = \log\beta + m\log T. \tag{2}$$

Finally, the crack proceeds to the unstable fatigue growth above the critical tearing energy, T_C, and the FCG rate becomes essentially infinite in region IV. Therefore, the FCG characteristic presents the complete fracture behaviour of material from crack initiation up to total rupture, where these fracture processes take place in the rubber during the abrasion process at different levels of tearing energy with respect to the loading boundary conditions acting on the tire from the road surface.

Although the abrasion mechanism theoretically is well known, evaluating the wear property of rubbery materials in a realistic way is still a challenging task. On the one hand, on-site wear tests are very time-consuming and costly and, on the other hand, on-site wear processes are very complex. Different road and/or climate conditions, and even the driving habit, can have large impact on the testing results. Therefore, predictive testing of wear resistance in laboratory using experimental laboratory equipment is of the highest importance. Although laboratory testers have been developed in the past, the tire industry still needs better laboratory solutions with more reliable results in characterisation of wear performance of rubber materials. A very rudimentary projection to quantify the wear is the determination of the abraded volume with respect to mass loss. In addition to well-known standard DIN-Abrader measuring in accordance with the standards ISO 4649 and ISO 20871, the most widespread device determining the mass loss is the so-called

Laboratory Abrasion Tester (LAT100), which originally was developed by Grosch [27] to measure the side force and the abrasion loss of a sample in the form of a solid rubber wheel as a function of various slip angles, loads, speeds and abrasive surfaces [28, 29]. Both mentioned equipment quantitatively simulate the fatigue wear process in the range of very low tearing energy level. In the past a simple laboratory equipment has been developed for measuring CC wear mechanism [30, 31]. However, this simple device is based on lab CC test methods with a small rotating rubber wheel, which is penetrated with a dead weight loaded impactor mounted on a lever. The loading conditions cannot be applied sufficiently and reproducible, which is not comparable with the real loading conditions of a tire. Due to the limited appliable impact energy, this simple traditional CC method did not target the high level of tearing energy close to ultimate strength. Various methods were developed and used for characterising the fatigue behaviour of elastomers in the past. The de Mattia test is a well-known and widely used method in the rubber industry to characterise the fatigue behaviour of rubbers [32]. Because the loading mechanism of the de Mattia appliance results in uncontrolled large strain in the sample by bending deformations and causes high strain concentration at the notch tip too, further on, from the practice it is assumed that this type of deformation does not correspond to that in tires during service, it can be taken as a sort of ranking under some rubbers, not as reliable physical results. As there is no instrumentation, visual observation of the test and classification of the crack initiation and its growth are necessary and consequently a high data scatter is usual. Contrary to qualitative information of the abrasion based on the determination of the mass loss and visual observation and classification of the crack initiated, the most proper and very promising way is to determine the abraded volume per unit energy dissipation or in-situ measurement of crack growth and associated these with the tearing energy level and furthermore with the complex FCG characteristics. The very important question is: How the knowledge of complex fracture behaviour over a broad range of tearing energies from fatigue threshold up to the ultimate strength could be beneficial for predicting the fracture behaviour with respect to lifetime of tires? The answer is very simple: When the FCG of a given specific rubber compound would be known in the complete range of applicable tearing energies, this compound can then be used in tires for the specific application for which the complex or selected range of fracture properties are best suited. For this purpose, efficient laboratory testing devices should be able to perform these tests under well-controlled testing conditions as close as possible to the real in-field situation.

With this approach, a unique set of testing devices had been developed by Coesfeld GmbH & Co. KG, Germany, to analyse the complete FCG characteristics. As reported in some publications, e.g. [33, 34], an important step in prediction of fatigue wear of rubber under defined loading conditions was introduced in the mid of 1990. The so-called **T**ear and **F**atigue **A**nalyser (**TFA©**) has the capability to load up to 10 notched single strain samples simultaneously with selectable strain and different cyclical load profiles [33–40]. In addition, this equipment is the only one that allows to analyse up to 10 plane strain test pieces, more commonly known as pure shear (PS). PS samples are preferably used for two reasons. Firstly: The tearing energy required to propagate a crack is independent from the crack length compared to the use of single edge notched tensile (SENT) geometry where it is not the case. SENT was used previously [41–45]. Secondly: Because of its width only one sample is necessary to measure the crack growth rate (FCGR) under appliance of numerous stain energies. Consequently the complete fatigue crack growth behaviour (FCG) can be determined from fatigue threshold, T_0 to ultimate strength, T_C quantitatively [46]. The TFA, compared to existing other appliances is fully instrumented. It supports tire developers with a power full tool to rank different rubbers automatically with respect to their fatigue crack growth rates. The crack length and the applied strain energy per strain cycle are synchronously measured. The crack length is measured by a CCD camera, strain and force are measured by transducers from which the total energy, elastic energy and the dissipated energy per load cycle are calculated. This feature saves operator attention and start-up time. The unique study performed in [46] evidenced that the accurate determination of the fatigue threshold value using the TFA device works well. But it is very time consuming. The time to determine the fatigue threshold is approximately 50 times longer than the time elapsed to determine the remaining regions of the FCG characteristic.

Therefore, it seemed likely to develop a methodology and related test equipment with whom it is possible to determine the fatigue threshold in a minimum amount of time. It is clear that the time-consuming phase of this measurement is to reach the lower limit for the breakage of rubber chains based on the strain energy alone. It was therefore necessary to accelerate the breaking of the chains at this very low strain. The pioneering work has been done by Lake and Yeoh [47]. In the experiment conducted, a PS sample is used that is lightly loaded and simultaneously a sharp razor blade cuts orthogonally to the principal stress of the loaded sample to cut the chains. With this method, it was possible for the first time to experimentally relate the measured intrinsic cutting energy to the fatigue threshold or intrinsic strength. Based on this concept, the theoretical background of the method was mainly elaborated by Mars [48]. The method described by Lake and Yeoh [47] was redefined and revised to the current state of the art by Mars and patented in 2017 [49]. Finally, for this method, a device called the Infrared Inverter was developed by Coesfeld GmbH & Co. KG, Germany, a device called Intrinsic Strength Analyser (ISA©), which is able to determine the value of the fatigue threshold very efficiently in a time of about 1 h [50–54]. With these two devices, TFA and ISA, it is thus possible to determine the total FCG properties in a convenient, reliable and efficient way. After a few hours the complete description of crack propagation from initiation, T_0 to critical strain energy, T_c for up to 10 rubber compounds is determined with only one set of PS sample each. However, there is still a gap in knowledge when assessing the overall resistance to wear. Due to contact with the various road surfaces, the tire is subjected to multiaxial loads and the resulting fracture processes are complicated, cracks develop and propagate very differently in each element, but their behaviour is always predetermined by the acting tearing energy and its effect as determined in the FCG characteristic curve. In order to characterise realistic wear processes on the tire tread from fatigue abrasion to CC wear, the load that takes place on a tire under realistic service conditions has to be adjusted in a controlled and reproducible way.

For this purpose, a fully instrumented device named **I**nstrumented **C**hip & **C**ut **A**nalyser (**ICCA©**) manufactured by Coesfeld GmbH & Co. KG, Germany, e.g. Refs. [14–17, 55], has been developed. The measuring principle is based on the above made assumption that wear with increasing load or increasing severity conditions of road produces an increasing wear profile or roughness on the rubber surface as already seen in Fig. 1. Therefore, the ICCA penetrates a rotating small sample wheel radial with a selectable force at an adjustable frequency. The rotating speed of the test wheel is also selectable. All of these parameters are controlled. Even the Heat build-up process is recorded by using an IR camera. The fatigue abrasion at low tearing energy level results in a smooth surface, abrasive wear in the range of middle tearing energy level results in a somewhat rougher surface and the CC wear which can be associated with the tearing energy close to ultimate strength results in larger volumes released from the rubber matrix. Changes in the surface roughness of the abraded rubber are used to quantify the surface damage. The impact force is controlled and the resulting tangential force is measured rectangular to the impact force every load cycle and is taken to define the degree of wear. The fluctuation of the tangential force is evaluated by a numerical algorithm. When the impact force increases the tearing energy increases accordingly and forces damaging the rubber surface following the rules determined by the FCG analysis. In other words, the key characterisation factor for the wear damage is basically the roughness and with it the fluctuation of the tangential force (in the methodology it is named "CC, Damage"). The CC damage is a physical parameter describing the kinetic of wear surface development.

Figure 3 therefore represents the FCG characteristic with the assignment of the above devices to specific regions in order to characterise the complete fracture behaviour curve under cyclic dynamic of tires. It is evident from the scheme that ISA is focused on the effective characterisation of only the fatigue threshold value, T_0, with the two remaining devices capable of analysing fracture behaviour over a wide region of tearing energy, where they efficiently analyse region III with minimal time requirements. However, whereas TFA analyses fracture behaviour quantitatively, directly characterisation tearing energy in relation to the crack propagation, ICCA quantitatively characterises the level of damage caused by fracture processes without directly quantifying the value of tearing energy and crack growth rate.

In the next part of this chapter, recent experiments are described to relate FCG characteristics to the wear phenomena found with the three above-mentioned instruments. The experiments were done on NR, SBR and BR which are typical polymers widely used for tire tread applications.

2 Experimental

Elastomers used in this study were natural rubber (SMR 20 CV/BP1, Lee Rubber Co. Pte Ltd., Malaysia), emulsion polymerised styrene-butadiene rubber (E-SBR 1500, Trinseo, Germany), and neodymium-catalysed butadiene rubber (High-Cis

Table 1 Rubber formulations

Batch	Raw material	Content [phr] SBR	NR	BR
Master batch	E-SBR 1500	100	–	–
	SMR20 natural rubber	–	100	–
	High-Cis Nd-BR	–	–	100
	N330 carbon black	50	50	50
Final batch	Zinc oxide	2	2	2
	Stearic acid	1	1	1
	Sulphur	2	2	2
	TBBS	2	1	2
	MTBS	–	–	0.5
	6PPD	1	1	1
Total content [phr]		158	157	158.5

Nd-BR, Trinseo, Germany). These polymers were filled with 50 phr of N330 type carbon black (CB) to obtain a filler volume fraction (ϕ) of 0.19. Table 1 lists the complete formulations for the rubber compounds.

Rubber compounds were prepared using a procedure with two mixing stages, where both steps were performed in an internal mixer (SYD-2L, Everplast, Taiwan). In the first step, the master batch was prepared with a rotor speed of 50 rpm, mixing chamber temperature of 100°C, and fill factor of 0.7. First the polymer (NR, SBR or BR) was mixed for 1 min followed by adding the CB and mixing for four additional minutes to reach a drop temperature of 140 to 150°C. The master batch was afterwards milled and sheeted with a two-roll mill at 60°C. In the second step, the final batch was prepared using a rotor speed of 35 rpm with a mixing chamber temperature of 60°C, wherein first the master batch compound was mixed for 1 min followed by adding the complete curing system and mixing for two more minutes, with a drop temperature in the 100 to 110°C range. The final batch was then milled and sheeted with a two-roll mill at 60°C. The final batches were stored for 24 h before curing. The curing characteristics – including time to reach 90% of final cure state (t90) – were evaluated at 160°C using a moving die rheometer (MDR 3000 Basic, MonTech, Germany) according to ASTM D 6204. The rubber sheets of a thickness 1 mm for preparation of plane strain samples as well as the specific disc geometry for CC analysis (details about the sample disc geometry can be found e.g. in [10–13]) were cured in a compression mould using a heated press (LaBEcon 300, Fontijne Presses, the Netherlands) at 160°C according to t90 for 1 mm thick sheets and t90 + 13 min (13 min = 1 min per 1 mm of the thickness) for the disc geometry.

2.1 Fatigue Threshold

The fatigue threshold with respect to intrinsic strength, T_0, has been analysed by using the ISA in accordance with the testing methodology developed by Endurica (LLC, USA) [49]. A specimen of plane strain geometry, which is pre-cracked on one edge, is exposed to several different strain levels in a range from 0% to 0.5%. The specimen is held under each defined strain and allowed to equilibrate for 10 min. Afterwards, the stress is determined from the normal force before cutting. The stress-strain curve is generated by combining the results from the different strain levels. When equilibration is reached (at each given strain), the tearing energy is computed as the product of the strain energy density, w, and the unstrained section gauge height, L_0,

$$T = wL_0 \tag{3}$$

Then a very sharp blade is brought into contact with the crack tip and is moved to slice the specimen at varied constant sequential rates. During cutting by the blade, an applied motive force, f, is required to maintain a constant rate of cutting speed. The cutting energy (3) contributes to the total energy release rate and drives the crack tip.

$$F_c = \frac{f}{B} \tag{4}$$

where B is the thickness of the specimen. When the crack tip dissipation is sufficiently small, the cutting energy, S, for strained PS specimen, undergoing the cutting process, may be written as the sum of individual energy release rates for tearing and cutting:

$$S = T + F_c, \tag{5}$$

where T is the tearing energy, and F_c is the cutting energy.

The intrinsic cutting energy $S_{0,C}$ is a calculated parameter from the consecutive measurements applying varied tearing and cutting energies. The intrinsic strength T_0 is proportional to the intrinsic cutting energy $S_{0,C}$ with a constant of proportionality b, where b varied between 0.1 and 0.2 in our experiments:

$$T_0 \;=\; b\; S_{0,C}. \tag{6}$$

2.2 FCG Analysis

The stable crack growth as well as the ultimate strength T_C, which represents region III in the FCG characteristic shown in Fig. 3, was determined with the TFA. Three double-notched mini-plane-strain samples were used to investigate each compound. The TFA experiments were performed with a Gaussian pulse of a single loading frequency with a constant pulse-width (Table 2). The load amplitude was stepwise increased. The complete applied loading conditions are listed in Table 2. The loading conditions have been chosen close to the loading frequency of truck tires. All three materials have been tested simultaneously using a TFA equipment with three electrically driven actuators . The complete TFA analyses have been performed in accordance with the testing protocol described in detail previously in Ref. [46].

2.3 CC Analysis

The cut and chip behaviour of the compounds was evaluated at room temperature with the ICCA. Testing with the ICCA involves rotating the rubber sample at a selected rotation speed and impacting the sample with a stainless-steel tool with specified frequency. The most important value calculated is a rate of damage. As the tangential force is the resulting answer on normal force which is assumed to grow proportional with increasing roughness of the penetrated surface, it is taken to calculate the degree of damage in every load cycle. A CC damage-parameter "P" is calculated, representing the reciprocal equivalent to CC resistance. To calculate P, a numerical algorithm which is based on the description of the tangential force over time, because of the data of tangential force over time shows an increasing scattering according to increasing roughness. For this reason an enwrapping curve is calculated which is numerically integrated from a certain starting point C_0 to the cycle count of interest, C_n. Each sum will be divided by the corresponding cycle number C_k with $0 < k < n$. This value is a scalar to quantify the CC damage at cycle count k and is called P. Therefore, a lower value of P is associated with greater CC resistance and vice versa.

The following test conditions have been applied for the ICCA analysis:

- Rotational speed: 150 rpm
- Normal (impact) forces: 70 N, 90 N, 110 N, 130 N and 150 N
- Sliding time: 50 ms.

Table 2 The loading condition for TFA experiment

Loading mode	Pulse-width frequency, f_1 [Hz]	Loading frequency, f_2 [Hz]	Temperature, T_e [°C]	Pre-force, F_{pre} [N]
Gaussian pulse	10	5	28	1

The complete ICCA analyses have been performed in accordance with the testing protocol described in detail previously in Refs. [17–19]. If different loads or normal forces are applied, it must be remembered that the low values of load correspond to very low abrasion, called fatigue abrasion, where medium values correspond to abrasive wear and high values correspond to cutting wear.

3 Results and Discussions

3.1 Fatigue Threshold

The relationship between cutting energy F_c and tearing energy T for the analysed compounds using the ISA is plotted in Fig. 4. In case of analysed rubbers, the lowest intrinsic cutting energy $S_{0,c}$ = 482.1 J/m^2 was found for the compound based on NR. The highest intrinsic cutting energy $S_{0,c}$ = 949.2 J/m^2 exhibits the rubber, based on BR. Finally, an intrinsic cutting energy, $S_{0,c}$ = 851.4 J/m^2 was determined for SBR.

To estimate the intrinsic strength T_0, it is necessary to determine the proportionality constant b, using a reference material. The details about the composition and performing of the reference ISA test are presented in Ref. [56]. In our experiment an unfilled SBR was used, for which the value of the intrinsic strength T_0 = 60 J/m^2 is known. From the intrinsic cutting energy $S_{0,c}$ = 571.6 J/m^2 which was measured and the known intrinsic strength, T_0 = 60 J/m^2 (endurance limit), the proportionality constant b was evaluated using Eq. 2. The proportionality constant was determined to be b = 0.105. Using this proportional factor, the intrinsic strength values T_0 for the

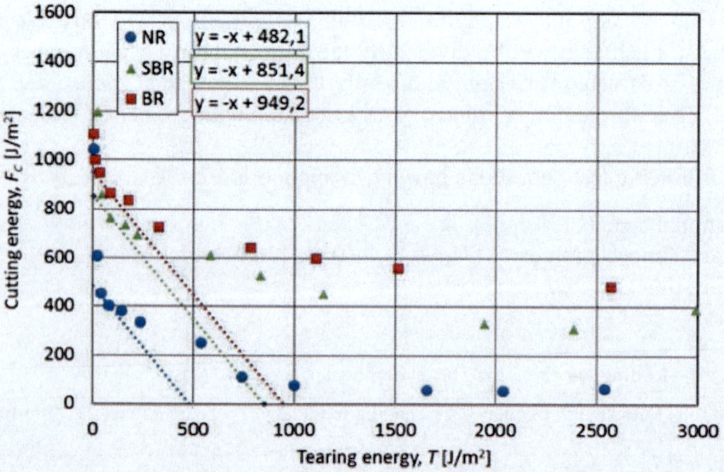

Fig. 4 ISA results for the analysed compounds

Table 3 Intrinsic strength for all analysed materials

Rubber	Intrinsic cutting energy $S_{0,c}$ [J/m^2]	Proportionality constant b [−]	Intrinsic strength T_0 [J/m^2]
NR	482.1	0.105	45.0
SBR	851.4		89.4
BR	949.2		99.7

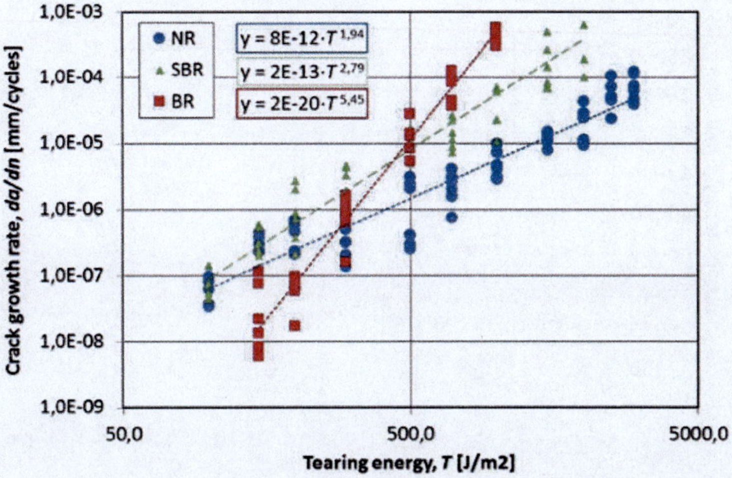

Fig. 5 FCGR, *da/dn* as a function of tearing energy, *T* for the analysed compounds

analysed materials could be calculated and they are listed in Table 3. The estimated T_0 values verify the expected trend, showing the highest intrinsic strength for BR followed by SBR and the lowest value for NR. This is in a full agreement with the resistance against fatigue abrasion of these materials as well known from operating tires.

3.2 FCG Analysis

The FCG characteristics of the compounds studied by TFA are shown in Fig. 5, using the power-law defined in Eq. 1, with the associated material constants β and m listed in Table 4. The exponent m indicates the slope of the crack propagation curve with respect to the exponent of the power-law. It can be clearly seen that NR rubber is highly resistant to FCG. The reason for this is the strain-induced crystallinity already mentioned above, which causes improved resistance to crack growth with increasing strain with respect to tearing energy. The highest value of m was determined for the BR compound, i.e. it has the lowest resistance to fatigue crack growth with increasing tearing energy. The exponent of power-law for the SBR

Table 4 List of Power-law for all compounds representing the stable crack growths of FCG characteristics

Material	*da/dn*	β	*m*
NR	8E-12x$^{1.94}$	8E-12	1.94
SBR	2E-13x$^{2.79}$	2E-13	2.79
BR	2E-20x$^{5.45}$	2E-20	5.45

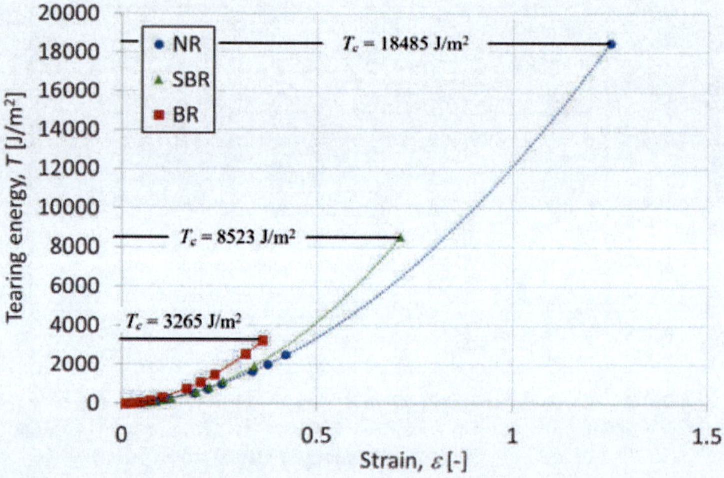

Fig. 6 Strain ε as a function of tearing energy T for the all analysed compounds

rubber is slightly higher than for NR, but significantly lower than for BR. It is remarkable that there is a crossing between all materials at a medium strain energy level. The FCG curve of BR crosses both curves for NR approximately at a tearing energy of $T \sim 300$ J/m^2 and for SBR approximately at $T \sim 400$ J/m^2. Thus, BR is more resistant to FCG compared to NR or SBR below the indicated crossing energies. This observation is exactly in line with the data previously obtained by Gent et al. [25].

The critical tearing energy T_c is defined as the tearing energy where the crack grows without any further energy input. In Fig. 6, the critical tearing energy, T_c defines the most right point of the strain vs. tearing energy function. The highest critical tearing energy $T_c = 18{,}485$ J/m^2 exhibits the rubber based on NR, whereas the lowest critical tearing energy $T_c = 3{,}265$ J/m^2 was determined for BR. A critical tearing energy $T_c = 8{,}523$ J/m^2 was determined for SBR.

Finally, from the data obtained with the ISA and TFA methods, Fig. 7 plots the FCG curves for the compounds investigated over all ranges of the fracture curve. It can be deduced from the strength values for these compounds that NR should be the most resistant material to cut wear in very rough terrain compared to SBR, while BR will have a very low resistance.

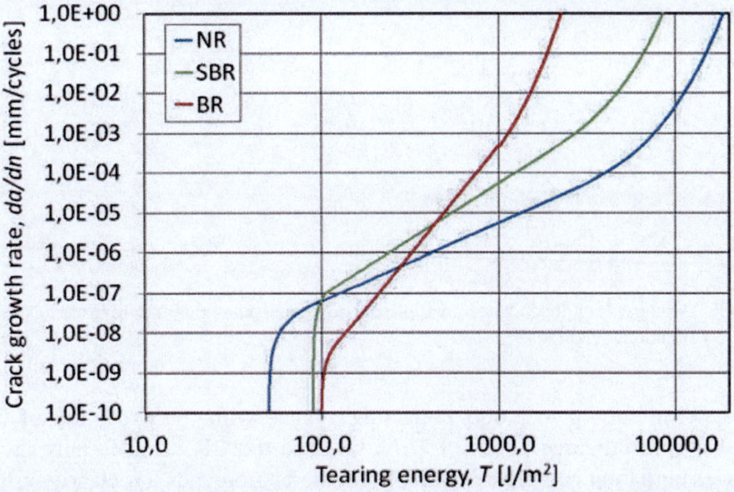

Fig. 7 Complete FCG characteristics for all analysed compounds containing all regions of fracture processes represented the average values

Table 5 Topographies of the abraded surfaces after final cycle count

Normal force, F_N [N]	NR	SBR	BR
70	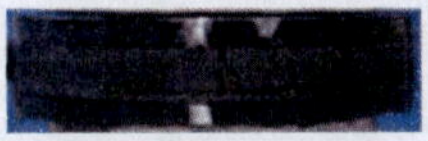		
90			
110		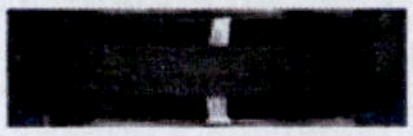	
130			
150			

3.3 CC Analysis

Table 5 shows the topography of the damaged surfaces after the last cycle number for all three investigated materials in dependence on the varied normal forces by ICCA. In this table, one sample for each compound is shown as an example, and 3 samples of each compound were analysed to check repeatability. The materials

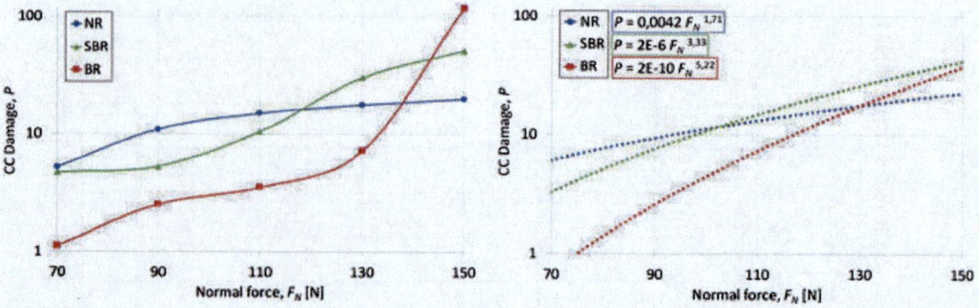

Fig. 8 CC Damage, *P* in dependence on applied loading (left) and their approximations by power-law (Eq. 7) functions (right)

were analysed for up to 3,000 impacting cycles, with the exception of the BR material at normal forces $F_N = 150$ N, because the BR samples only last up to 300 cycles until their complete rupture. From the photographs it is clearly visible that the samples based on BR are completely free of any surface damage for 70 and 90 N, while for 110 and 130 N they contain only one or three visible cracks, but their surface is still completely smooth. At an increased load to 150 N, the sample is very quickly and completely destroyed, as large pieces of rubber being cut and chipped out of the surface. The surface topography of the SBR samples shows a high wear resistance in the region from 70 to 110 N. But a small sliding mark is visible on the surface. This can be attributed to a smearing effect caused by the development of a significant heat build-up during the wear process, as SBR shows the highest heat build-up under the analysed materials. It can therefore be assumed that the SBR material has the worst resistant against smearing wear. For load ranges from 130 to 150 N, rapid abrasion can be observed, which is characterised by the chipping of larger particles from the surface of the sample, but the particles are clearly smaller in size compared to BR material. Finally, for the NR material, cracks are visible on the sample already at a load of 90 N, while the topography of abrasion on the surface increases systematically with increasing load. From an optical point of view, the surface damage increases almost constantly over the increased load levels. In terms of the size of the particles that chipped off the surface of the samples, the NR rubber shows the smallest particles compared to SBR and BR.

The diagram in Fig. 8, left, shows the dependence of CC damage, *P* on load and the measured data are fully corresponding with the surface topography discussed above and shown in Table 5. If we consider that CC damage *P* below 10 indicates no or minimal abrasion, then it is obvious that up to a load value of 130 N the BR material is more or less fully resistant to abrasion, while above this value it becomes rapidly vulnerable. The SBR based material is resistant up to an average value of 110 N, after which surface abrasion accelerates significantly, but with a smaller increase in damage than with the BR material. Thus, the curves cross and the SBR material becomes more resistant than the BR material beyond 140 N. NR shows CC damage, P over the entire applied load range, but this value of CC damage is very stable. Thus, for low loads this material is the least resistant to abrasion, but in the

Table 6 List of power-law for all compounds representing the CC damage, *P* in dependence on normal force, F_N

Material	P	β_{cc}	m_{cc}
NR	$0.0042x^{1.71}$	0.0042	1.71
SBR	$2E\text{-}06x^{3.33}$	2E-06	3.33
BR	$2E\text{-}10x^{5.22}$	2E-10	5.23

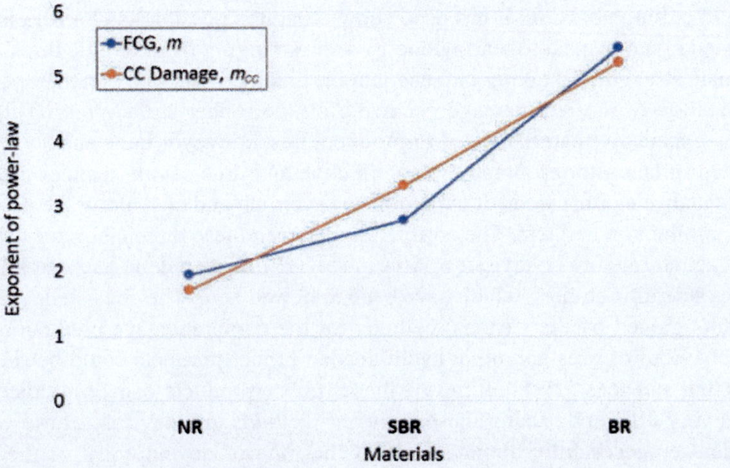

Fig. 9 Comparison of material constant, *m* and m_{cc} representing the exponents of power-law functions for FCG and CC damage behaviour of all materials

middle region and for high loads it becomes the most resistant material. In addition, Fig. 8, right, shows the approximation of these curves by power-law functions defined as follows:

$$P = \beta_{cc} F_N{}^{m_{cc}}, \tag{7}$$

where the related material constants β_{cc}, m_{cc} are listed in Table 6.

The diagram in Fig. 9 compares the material constants *m* (TFA) and m_{cc} (ICCA) representing the exponents of the power-law functions, between the FCG characteristic and the CC damage, respectively. From this plot, it can be seen that Paris-(FCG)-exponent, *m* and exponent of CC damage, m_{cc} show obviously similar values.

Last but not least, it is quite evident that for the region of low tearing energies or normal forces the trend of the material resistance to fatigue wear in order NR < SBR < BR from lowest to highest resistance is identical to the trend for the fatigue threshold, T_0, measured by ISA.

4 Conclusion

It could be shown that with the help of the three measuring appliances (TFA©, ICCA©, ISA©) a complete characterisation of rubber with respect to their wear behaviour under cyclical load can be predicted. The study demonstrates with basic materials that by using the above-described experimental methodologies, rubber wear can be characterised very effectively in laboratory conditions, ranging from fatigue to cutting abrasion. It has been shown that it is possible to characterise these phenomena with respect to tearing energy values ranging from fatigue threshold, T_0 to ultimate strength, T_c, taking into account the crack growth rate parameter as in the quantification of the real damage occurring on the rubber surface during the wear process. Especially materials used to produce tires, conveyor belts and so forth can be ranked in laboratories already in an effective and time-saving manner predicting their resistance against wear. It is helpful to save time and cost along the process of design, production and test. The correlation between these three laboratory methods and the actual fracture behaviour of tires in service has been demonstrated in several previous scientific studies, which have been reviewed here. The consistency between the results gained by the various methods for the comprehensive prediction of the wear behaviour of tires and other cyclic loaded rubber products could be shown.

Last but not least, the lifetime of these rubber products can be predicted and verified very efficiently using these complex methods and devices. These verifications can be successfully applied early at the time of development of the rubber compound with regard to its wear and fracture resistance. This lowers overall costs, reduces air pollution and improves environmental protection, leading to enhanced sustainability and savings in raw material resources.

Funding This work was supported by the Ministry of Education, Youth and Sports of the Czech Republic – DKRVO (RP/CPS/2022/006) and within the project IGA/CPS/2021/006.

References

1. Pacejka HB, Sharp RS (1991) Shear force development by pneumatic tyres in steady state conditions: a review of modelling aspects. Veh Syst Dyn 20(3–4):121–175. https://doi.org/10.1080/00423119108968983
2. Pacejka HB (2005) Tyre and vehicle dynamics.2nd edn. Elsevier. ISBN: 9780080543338
3. Persson BNJ (2009) Theory of powdery rubber wear. J Phys Condens Matter 21:485001. https://doi.org/10.1088/0953-8984/21/48/485001
4. Stoček R, Stěnička M, Kipscholl R (2019) Heat build-up characterization under realistic load. In: Constitutive models for rubber XI – proceedings of the 11th European conference on constitutive models for rubber. CRC Press/Balkema, Nantes, pp 157–162. https://doi.org/10.1201/9780429324710-28
5. Peter O, Stoček R, Kratina O (2022) Experimental and numerical description of the heat build-up in rubber under cyclic loading. In: Advances in polymer science. Springer, Berlin. https://doi.org/10.1007/12_2021_108

6. Muhr AH, Roberts AD (1992) Rubber abrasion and wear. Wear 158(1992):213–228. https://doi.org/10.1016/0043-1648(92)90040-F
7. Veith AG (1992) A review of important factors affecting treadwear. Rubber Chem Technol 65(3):601–659. https://doi.org/10.5254/1.3538631
8. Schallamach A (1952) Abrasion of rubber by a needle. J Polym Sci 9:385. https://doi.org/10.1002/POL.1952.120090501
9. Gent AN, Nah C (1996) Abrasion of rubber by a blade abrader: effect of blade sharpness and test temperature for selected compounds. Rubber Chem Technol 69(5):819–833. https://doi.org/10.5254/1.3538405
10. Klüppel M (2014) Rubber friction. In: Kobayashi S, Müllen K (eds) Encyclopedia of polymeric nanomaterials. Springer, Berlin. https://doi.org/10.1007/978-3-642-36199-9_305-1
11. Heinrich G, Klüppel M (2022) Basic mechanisms and predictive testing of tire-road abrasion. In: Advances in polymer science. Springer, Berlin. https://doi.org/10.1007/12_2022_113
12. Kienle RN, Dizon ES, Brett TJ, Eckert CF (1971) Tread wear and wet skid resistance of butadiene-styrene elastomers and blends. Rubber Chem Technol 44(4):996–1014. https://doi.org/10.5254/1.3547400
13. Grosch KA, Schallamach A (1961) Tyre wear at controlled slip. Wear 4(5):356–371. https://doi.org/10.1016/0043-1648(61)90003-5
14. Stoček R, Mars WV, Kipscholl R, Robertson CG (2019) Characterisation of cut and chip behaviour for NR, SBR and BR compounds with an instrumented laboratory device. Plast Rubber Compos 48(1):14–23. https://doi.org/10.1080/14658011.2018.1468161
15. Kipscholl R, Stoček R (2019) Quantification of chip and cut behaviour of basic rubber (NR, SBR). RFP Rubber Fibres Plastics 2:88–91. ISSN 1863-7116
16. Stoček R, Ghosh P, Machů A, Chanda J, Mukhopadhyay R (2021) Fatigue crack growth vs. chip and cut Wear of NR and NR/SBR blend-based rubber compounds. In: Advances in polymer science. Springer, Berlin. https://doi.org/10.1007/12_2020_67
17. Stoček R, Heinrich G, Kipscholl R, Kratina O (2021) Cut & chip wear of rubbers in a range from low up to high severity conditions. Appl Surf Sci Adv 6(1):100152. https://doi.org/10.1016/j.apsadv.2021.100152
18. Stoček R, Mars WV, Robertson CG, Kipscholl R (2018) Characterizing rubber's resistance against chip and cut behaviour. Rubber World 257:38–40
19. Stoček R, Heinrich G, Schulze A, Wunde M, Klüppel M, Vatterott C, Tschimmel J, Lacayo-Pineda J, Kipscholl R (2020) Chip & cut wear of truck tire treads: comparison between laboratory and real tire testing. Kautschuk Gummi Kunststoffe 73(6):51–55
20. Irwin GR (1957) Analysis of stresses and strains near the end of a crack traversing a plate. J Appl Mech 24(3):361–364. https://doi.org/10.1115/1.4011547
21. Paris PC, Gomez MP, Anderson WE (1961) A rational analytic theory of fatigue. Trend Eng 13: 9–14
22. Gent AN, Mars WV (2013) Chapter 10. In: Mark JE, Erman B, Roland M (eds) The science and technology of rubber4th edn. Academic Press, Boston, pp 473–516. https://doi.org/10.1016/B978-0-12-394584-6.00010-8
23. Paris PC, Erdogan F (1963) Critical analysis of crack propagation laws. J Basic Eng 85:528–534. https://doi.org/10.1115/1.3656900
24. Paris PC (1998) Fracture mechanics and fatigue: a historical perspective, fatigue fract. Eng Mater Struct 21:535–540. https://doi.org/10.1046/j.1460-2695.1998.00054.x
25. Gent AN, Lindley PB, Thomas AG (1964) Cut growth and fatigue of rubbers. I. The relationship between cut growth and fatigue. J Appl Polym Sci 455-466. https://doi.org/10.1002/app.1964.070080129
26. Lindley PB (1974) Non-relaxing crack growth and fatigue in a non-crystallizing rubber. Rubber Chem Technol 47(5):1253–1264. https://doi.org/10.5254/1.3540497

27. Grosch KA (1997) A new way to evaluate traction-and wear properties of tire tread compounds. In: 152th meeting rubber division, American Chemical Society. Oct 21–24, 1997, Cleveland, OH, USA
28. Salehi M, Noordermeer JWM, Reuvekamp LAEM, Dierkes WK, Blume A (2019) Measuring rubber friction using a laboratory abrasion tester (LAT100) to predict car tire dry ABS braking. Tribol Int 131:191–199. https://doi.org/10.1016/j.triboint.2018.10.011
29. Salehi M, Noordermeer JWM, Reuvekamp LAEM, Blume A (2022) Characterization of counter-surface substrates for a laboratory abrasion tester (LAT100) compared with asphalt and concrete to predict car tire performance. Lubricants 10:8. https://doi.org/10.3390/lubricants10010008
30. Beatty JR (1979) Testing apparatus and method for measuring cutting, chipping and abrasion resistance, U.S. Patent 4, 144, 740
31. Beatty JR, Miksch B (1982) A laboratory cutting and chipping tester for evaluation off-the-road and heavy-duty tire treads. Rubber Chem Technol 55:1531–1546. https://doi.org/10.5254/1.3535947
32. ASTM D 430–95 and ASTM D813–95, Standard test method for rubber deterioration. Fatigue and crack growth
33. Eisele U, Kelbch SA, Engels H-W (1992) The tear analyzer – a new tool for quantitative measurements of the dynamic crack growth of elastomers. Kautschuk-Gummi-Kunststoffe 45: 1064–1069
34. Heinrich G, Struve J, Gerberb G (2002) Mesoscopic simulation of dynamic crack propagation in rubber materials. Polymer 43(2):395–401. https://doi.org/10.1016/S0032-3861(01)00505-5
35. Stoček R, Gehde M, Heinrich G (2009) Analyse des dynamischen Risswachstums von Elastomeren-simultaner Zug-(tensile) und Pure-shear-Prüfmodus bei optimierter Risserfassung. Kautschuk Gummi Kunststoffe 62:170–176
36. Andreini G, Straffi P, Cotugno S, Gallone G, Polacco G (2013) Crack growth behaviour of styrene-butadiene rubber, natural rubber and polybutadiene rubber com-pounds: comparison of pure-shear versus strip tensile test. Rubber Chem Technol 86:132–145. https://doi.org/10.5254/rct.13.88957
37. Stoček R, Heinrich G, Gehde M, Kipscholl R (2013) Analysis of dynamic crack propagation in elastomers by simultaneous tensile- and pure-shear-mode testing. In: Grellmann W et al (eds) Fracture mechanics & statistical mechanics, LNACM 70, 269–301. https://doi.org/10.1007/978-3-642-37910-9_7
38. Ghosh P, Mukhopadhyay R, Stocek R (2016) Durability prediction of NR/BR and NR/SBR blend tread compounds using tear fatigue analyser. Kautschuk-Gummi-Kunststoffe 69(6): 53–55
39. Ghosh P, Stoček R, Gehde M, Mukhopadhyay R, Krishnakumar R (2014) In-vestigation of fatigue crack growth characteristics of NR/BR blend based tyre tread compounds. Int J Fract 188:9–21. https://doi.org/10.1007/s10704-014-9941-9
40. Stoček R, Kratina O, Ghosh P, Maláč J, Mukhopadhyay R (2017) Influence of thermal ageing process on the crack propagation of rubber used for tire application. Springer Ser Mater Sci 247: 351–364. https://doi.org/10.1007/978-3-319-41879-7_24
41. Stoček R, Stěnička M, Maloch J (2020) Determining parametrical functions defining the deformations of a plane strain tensile rubber sample. In: Heinrich G, Kipscholl R, Stoček R (eds) Fatigue crack growth in rubber materials. Advances in polymer science, vol 286. Springer, Cham. https://doi.org/10.1007/12_2020_78
42. Rivlin RS, Thomas AG (1953) Rupture of rubber. I. Characteristic energy for tearing. J Polym Sci 10:291–318. https://doi.org/10.1002/pol.1953.120100303
43. Stoček R, Gehde M, Heinrich G (2009) Analyse des dynamischen Risswachstums von Elastomeren in simultaner Zug-(tensile) und Pure-shear Prüfmodus bei optimierter Risserfassung. Kautschuk Gummi Kunststoffe 62:170–176
44. Stoček R, Horst T, Reincke K (2017) Tearing energy as fracture mechanical quantity for elastomers. In: Stöckelhuber KW, Das A, Klüppel M (eds) Designing of elastomer

nanocomposites: from theory to applications. Advances in polymer science, vol 275. Springer, New York, pp 361–398. https://doi.org/10.1007/12_2016_10
45. Stoček R, Heinrich G, Gehde M, Kipscholl R (2012) A new testing concept for determination of dynamic crack propagation in rubber materials. Kautschuk Gummi Kunststoffe 65:49–53
46. Stoček R (2020) Some revisions of fatigue crack growth characteristics of rubber. In: Heinrich G, Kipscholl R, Stoček R (eds) Fatigue crack growth in rubber materials. Advances in polymer science, vol 286. Springer, Cham. https://doi.org/10.1007/12_2020_72
47. Lake GJ, Yeoh OH (1978) Measurement of rubber cutting resistance in the absence of friction. Int J Fract 14:509–526. https://doi.org/10.1007/BF01390472
48. Mars WV (2007) Fatigue life prediction for elastomeric structures. Rubber Chem Technol 80(3):481–503. https://doi.org/10.5254/1.354817
49. Mars WV (2017) Instrument for measuring the intrinsic strength of polymeric materials via cutting. No.: US 2017/0003207 A1, USA
50. Stocek R, Kratina O, Kipscholl R (2015) A new experimental approach to rubber resistance against cutting by sharp objects. In: Constitutive models for rubber IX. Proceedings of the 9th European conference on constitutive models for rubbers, ECCMR, pp 357–362
51. Stoček R, Mars WV, Kratina O, Machů A, Drobilík M, Kotula O, Cmarová A (2017) Characterization of ageing effect on the intrinsic strength of NR, BR and NR/BR blends. In: Constitutive models for rubber X – proceedings of the 10th European conference on constitutive models for rubber, pp 371–374
52. Mars WV, Robertson CG, Stoček R, Kipscholl C (2019) Why cutting strength is an indicator of fatigue threshold. In: Constitutive models for rubber XI – proceedings of the 11th European conference on constitutive models for rubber, pp 351–356
53. Robertson CG, Stoček R, Kipscholl C, Mars WV (2019) Characterizing the intrinsic strength (fatigue threshold) of natural rubber/butadiene rubber blends. Tire Sci Technol 47(4):292–307. https://doi.org/10.2346/tire.19.170168
54. Robertson CG, Stoček R, Mars WV (2020) The fatigue threshold of rubber and its characterization using the cutting method. In: Heinrich G, Kipscholl R, Stoček R (eds) Fatigue crack growth in rubber materials. Advances in polymer science, vol 286. Springer, Cham. https://doi.org/10.1007/12_2020_71
55. Robertson CG, Suter JD, Bauman MA, Stoček R, Mars WV (2020) Finite element modeling and critical plane analysis of a cut-and-chip experiment for rubber. Tire Sci Technol 49(2): 128–145. https://doi.org/10.2346/tire.20.190221
56. Stoček R, Stěnička M, Zádrapa P (2021) Future trends in predicting the complex fracture behaviour of rubber materials. Contin Mech Thermodyn 33:291–305. https://doi.org/10.1007/s00161-020-00887-z

Adv Polym Sci (2023) 289: 209–238
https://doi.org/10.1007/12_2021_108

Published online: 20 February 2022

Thermal and Thermo-Oxidative Degradation of Rubbers: Some Recent Studies

T. Saha and A. K. Bhowmick

Contents

Abstract Elastomers or rubbers are a special kind of material with inherent elasticity, i.e. the ability to retain its structure upon deformation. Due to their unique properties, elastomers are used in different applications starting from automotive tires, seals, gaskets, conveyor belts, V-belts to musical instruments, toys, and even in garments. However, there is a much-awaited need for an elastomer with a higher degree of heat resistance property due to the stringent industrial requirements. The heat resistance of an elastomer is extremely desirable in automotive, aerospace, and off-shore industries where several rubber parts experience much higher temperature. Thus, it is very important to understand the degradation of different elastomers over

T. Saha
Rubber Technology Centre, Indian Institute of Technology, Kharagpur, India

A. K. Bhowmick (✉)
Department of Chemical and Biomolecular Engineering, The University of Houston, Houston, TX, USA

Formerly with Indian Institute of Technology, Kharagpur, India
e-mail: akbhowmick@uh.edu

a range of temperatures to prevent it. This chapter summarizes the high-temperature degradation of various elastomers with the help of the reactive molecular dynamics simulation technique. Different fragmented products were identified with the help of the simulation method and experimentally verified. The simulation technique was also used to calculate the activation energy for the degradation of different elastomers. The final section of this chapter describes the thermo-oxidative degradation of various elastomers and their nanocomposites.

Keyword Degradation · Elastomer · Heat resistance · Lifetime · Molecular simulation · Thermal stability

Abbreviations

ACM	Polyacrylic ester
ADF	Amsterdam density function
BR	Polybutadiene rubber
DSC	Differential scanning calorimetry
EA	Ethyl acrylate
EPDM	Ethylene propylene diene monomer
FTIR	Fourier transform infrared spectroscopy
FWO	Flynn-Wall-Ozawa
GUI	Graphical user interface
HDPE	High-density polyethylene
HNBR	Hydrogenated acrylonitrile-butadiene rubber
HSN	Highly saturated nitrile rubber
IR	Infrared spectroscopy
KAS	Kissinger-Akahira-Sunose
MDs	Molecular dynamics simulation
nBA	n-Butyl acrylate
NBR	Acrylonitrile-butadiene rubber
NMR	Nuclear magnetic resonance
NR	Natural rubber
PC	Polycarbonate
PE	Polyethylene
PIB	Polyisobutylene
PMMA	Polymethylmethacrylate
PMS	Poly(α-methyl styrene)
PP	Polypropylene
PS	Polystyrene
PSD	Power spectral density
PYGCMS	Pyrolysis-gas chromatography-mass spectrometry
QM	Quantum mechanical
ReaxFF	Reactive force field

SBR	Styrene-butadiene rubber
TGA	Thermogravimetric analysis
TTS	Time-temperature superposition
UFF	Universal force field

1 Introduction

Elastomers are long-chain macromolecules with inherent viscoelasticity. The long macromolecular chains interact together with weak intermolecular forces of interactions. As a result, they have high elongation [1, 2]. They exhibit the unique property of retaining their original shape and size when subjected to a uniform deformation [3, 4]. Elastomers can either be naturally occurring (e.g., Natural rubber) or can be synthesized in a chemical plant (e.g., Styrene-butadiene rubber). Elastomers are composed of long polymeric chains where thousands of chemical units (known as monomer) are connected by covalent bonds [5–7]. The monomer units are generally comprised of carbon, oxygen, and hydrogen. In some special cases, silicon, fluorine, chlorine, and/or sulfur (in Thiokol rubber) may be present [1, 8]. For a polymeric material to act as an elastomer, it must possess high molecular weight along with flexible molecular chains (glass transition temperature lower than room temperature) and low intermolecular forces of interaction.

The long molecular chains and flexible structure of an elastomer are sufficient to exhibit the properties of elasticity and extensibility. But to produce a useful rubbery product, the polymeric chains need to be crosslinked to form a three-dimensional structure with the help of external agents (known as curatives) in the presence of heat and pressure [9]. This method is called curing or vulcanization of elastomer. Only then rubber products are used in different fields of applications such as automobile tires, wire and cables, conveyor belts, consumable products like shoe soles, toys, musical equipment, etc., adhesives and sealants, industrial tools, lubricants, seals, gaskets, and even in the medical field [10]. Due to its diversified applications, elastomer accounts for the highest amount of production and usage of polymers.

But these elastomers encounter undesirable changes in properties like modulus, tensile strength, hardness, etc. under extreme environmental conditions [11–13]. In many industrial applications, any change in these properties of an elastomer under a stringent environment is highly unacceptable. Thus, it is very essential to understand their degradation behavior and prevent them [14, 15]. Bhowmick et al. have studied the degradation of different elastomers for nearly three decades [13–19]. But the relationship between the change in chemical structure of an elastomer with the change in properties under the influence of heat or oxygen is still an active area of research. Thus, recently they have made an effort to investigate the degradation of different elastomers at a molecular level and tried to establish a structure-degradation correlation for different elastomers [20–23]. This can be useful to prevent the degradation of the elastomers.

Molecular modeling can be an effective way to understand the degradation of various elastomers. Recently, molecular dynamics simulation appears as one of the promising tools to study the structure-property relationship of different polymers and their allied materials [24–26]. Various researchers around the world have worked extensively for over a decade to simulate long macromolecular chains [27–30]. But the major question is whether the MD simulation is capable enough to predict degradation pattern of an elastomer correctly? The answer lies in a new kind of force field termed as the "Reactive Force Field." It was invented by van Duin et al. in the year 2001 [31]. Reactive force field simulation technique can be used to identify major decomposition fragments of an elastomer when it is heated at a high temperature [32, 33]. This chapter emphasizes the use of the reactive force field method to predict the degradation pattern of various elastomers. Both the thermal and the thermo-oxidative behaviors are covered in this chapter.

In addition, the mechanical properties of a few elastomers before and after degradation, recently studied in our laboratory have been highlighted in this chapter.

2 Principles of Elastomer Degradation

The degradation of rubber is a topic of interest for many researchers for the last two decades. It is mainly associated with the change in molecular characteristics of different elastomers along with the change in physical properties like modulus, tensile strength, tear strength, etc. Different elastomers undergo different types of decomposition like end group scission, random chain scission, crosslinking due to the action of heat, light, and/or chemicals [34, 35]. In the case of crosslinked systems, there is a possibility of breakage of weaker crosslinks or crosslink exchange. The degradation is also influenced by the compounding ingredients used in a rubber formulation. Diffusion limited oxidation contributes to changes in structure. All these create an adverse effect on the properties and service life of the finished products. Degradation of any elastomer is a complex process and without thorough understanding, it is very difficult to prevent this process. Different researchers have utilized different analytical methods like Thermogravimetric analysis (TGA), Pyrolysis-gas chromatography-mass spectrometry (Py-GC-MS), Fourier-transform infrared spectroscopy (FTIR), etc. to analyze the degradation of various elastomers [36–39]. Determination of various properties before and after degradation at specified temperatures is common in the rubber industry.

3 Thermal and Thermo-Oxidative Degradation of Elastomers

The degradation of elastomers is a very complicated phenomenon and hence extensive research has been done by various researchers for different rubbers and rubber blends. Bhowmick et al. studied the degradation of guayule rubber in presence of stearic, oleic, linoleic, and linolenic acid. Guayule rubber degradation started at around 230°C in air and 330°C in nitrogen, but in the presence of an acid, the onset point was reduced due to the low decomposition temperature of the acids [40]. The most abundant fragmented product obtained during the pyrolysis of guayule rubber in an inert atmosphere was isoprene. Deuri et al. investigated the aging of EPDM-based rocket insulator compounds filled with cork, asbestos fiber, and iron oxide in nitrogen and air. They observed that the mechanical properties of the insulator such as tensile strength, tear strength, and hardness increased initially with the increase in time and temperature of aging [41].

Bhattacharjee et al. analyzed low-temperature degradation of nitrile rubber (NBR), highly saturated (HSN) nitrile rubber, and hydrogenated nitrile rubber (HNBR) [16]. They reported that the molecular weight of highly saturated (HSN) nitrile rubber was first increased and then decreased after an optimum aging time. For nitrile rubber, the molecular weight was first decreased and then increased on prolonged aging. IR and ESCA analysis suggested that the degradation of nitrile rubber occurred through the attack of the double bond, whereas, in the case of HSN and HNBR, the degradation took place through the attack on -C ≡ N group and free radical decomposition.

De Sarkar et al. studied the degradation of hydrogenated styrene-butadiene rubber (HSBR) with different levels of unsaturation under anaerobic and aerobic conditions using TGA, DSC, IR, and NMR spectroscopy [14]. IR and NMR studies showed that the degradation mainly occurred via thermal isomerization, oxidation, cyclization, depolymerization, and chain scission. Garbarczyk et al. used NMR spectroscopic technique to characterize the aging of NBR and observed that aging occurred mainly via additional crosslinking [42].

Various researchers have studied the effect of nanofillers on the thermal degradation of different elastomers. Maiti et al. investigated high- and low-temperature degradation of fluoroelastomers in presence of nanoclays [17]. The degradation of fluoroelastomer both in high and low temperatures occurred via breakdown of the main chain backbone. Even, the degradation mechanism remained the same after the addition of nanoclays. Similarly, Choudhury et al. studied the effect of different nanofillers (nanoclay and nanosilica) on high- and low-temperature degradation of hydrogenated nitrile rubber (HNBR) [19]. At higher temperatures, HNBR encountered allylic chain scission, producing di-radicals and alkyl radicals. On the other hand, FTIR analysis revealed that at low temperature under aerobic hot aging conditions, hydrogenated nitrile rubber (HNBR) undergoes crosslinking reactions that lead to embrittlement and ultimately failure. The incorporation of nanofillers significantly improved the degradation profile of the nanocomposites by increasing the degradation temperature along with significant reduction in the rate of decomposition.

4 Reactive Force Field Molecular Dynamics Simulation (ReaxFF)

The concept of reactive force field was first introduced by van Duin in the year 2001 to resolve the discontinuity between quantum mechanical (QM) and classical molecular dynamics simulation (MD) [31–33]. This new force field is less expensive and less time-consuming than the QM and classical MD simulation methods, respectively. Reactive force field can simulate the bond formation and bond breaking in a large molecular system with greater accuracy. The energy function used in ReaxFF is shown in Eq. (1).

$$E_{\text{system}} = E_{\text{bond}} + E_{\text{over}} + E_{\text{under}} + E_{\text{val}} + E_{\text{tors}} + E_{\text{conj}} + E_{\text{vdW}} + E_{\text{coulomb}} \quad (1)$$

Here, E_{system} is the total energy of the system and E_{bond} corresponds to the bond energy. Other terms like E_{over}, E_{under}, E_{val}, E_{tors}, E_{conj}, E_{vdW}, E_{Coulomb} are over-coordination energy, under coordination energy, valence energy, torsion energy, conjugation effects to molecular energy, non-bonded van der Waals interaction, and Coulomb interaction, respectively.

ReaxFF simulation permits the bonded (bond angle and torsion angle) and non-bonded (van der Waals, Coulombic) interactions to change simultaneously during the breaking and formation of bonds. The interatomic distances are used to evaluate the bond order and are revised with each step. This enables ReaxFF to identify the dissociation of existing bonds as well as the formation of new bonds. ReaxFF incorporates the valence terms (bond angle, torsion angle) so that it ensures that the energy terms related to them diminish upon dissociation of bonds. On the other hand, excessive close-range non-bonded energies are avoided by shielding. These energies are evaluated between every atom pair, irrespective of their connectivity. The flow diagram for the reactive force field (ReaxFF) method is shown in Fig. 1.

Visualization of molecular structure by molecular dynamics simulation is an active area of research recently. Reactive forcefield molecular dynamics simulation (ReaxFF) can be utilized in this respect to understand the degradation pattern of different elastomers along with the different products formed.

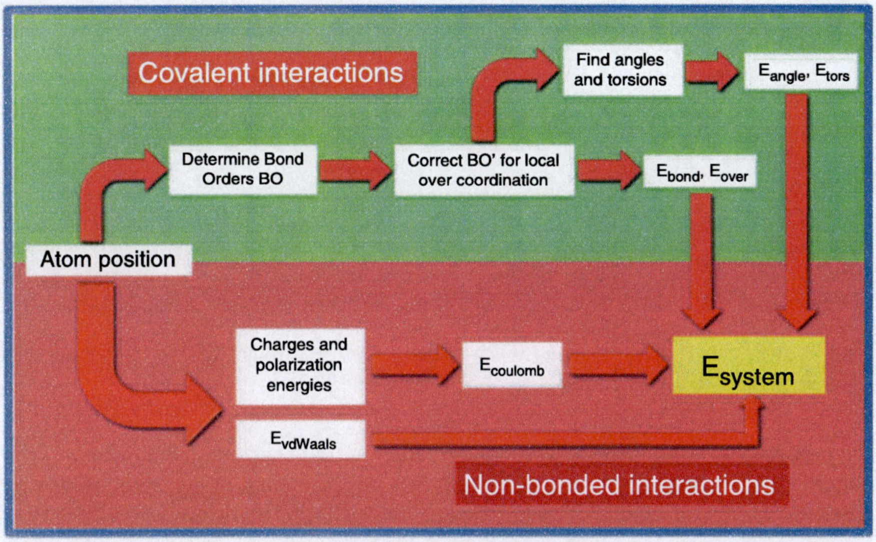

Fig. 1 Flow diagram of the reactive force field (ReaxFF) method

4.1 Polymer Degradation from ReaxFF Molecular Simulation

After the development of reactive force field molecular dynamics simulation (ReaxFF) in 2001, various researchers are currently working on the visualization of the pyrolysis phenomenon of different long polymeric chains. But still, there are very few literatures available in this area.

Nyden et al. utilized the reactive simulation for the first time on an aromatic polymer (polystyrene) and established that the degradation of polystyrene took place via β-scission reaction to form styrene monomer [43]. Later, they extensively performed the simulation on polyisobutylene (PIB), polypropylene (PP), and polyethylene (PE) [44]. They observed that in the case of polyisobutylene, the major degraded product was identified as isobutylene. On the contrary, in the case of PP and PE, low molecular weight hydrocarbons along with different cyclic products were obtained.

Stolialov in association with Nyden further extended the simulation study on polymethylmethacrylate (PMMA) and polyisobutylene (PIB) separately to understand their degradation characteristics [45–47]. PMMA exhibits the formation of its monomer via the unzipping mechanism of the long polymeric chain. On the other hand, the decomposition of polyisobutylene involved homolytic cleavage of polymeric chains followed by depolymerization of its monomer units.

Diao et al. investigated the effect of time, temperature as well as heating rate on the model compounds of epoxy resin consisting of 15 numbers of monomers [48]. They observed that as the simulation temperature and heating rate were

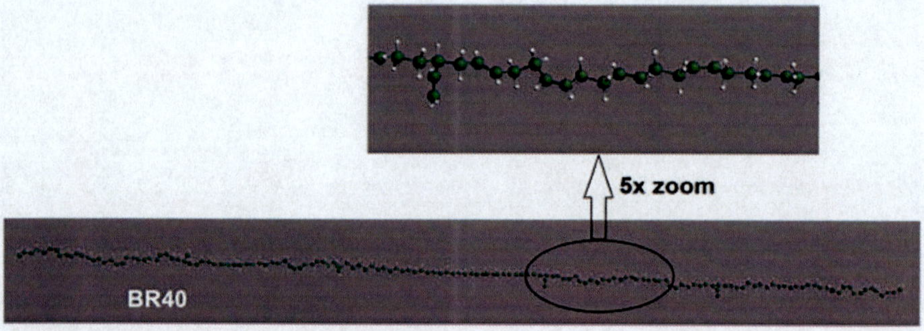

Fig. 2 Single model chain of BR40 (Carbon-green, Hydrogen-white) (Figure reproduced with permission from Ref. [20]. Copyright 2020 John Wiley & Sons)

increased, these shortened the onset of degradation. Liu et al. studied the degradation pattern of high-density polyethylene (HDPE) with the help of reactive simulation method. They reported that decomposition of HDPE occurred via β-scissions, H-abstractions, and intra-molecular H-shift [49].

Hu et al. calculated the activation energy for degradation of poly(α-methyl styrene) [PMS] by isothermal heating at five different temperatures using the simulation method [50]. Recently, Zhao et al. performed the reactive simulation on polycarbonate (PC) model compounds to analyze their decomposition characteristics at extremely high temperature [51]. The breaking of C-O bonds between the monomer units and also from the terminal position is mainly responsible for the degradation of polycarbonate.

Saha et al. performed the reactive dynamics simulation of different butadiene-based model elastomers like polybutadiene rubber (BR), styrene-butadiene rubber (SBR), acrylonitrile-butadiene rubber (NBR), and hydrogenated acrylonitrile-butadiene rubber (HNBR) [20–22].

Polybutadiene-based model elastomer was designed by connecting 40 different butadiene monomers with the help of "*Graphical User Interface (GUI)*" present in Amsterdam density function (ADF) software [31]. The structure was then optimized with the help of "*Universal Force Field (UFF)*" optimizer in ADF. The optimized structure is shown in Fig. 2 and designated as BR40.

The major degraded products identified from the simulation method were 1,3-butadiene (C_4H_6), i.e. the monomer of BR40 along with high molecular weight di, tri, and pentamer of butadiene. These were formed via a free radical mechanism With time, these high molecular weight fragments were further decomposed into smaller fragments of ethylene (C_2H_4), 1,3-butadiene (C_4H_6), and 2-butene (C_4H_8). The formation of 2-butene followed a proton transfer reaction between two allylic carbons [52, 53]. Along with these, a small amount of 4-vinylcyclohexene (a cyclic dimer of 1,3-butadiene) was also identified. Different fragmented products obtained from the reactive simulation of BR40 are represented in Fig. 3.

In the case of styrene-butadiene rubber, a copolymer of styrene and butadiene was designed by connecting 30 units of 1,3-butadiene (75 mol%) with 10 units of styrene

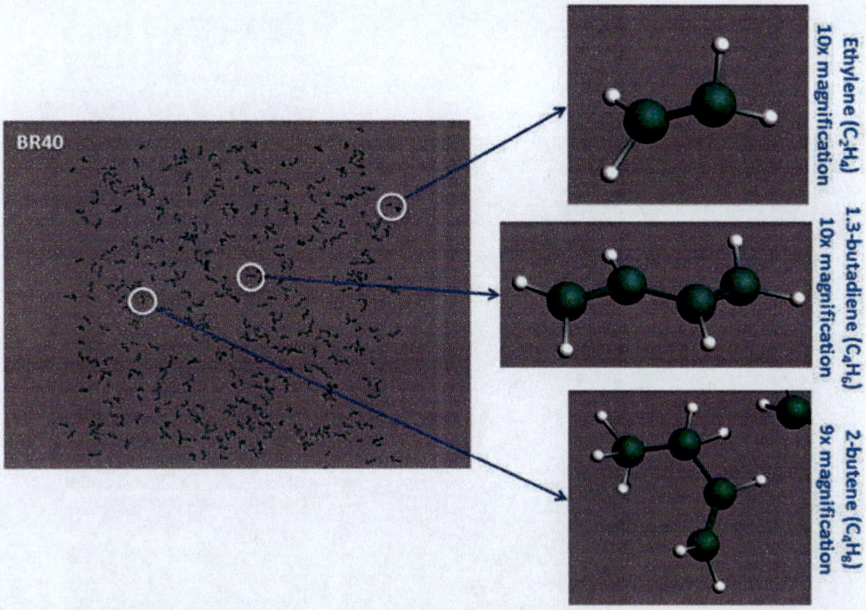

Fig. 3 Identification of different degraded products of BR40 from reactive simulation (Carbon-green, Hydrogen-white) (Figure reproduced with permission from Ref. [20]. Copyright 2020 John Wiley & Sons)

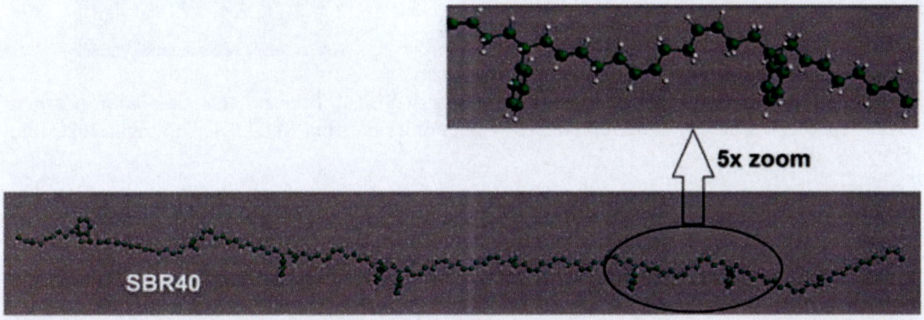

Fig. 4 Single model chain of SBR40 (Carbon-green, Hydrogen-white) (Figure reproduced with permission from Ref. [20]. Copyright 2020 John Wiley & Sons)

(25 mol%). The structure of the model chain is depicted in Fig. 4 and designated as SBR40.

1,3-Butadiene (C_4H_6), 2-butene (C_4H_8), and styrene (C_8H_8) were recognized as the major degraded products when SBR40 was subjected to high-temperature heating. Along with these, 4-vinylcyclohexene was again identified with a smaller amount. 1,3-butadiene, 2-butene, and 4-vinylcyclohexene were obtained from the degradation of the butadiene segment and styrene was formed from the styrene part

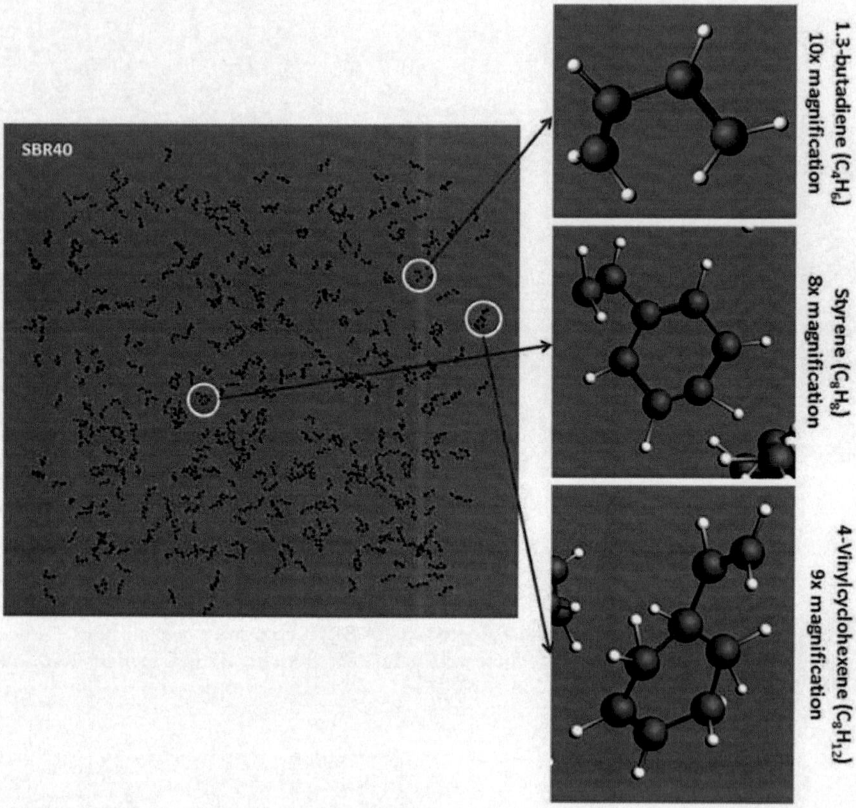

Fig. 5 Identification of different degraded products of SBR40 from reactive simulation (Carbon-green, Hydrogen-white) (Figure reproduced with permission from Ref. [20]. Copyright 2020 John Wiley & Sons)

of the SBR40. Moreover, a smaller fraction of different oligomeric products consisting of 1,3-butadiene and styrene ($C_{16}H_{20}$ and $C_{20}H_{26}$) were also detected from the simulation method. These products on further heating dissociated into the monomers. Different fragmented products obtained from the reactive simulation of SBR40 are shown in Fig. 5.

For acrylonitrile-butadiene rubber, acrylonitrile comonomer was used in place of styrene along with butadiene. The model of acrylonitrile-butadiene rubber chains (NBR40) by randomly connecting 27 units of 1,3-butadiene (67.5 mol%) with 13 units of acrylonitrile (32.5 mol%) was constructed, as displayed in Fig. 6.

Saha et al. identified acrylonitrile (C_3H_3N), butadiene (C_4H_6), and 2-butene (C_4H_8) as the dominant products during the reactive simulation of NBR40 at the initial stage of degradation. They proposed that the disintegration of NBR40 model chains took place due to the breaking of the main polymeric chain. When the model compound was heated further, linear chain radicals slowly converted into cyclic

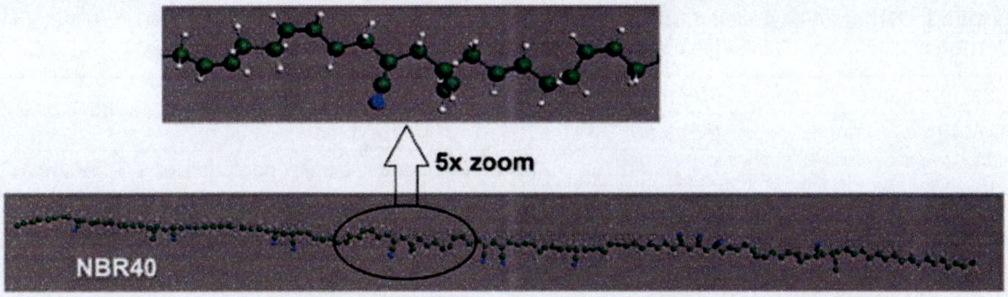

Fig. 6 Single model chain of NBR40 (Carbon-green, Hydrogen-white, Nitrogen-blue)

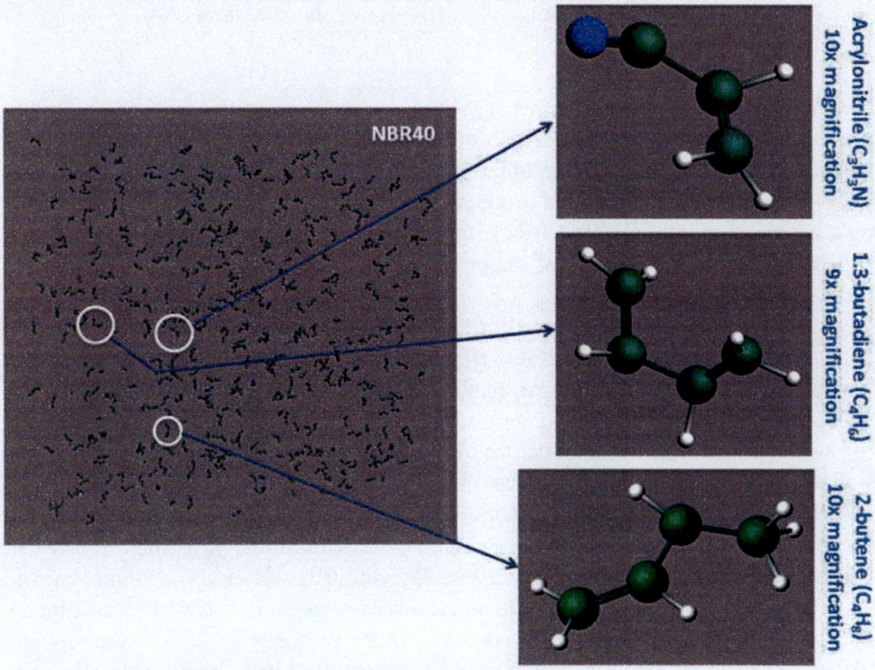

Fig. 7 Identification of different degraded products of NBR40 from reactive simulation (Carbon-green, Hydrogen-white, Nitrogen-blue)

structures; as a result, vinyl cyclohexene (dimer of 1,3-butadiene) along with cyclohex-3-enecarbonitrile was also found [21, 22]. It indicated that at a higher temperature, linear chain radicals slowly got converted to cyclic structures. The dominant degraded products obtained from the reactive simulation of NBR40 are shown in Fig. 7.

By studying the degradation patterns of butadiene-based homopolymer (BR40) as well as copolymers (SBR40 and NBR40), it was concluded that the butadiene portions present in all the structures followed the same free radical degradation

Table 1 Different degraded products obtained from the reactive simulation of BR40, SBR40, and NBR40

Sample name	Major products	Minor products
BR40	Ethylene	4-Vinylcyclohexene
	1,3-Butadiene	Dimer, trimer, pentamer of 1,3-butadiene
	2-Butene	
SBR40	1,3-Butadiene	Ethylene
	2-Butene	Propylene
	Styrene	4-Vinylcyclohexene
NBR40	1,3-Butadiene	Ethylene
	2-Butene	4-Vinylcyclohexene
	Acrylonitrile	Cyclohex-3-enecarbonitrile

mechanism to form the monomer 1,3-butadiene. Along with this, another major product, 2-butene (C_4H_8) was also observed. Additionally, in the case of copolymers such as SBR40 and NBR40, the other monomers, i.e. styrene and acrylonitrile were identified. Different fragmented products (both major and minor) obtained from the reactive simulation of BR40, SBR40, and NBR40 are given in Table 1.

The formation mechanism of major fragmented products of BR40, SBR40, and NBR40 is shown in Fig. 8.

But, in the case of hydrogenated acrylonitrile-butadiene rubber, the degradation pattern was different from the rest of the butadiene-based elastomer. To get an insight into its decomposition mechanism, a unit chain of HNBR40 consisting of 13 units of acrylonitrile (36 wt.%) and 27 units of hydrogenated butadiene units (64 wt.%) was utilized. The structure of model HNBR40 is shown in Fig. 9.

The major fragmented products observed during the reactive simulation of HNBR40 were acrylonitrile (a comonomer) along with different unsaturated hydrocarbons such as ethylene, propylene, butene, etc. Additionally, a small number of unsaturated nitrile products like 2-butylacrylonitrile ($C_7H_{11}N$), 3-butenenitrile (C_4H_5N) were also detected. The degradation mechanism of HNBR was slightly different than that of NBR. This was due to the presence of active α-methylene carbon. The higher electron-withdrawing effect of the nitrile group, as well as the electron withdrawal from the double bond, made it susceptible to degradation. As the concentration of these double bonds was reduced in HNBR, the degradation from α-methylene carbon was also reduced. The same phenomenon was observed by Bhattacharjee et al. from their earlier studies [16].

Another important oil and heat-resistant elastomer is Polyacrylicester (ACM). It is a random terpolymer comprised of ethylene and two different types of acrylic ester groups. 1 unit of ethylene (9.6 mol%), 3 units of n-butyl acrylate (27 mol%), and 6 units of ethyl acrylate (63 mol%) were connected in a randomly placed orientation constructing a model chain of ACM (termed as ACM10), following the molecular structure of the actual elastomer from DENKA Company Limited, JAPAN. The structure of the model polyacrylic ester elastomer (ACM) is given in Fig. 10.

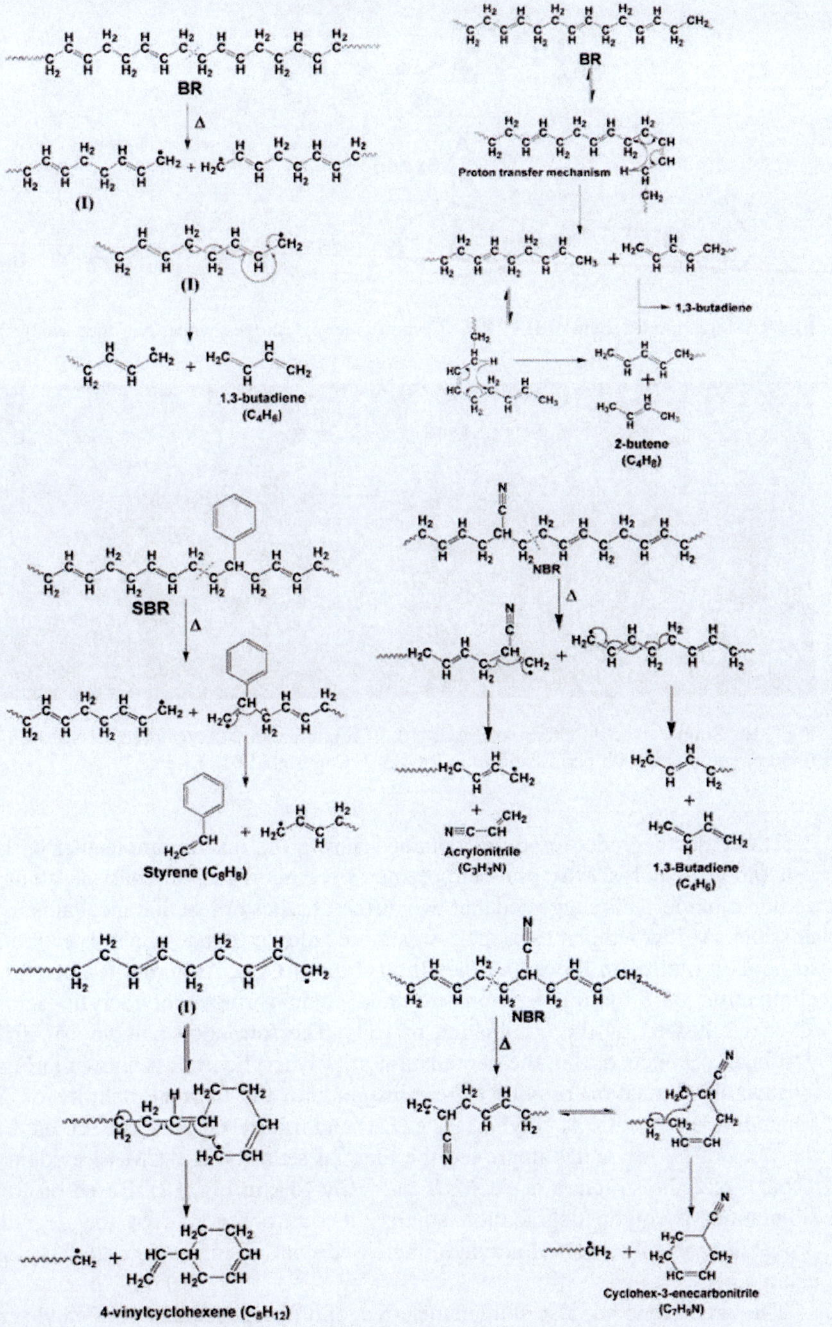

Fig. 8 The formation mechanism of major fragments of BR40, SBR40, and NBR40 (Figure reproduced with permission from References [20, 21]. Copyright 2017 Elsevier and 2020 John Wiley & Sons)

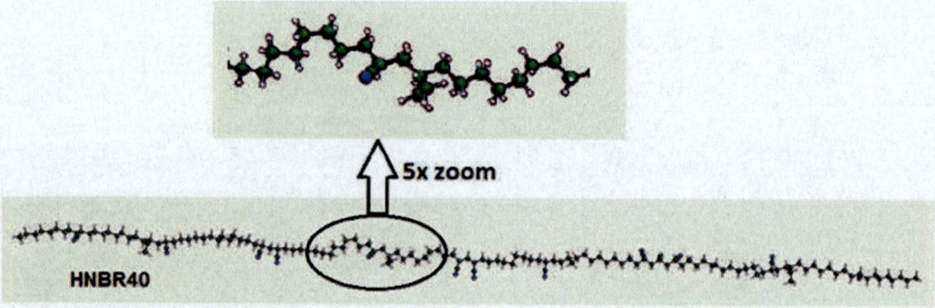

Fig. 9 Single model chain of HNBR40 (Carbon-green, Hydrogen-white, Nitrogen-blue)

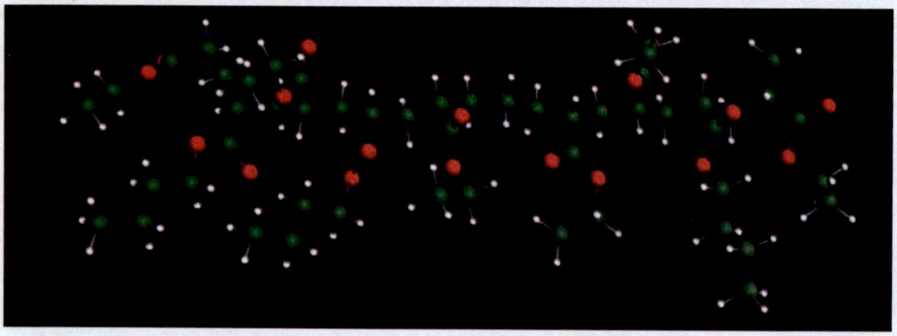

Fig. 10 Single model chain of ACM10 (Carbon-green, Oxygen-red, Hydrogen-white) (Figure reproduced with permission from Ref. [57]. Copyright 2018 Elsevier)

The major degraded products obtained during the reactive molecular dynamics simulation of polyacrylic ester terpolymer were alkyl radicals, alkene along with carbon dioxide. This suggested that two different types of reaction mechanisms were possible. At low temperature, there was a breakdown of side-chain ester groups to form alkyl radicals followed by the elimination of CO_2. At higher temperature, the elimination of β-hydrogens from the side chain formed poly(acrylic acid) and alkene, followed by the elimination of CO_2. The formation pathway of different dominant products during the degradation of polyacrylic ester is shown in Fig. 11.

ReaxFF simulation provided great insight into the thermal stability of ACM, depending on the ratio of ethyl acrylate (EA) and n-butyl acrylate (nBA). Increase in the ratio of EA/nBA has improved the thermal stability of ACM as evident from Table 2 (see the structure of ACM10 and ACM10A in Fig. 12). Based on ReaxFF simulation, pyrolytic degradation studies, it was observed that the degradation probability of poly (n-alkyl acrylate) increased with an increase in the alkyl group chain length.

The same group has also studied the effect of different sequences of ethyl acrylate and n-butyl acrylate in the pyrolysis simulation of ACM. Three separate single model chains with different sequence of ethyl acrylate and n-butyl acrylate

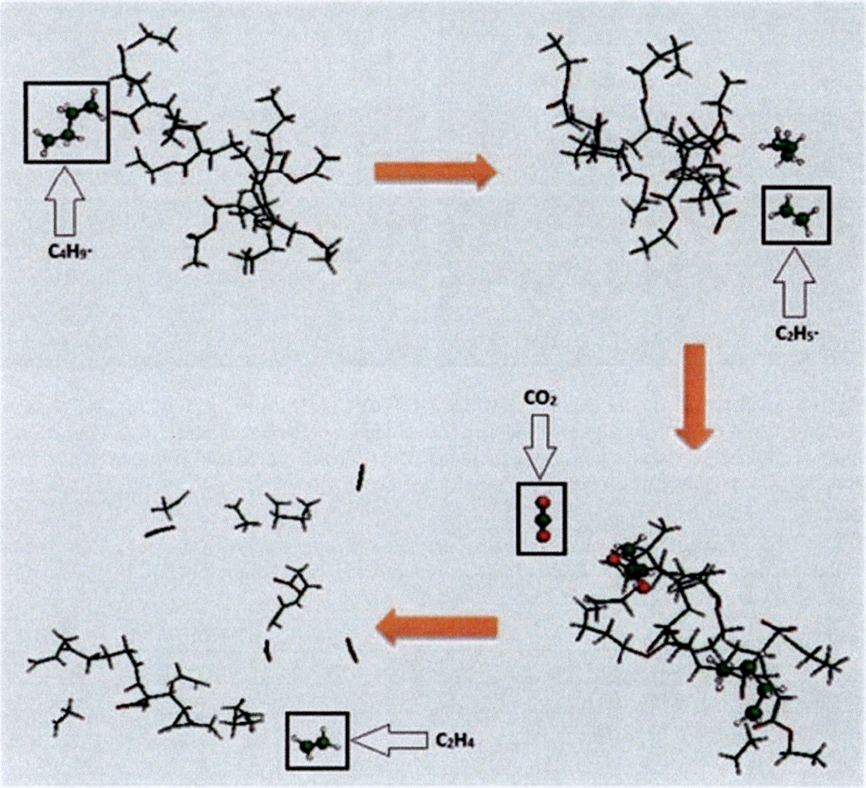

Fig. 11 Snapshot of formation pathway of dominant products at different stages of degradation (Carbon-green, Oxygen-red, Hydrogen-white) (Figure reproduced with permission from Ref. [23]. Copyright 2016 Elsevier)

Table 2 The effect of ethyl acrylate and n-butyl acrylate on thermal degradation of ACM (Table reproduced with permission from Ref. [23]. Copyright 2016 Elsevier)

Sample	PEA	ACM10	ACM10A	PnBA
Ethylene:EA:nBA	0:100:0	10:60:30	10:30:60	0:0:100
Total number of products	146	162	196	196
Initial decomposition time (ps)	20.5	19.5	18.5	17.5
Initial decomposition temperature (K)	1,320	1,225	1,150	990

(designated as ACM10, ACM10B, and ACM10C) were built. The structure of ACM10, ACM10B, and ACM10C is shown in Fig. 12.

The reactive simulation was conducted on each model chain individually. The effect of different sequence of ethyl acrylate and n-butyl acrylate on the thermal decomposition of ACMs is tabulated in Table 3.

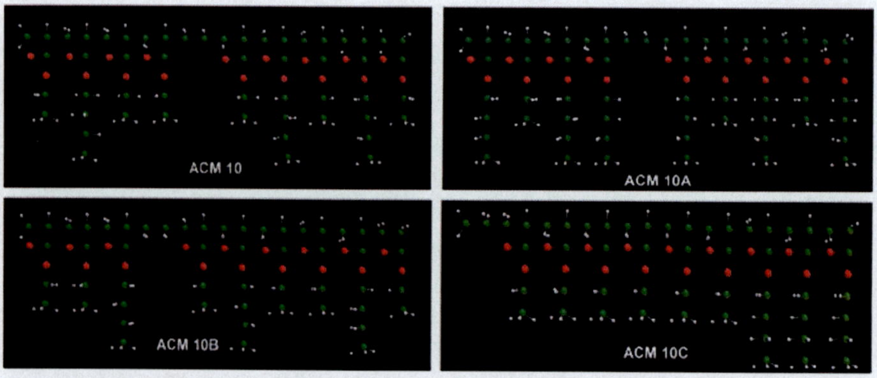

Fig. 12 Snapshot of model chains ACM 10, ACM10A, ACM 10B, and ACM 10C (Atoms: Carbon-green, Oxygen-red, Hydrogen-white). The models were not geometrically or energetically optimized for better visualization of the monomer units (Figure reproduced with permission from Ref. [23]. Copyright 2016 Elsevier)

Table 3 Effect of different sequences of ethyl acrylate and n-butyl acrylate in a single model chain on thermal decomposition of ACM10 (Table reproduced with permission from Ref. [23]. Copyright 2016 Elsevier)

Sample	ACM10	ACM10B	ACM10C
Total number of products	162	165	169
Initial decomposition time (ps)	19.50	19.56	19.44
Initial decomposition temperature (K)	1,225	1,230	1,220
Final decomposition time (ps)	23.96	24.05	24.02
Final decomposition temperature (K)	1,670	1,680	1,675
Rate (molecules/picosecond)	2.24	2.23	2.18

The results are almost similar. From these results, it is clear that the thermal decomposition pattern of ACM from the ReaxFF simulation was not affected by changing the sequence of the monomer units.

For a better understanding of the degradation phenomenon and to validate the results obtained from the simulation method, Saha et al. performed Py-GC-MS analysis on commercial BR and SBR samples [20]. The pyrogram of commercial BR and SBR samples is given in Fig. 13.

From the chromatogram of commercial BR, it was observed that 1,3-butadiene, 2-butene, 1,3-cyclohexadiene, and 4-vinylcyclohexene were identified at the retention times of 1.99, 2.86, 4.13, and 5.18 min, respectively (Fig. 13a). In the case of SBR sample, similar peaks were observed along with a new peak at 6.37 min (Fig. 13b) due to the formation of styrene monomer (C_8H_8) with m/Z value of 104. The major peaks obtained during the Py-GC-MS of commercial BR and SBR samples are tabulated in Table 4. These experimental results corroborate the findings from the theoretical simulations.

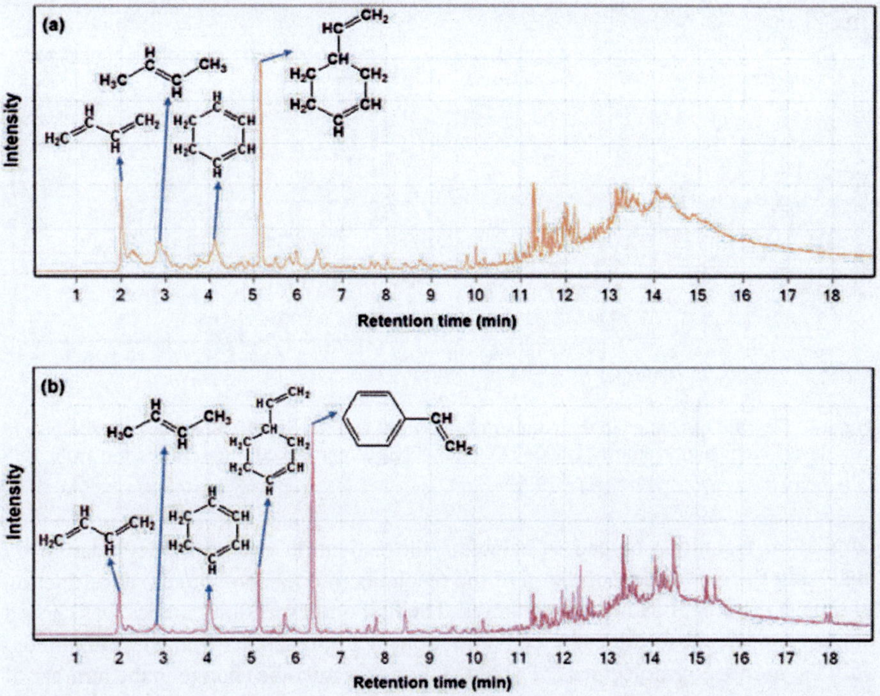

Fig. 13 Py-GC/MS chromatograms of commercial (**a**) BR and (**b**) SBR at 600°C (Figure reproduced with permission from Ref. [20]. Copyright 2020 John Wiley & Sons)

Table 4 Pyrolysis GC-MS of commercial BR and SBR at 600°C (Table reproduced with permission from Ref. [20]. Copyright 2020 John Wiley & Sons)

BR			SBR		
Retention time (min)	m/Z	Component	Retention time (min)	m/Z	Component
1.99	54	C_4H_6	1.99	54	C_4H_6
2.86	56	C_4H_8	2.90	56	C_4H_8
4.13	80	C_6H_8	4.03	80	C_6H_8
5.18	108	C_8H_{12}	5.16	108	C_8H_{12}
–	–	–	6.37	104	C_8H_8

4.2 *Degradation Kinetics from ReaxFF Molecular Simulation*

ReaxFF Molecular dynamics simulation was also used to study the kinetics of degradation of different model compounds. To study the kinetics, at first the model compounds were isothermally heated at different temperatures. The time evolution profiles of the reactants (in this case the model compounds) were utilized to investigate the degradation kinetics using first order reaction kinetics. The

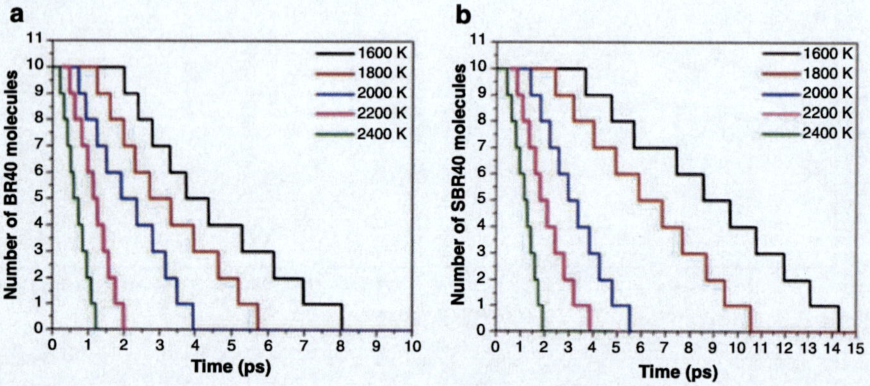

Fig. 14 The plot of change in the number of (**a**) BR40 and (**b**) SBR40 molecules with respect to time at different temperatures (1,600–2,400 K) (Figure reproduced with permission from Ref. [20]. Copyright 2020 John Wiley & Sons)

simulation technique helped to calculate the activation energy of degradation by analyzing the rate of consumption of the reactants (model compounds) as a function of time at different heating temperatures. The first order reaction kinetics was carried out by simply substituting the concentration of the reactants (model compounds) with the number of molecules [21, 32, 45]. For example, the change in the number of reactant molecules for BR40 and SBR40 model compounds was plotted in Fig. 14 as a function of time at different heating temperatures.

The rate constants (k_T) for each particular temperature were evaluated from the slope of the natural logarithm of the number of molecules (N_t) against the simulation time t, as expressed by Eq. (2).

$$\ln N_t - \ln N_0 = -k_T \times t \tag{2}$$

Here, N_t signified the number of BR40/SBR40 molecules at simulation time t and N_0 defined the number of BR40/SBR40 molecules at the start of the simulation.

The activation energy of degradation (E_a) was calculated by fitting the rate constants (k_T) obtained at each temperature into the Arrhenius equation, as demonstrated by Eq. (3).

$$\ln k_T = \ln A - {}^{E_a}\!/_{RT} \tag{3}$$

Here, A represented the pre-exponential factor and R characterized the universal gas constant (8.314 J mol^{-1} K^{-1}). The value of E_a was calculated from the slope of lnk_T vs. 1,000/T. The results for BR40 and SBR40 are shown in Fig. 15.

The activation energy of degradation for different model compounds obtained from the simulation method was further correlated with the experimental methods via TGA analysis. Two different techniques, Flynn-Wall-Ozawa (FWO) [54] and Kissinger-Akahira-Sunose (KAS) [55] were used in this aspect. The activation

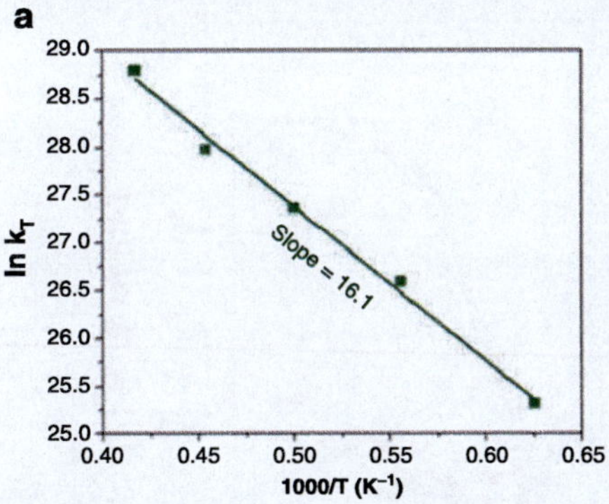

Fig. 15 Linearly fitted lnk_T vs. 1,000/T curve obtained from reactive simulation of (**a**) BR40 and (**b**) SBR40 (Figure reproduced with permission from Ref. [20]. Copyright 2020 John Wiley & Sons)

Table 5 Activation energies for degradation of different elastomers obtained from TGA and reactive simulation methods (Table reproduced with permission from Ref. [20]. Copyright 2020 John Wiley & Sons)

	Theory	Experimental	
Sample	ReaxFF MDs	Flynn-Wall-Ozawa	Kissinger-Akahira-Sunose
BR	134 ± 4	148 ± 3	146 ± 4
SBR	153 ± 5	166 ± 5	163 ± 4
NBR	180 ± 4	208 ± 15	206 ± 14

energies for degradation of different elastomers calculated both from simulation and experimental techniques are tabulated in Table 5. The trends in the theoretical and experimental results are in good agreement.

5 Thermo-Oxidative Degradation of Elastomers

The previous studies were focused on the thermal degradation of elastomers in an inert atmosphere. But in a practical scenario, it is very important to understand the thermo-oxidative degradation of different elastomers [15, 56]. Bhowmick et al. had extensively analyzed the thermo-oxidative degradation of different elastomers [17–19]. The degradation of fluoroelastomer and its nanocomposites was studied by Maiti et al. in oxygen as well as nitrogen atmospheres [17]. Sadhu et al. investigated the thermo-oxidative degradation of different elastomer-based nanocomposites [18]. Celina et al. correlated the chemical structure and thermal property of unfilled neoprene rubber during oxidative degradation up to 140°C [57].

Recently, Saha et al. explored the degradation of polyacrylic ester elastomer under thermo-oxidative conditions with the help of reactive molecular dynamics

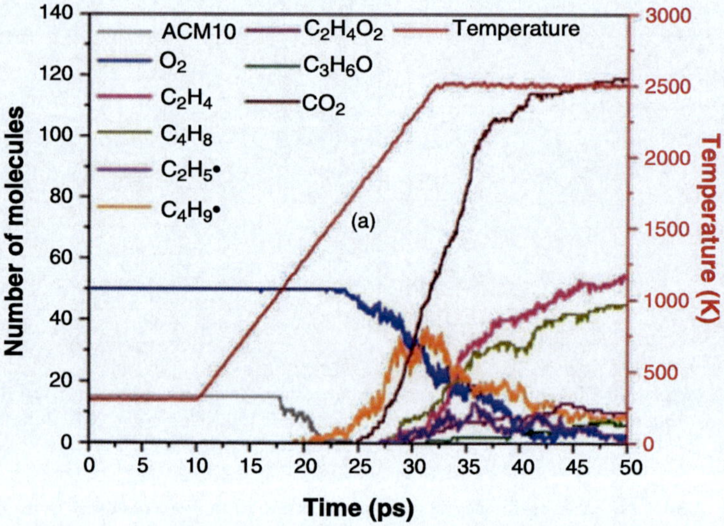

Fig. 16 Formation of different degraded products of ACM10 in presence of oxygen (Figure reproduced with permission from Ref. [58]. Copyright 2021 Elsevier)

simulation [58]. They performed the reactive simulation on 15 numbers of ACM model chains in the presence of 50 O_2 molecules inside a periodic box at a higher temperature. A large number of different fragmented products were identified during the simulation. The evolution of both major and minor fragmented products (aldehyde and acid) as a function of time is shown in Fig. 16.

Ethylene (C_2H_2), butene (C_4H_8), ethyl radicals (C_2H_5•), butyl radicals (C_4H_9•) along with carbon dioxide (CO_2) were identified as the major degraded products during the simulation. The formation mechanism of these products has already been discussed in the previous section of this chapter. Moreover, along with these, different low molecular weight aldehydes and acids such as acetic acid ($CH_3COOH \equiv C_2H_4O_2$) and propionaldehyde ($CH_3CH_2CHO \equiv C_3H_6O$) were also detected during the simulation. These low molecular weight aldehydes and acids were formed due to partial oxidation of the side chain of ACM. Similar results were obtained from the Py-GC-MS analysis. The results are given in Table 6.

The results were in line with the simulation method. But interestingly some higher carbon fraction aldehydes and acids which were also detected in the Py-GC-MS analysis were not identifiable from the simulation. This was due to higher simulation temperature where the higher carbon fraction products were broken down into smaller fractions. Hence, they were not detectable from the simulation method. A comparison of different degraded products obtained from the reactive simulation of ACM under thermal as well as thermo-oxidative conditions is shown in Table 7.

The above discussion was based on the high-temperature thermo-oxidative degradation of elastomers. But, along with the progress of modern technologies, it is very important to examine the performance of an elastomer in an outdoor condition,

Table 6 Pyrolysis GC-MS of pristine ACM in oxygen atmosphere (Table reproduced with permission from Ref. [58]. Copyright 2021 Elsevier)

Component	Molecular formula	m/Z	Retention time (min)
Carbon dioxide	CO_2	44	0.98
Ethylene	C_2H_4	28	1.96
Butene	C_4H_8	56	2.86
Propionaldehyde	C_2H_5CHO	58	5.22
Acetic acid	CH_3COOH	60	7.18
Butanaldehyde	C_3H_7CHO	72	8.31
Propionic acid	C_2H_5COOH	74	9.23

Table 7 Comparison of different degraded products obtained from the reactive simulation of ACM under thermal and thermo-oxidative conditions

Sample name	Thermal	Thermo-oxidative
	Ethylene (C_2H_2)	Ethylene (C_2H_2)
Commercial ACM	Butene (C_4H_8)	Butene (C_4H_8) Ethyl radical ($C_2H_5\bullet$)
	Ethyl radical ($C_2H_5\bullet$)	
	Butyl radical ($C_4H_9\bullet$)	Butyl radical ($C_4H_9\bullet$)
	Carbon dioxide (CO_2)	Carbon dioxide (CO_2)
		Low MW aldehydes (propionaldehyde, butanaldehyde, etc.
		Low MW acids (acetic acid, propionic acid, etc.)

i.e. under low-temperature thermo-oxidative conditions. The behavior of an elastomer during the outdoor application is very crucial for the commercialization of end products. Saha et al. had investigated the lower temperature degradation behavior of unvulcanized polyacrylic ester and its silica-filled nanocomposites with the help of change in intrinsic viscosity and number average molecular weight [59]. They reported that the values of number average molecular weight (M_n) of pristine ACM reduced drastically with the increase in aging temperature. This indicated the breaking of main molecular chain for the uncrosslinked system. On the other hand, for silica-filled nanocomposites, the change in M_n was greatly reduced due to its enhanced thermal stability in the presence of nanosilica.

Usually, crosslinked samples are used in practical tests. Uncrosslinked samples deform under severe aging conditions and are not suitable for such experiments. Elastomers that are heat resistant because of negligible or no double bonds or having special microstructure are still heat resistant in their crosslinked state. For example, fluorocarbon elastomers are more heat resistant than natural rubber, both for uncrosslinked and crosslinked states. The nature or amount of crosslinking agents in natural rubber cannot place this elastomer above fluorocarbon elastomers. While aging of crosslinked samples has been routinely done for many samples having

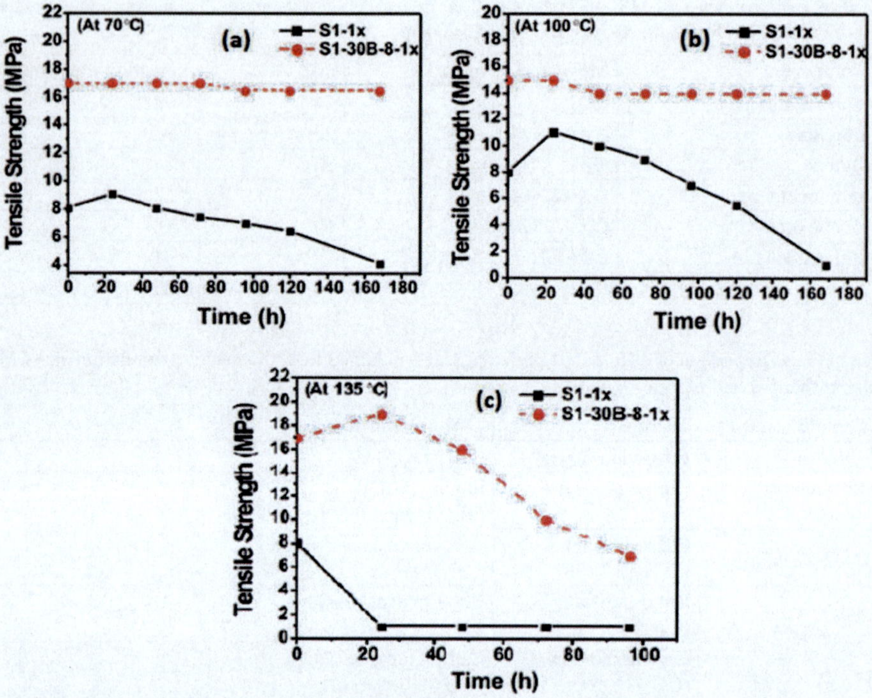

Fig. 17 Tensile strength vs. time of aging at three different temperatures (**a**) 70°C (**b**) 100°C (**c**) 135°C (Figures reproduced with permission from Ref. [19]. Copyright 2010 Elsevier)

variation of molecular structure, or nature and amount of crosslinks or additives in our laboratory, we are discussing below two recent examples.

Choudhury et al. studied the physicomechanical properties of pristine HNBR (S1-1x) and its nanocomposite counterpart (S1-30B-8-1x with 8 phr loadings of Cloisite 30B) under accelerated heat aging testing at 70, 100, and 135°C over different periods of time [19]. The retention of tensile strength was found to be much better in presence of organically modified clay (Cloisite 30B) as shown in Fig. 17. The symbol "1x" in Fig. 17 means the sample cured with 1 phr (parts per 100 g of rubber) of the curing agent.

Choudhury et al. have used the power spectral density (PSD) technique in AFM to understand the aging behavior of both the pristine HNBR and its nanocomposite. The 2D power spectral density diagram of pristine HNBR and its nanocomposite before and after aging is shown in Fig. 18. The variation of height in real space can be considered in terms of power spectrum in frequency space through the use of the Fourier transform in PSD. The surfaces were analyzed by a series of line scans of N_x steps, yielding a profile, $h(x)$, for each value from 1 to N_y in the y direction. The PSD was then calculated by taking the Fourier transform of each of these line scans, squaring the result to determine the power, and averaging the power calculated for

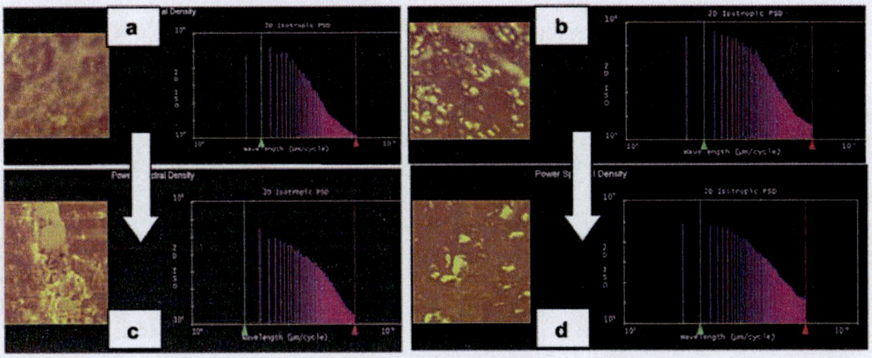

Fig. 18 Power spectral density of (**a**) S1-1x; (**b**) S1-30B-8-1x (before aging); (**c**) S1-1x; and (**d**) S1-30B-8-1x (after aging) (Figures reproduced with permission from Ref. [19]. Copyright 2010 Elsevier)

Table 8 Quantitative parameters determined from roughness and power spectral density analysis of unfilled elastomer and its nanocomposite before and after aging (Table reproduced with permission from Ref. [19]. Copyright 2010 Elsevier)

Sample		Average roughness R_a (nm)	RMS roughness R_q (nm)	Total power (nm^2)	Equivalent RMS (nm)
S1-1x	Before aging	3.4	4.2	18.9	4.3
	After aging	16.1	18.8	368.0	19.2
S1-30B-8-1x	Before aging	8.7	12.9	123.0	11.0
	After aging	9.9	16.2	280.0	16.6

each line scan to generate a single 1DPSD for the surface. Two-dimensional PSDs (2DPSD) typically use an annular average of the square of two-dimensional Fourier transforms of the surface rather than one-dimensional Fourier Transform.

The quantitative parameters obtained from the PSD technique such as average roughness (R_a), root mean square (RMS) roughness (R_q), total power (i.e., the sum of the power contained in the entire spectrum), and the equivalent RMS (square root of the total power) values are tabulated in Table 8.

From Table 8, it was observed that the values of *Rq* and *Ra* for the pristine elastomer increased considerably from 4.2 nm and 3.4 nm to 18.8 nm and 16.1 nm, respectively, after aging. It suggested that there was a drastic change in roughness upon aging for the pristine elastomer due to oxidative attack at the surface. On the contrary, only a slight change in roughness was observed on aging of the nanocomposite (*Rq* and *Ra* values were increased from 12.9 nm and 8.7 nm to

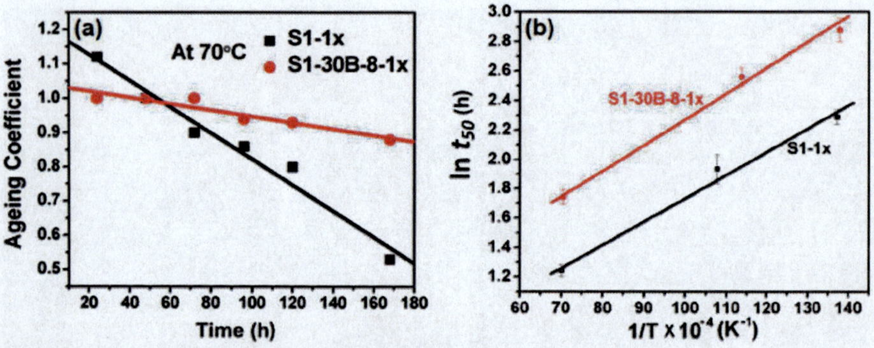

Fig. 19 (**a**) Aging coefficient vs. time of aging at 70°C and (**b**) Arrhenius plot of ln t_{50} vs. *1/T* for S1-1x and S1-30B-8-1x (Figures reproduced with permission from Ref. [19]. Copyright 2010 Elsevier)

16.2 nm and 9.9 nm, respectively). These results confirmed that the presence of nanofillers protects the surface of the matrix from oxidative attack [19, 60].

Furthermore, they estimated also the life cycle of pristine HNBR and its nanocomposites by measuring the time required when the tensile strength value reaches 50% of its original values [19]. They calculated the aging coefficient (ratio of tensile strength of aged samples over tensile strength of unaged sample) and plotted it as a function of aging time (in hours) as represented in Fig. 19a. To predict the life of nanoclay-filled and unfilled HNBR, the Arrhenius equation was used as given in Eq. (4).

$$\ln t_{50} = {}^{E}/_{RT} + \text{Constant} \tag{4}$$

Here, E represents the activation energy (J mol^{-1}) and R is the universal gas constant (8.314 J mol^{-1} K^{-1}), and T is the absolute temperature (K). t_{50} is the time required in hour to reach 50% of the original tensile strength value (aging coefficient) at a particular temperature (T). The value of t_{50} was calculated either directly or by extrapolation.

Figure 19b illustrated the plot of ln t_{50} vs. $1/T$ for unfilled and nanoclay-filled HNBR. The lifetime of the compounds at 40°C was evaluated from the slope of the straight line by the extrapolation method. The lifetime of the unfilled and the nanoclay-filled HNBR was 15 and 45 years, respectively, at an operating temperature of 40°C.

Saha et al. analyzed the physicomechanical behavior of pristine ACM and its silica-filled nanocomposite under accelerated aging at 175, 188, and 200°C over different periods of time. Both the pristine ACM and its nanocomposite exhibited excellent retention in tensile strength at 175°C. But as the heating time and temperature were increased, the tensile strength of the compounds exhibited a decreasing trend. ACM nanocomposite with 4 phr nanosilica displayed excellent retention in tensile strength as compared to the pristine ACM, which is evident from Fig. 20.

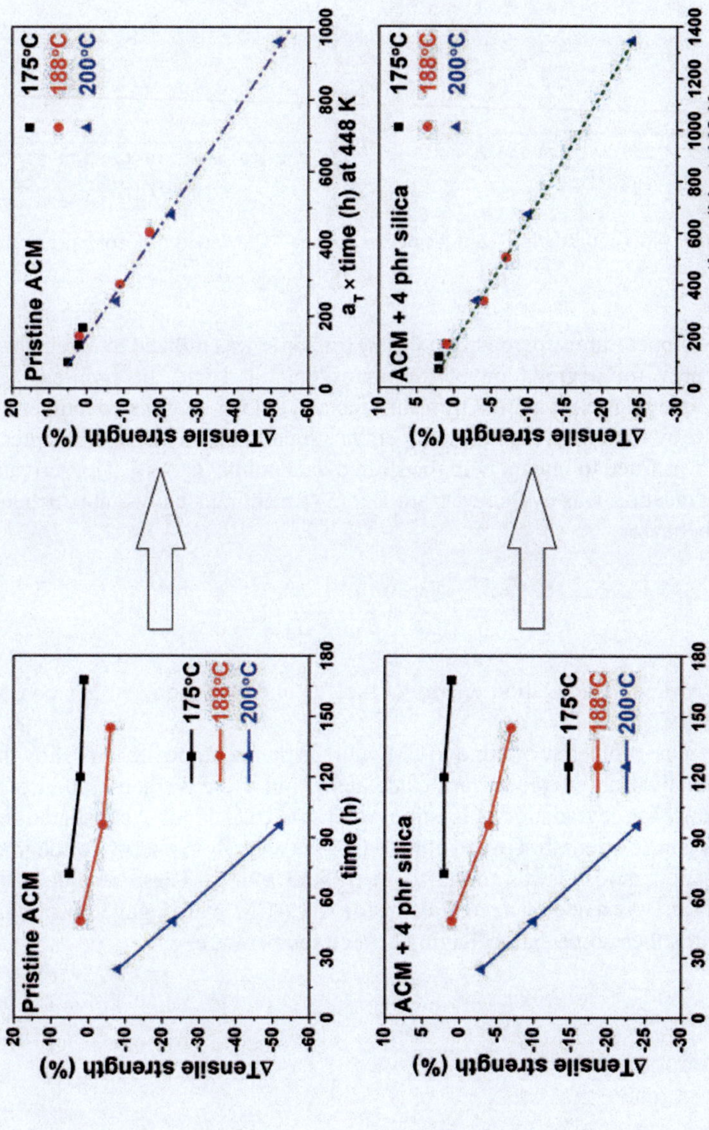

Fig. 20 Aging and accompanying time-temperature superposition (TTS) plots for change in tensile strength of pristine ACM and ACM filled with 4 phr nanosilica

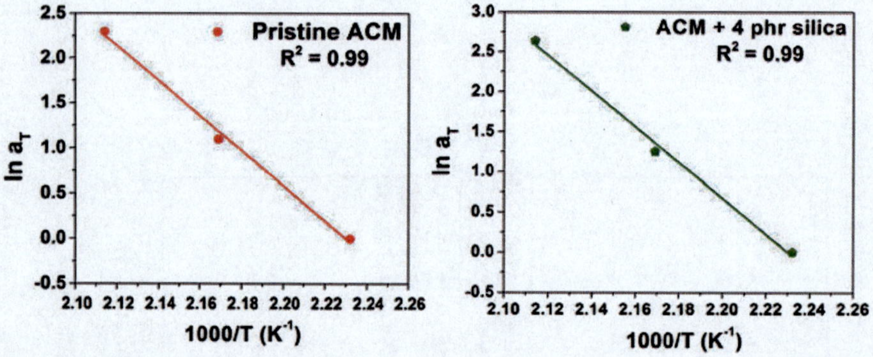

Fig. 21 Arrhenius plots of the derived shift factors for pristine ACM and its nanocomposite using a reference temperature (T_{ref}) of 175°C

The Time-temperature superposition (TTS) principle was utilized to calculate the activation energy for degradation of the nanosilica filled and the unfilled ACM [61, 62]. The aging data was shifted by a shift factor a_T along a reference temperature (T_{ref}) of 175°C by assuming the time-temperature superposition (TTS) principles. A master curve was fitted to line up with the shifted data points [63–65]. The activation energy of degradation was evaluated from Eq. (5) assuming steady-state Arrhenius degradation behavior.

$$a_T = \exp\left[\frac{E_{a'}}{R}\left(\frac{1}{T_{ref}} - \frac{1}{T}\right)\right] \tag{5}$$

$E_{a'}$ is the Arrhenius activation energy (J mol^{-1}) and R is the ideal gas constant (8.314 J mol^{-1} K^{-1}).

The Arrhenius plots for pristine ACM and its nanocomposite are shown in Fig. 21. The activation energies were calculated from these Arrhenius plots. The activation energy for degradation of pristine ACM was 162 kJ mol^{-1} calculated from the change in tensile strength. On the other hand, the activation energy for degradation of the nanocomposite was found to be 185 kJ mol^{-1}. These results clearly suggested that the presence of nanosilica inside the ACM matrix gave outstanding stability towards thermomechanical aging at high temperature.

6 Conclusions

In this chapter, a novel molecular dynamics simulation approach was used to understand the degradation pattern of different elastomers. The study first discussed the degradation of different polymeric model compounds at an extremely higher temperature, followed by the study of various synthetic rubbers like BR, SBR, NBR,

HNBR, and ACM. The chapter further highlighted different fragments that were formed during the degradation of different elastomers and attempted to visualize the formation pathway of these fragments. It was observed that the butadiene component present in both the homopolymer (BR) and the copolymers (SBR and NBR) followed the same free radical pathway to form the monomer 1,3-butadiene along with 2-butene (C_4H_8). But in the case of HNBR different unsaturated hydrocarbons such as ethylene, propylene, and butene were identified due to their saturated backbone. On the other hand, the degradation of ACM took place mainly via β-hydrogen elimination from the side-chain ester group. These results were further verified with the help of Py-GC-MS. Furthermore, this chapter also delineates a method to calculate the activation energy for degradation of these elastomers via the simulation technique. The results acquired from the reactive simulation method are in line with the results calculated from the TGA.

Furthermore, the chapter also described the thermo-oxidative degradation of different elastomers at high temperature. The reactive simulation was further utilized in presence of oxygen molecules to understand the degradation of ACM under thermo-oxidative conditions. The degradation pathway was very similar to that of the inert atmosphere discussed earlier. The only difference was the formation of low molecular weight aldehydes and acids due to partial oxidation of the side chain of ACM. Finally, the last section of this chapter deals with the lower temperature thermo-oxidative behavior of different elastomers and their nanocomposites based on accelerated aging tests followed by measuring the tensile strength at different time intervals. The first method involved measuring the aging coefficient (ratio of tensile strength of aged samples over tensile strength of unaged samples) and putting it into the Arrhenius equation. The lifetime of the unfilled and the nanoclay-filled HNBR calculated from this approach was 15 and 45 years, respectively, at an operating temperature of 40°C. On the other hand, the second method followed the Time-temperature superposition (TTS) principle to evaluate the activation energy for degradation of the pristine ACM and nanosilica filled ACM. The activation energies for degradation of pristine ACM and nanosilica filled ACM were 162 and 185 kJ mol^{-1} calculated from the change in tensile strength.

Acknowledgements The authors would like to acknowledge Denka Company Limited, Japan for their wholehearted support, cooperation and funding for the work on polyacrylic ester terpolymer. The research work on degradation of HNBR was supported by Lanxess, Germany. Funding by various Indian Government organizations (CSIR, ISRO, BARC, DRDO, IIT Kharagpur) to Professor Anil Bhowmick for carrying out systematic studies on degradation of rubbers is gratefully acknowledged. The authors are grateful to Indian Institute of Technology Kharagpur for providing experimental facilities for this work. Professor Bhowmick served the Institute during 1984–2019 as Assistant Professor, Professor, and Professor of Eminence.

References

1. Stephens HL, Bhowmick AK (2001) Handbook of elastomers.2nd edn. Marcel Dekker, New York
2. Morton M (1973) Rubber technology.2nd edn. Van Nostrand, New York
3. Blow CM, Hepburn C (1982) Rubber technology & manufacture.2nd edn. Butterworths, London
4. Bhowmick AK, Hall MM, Benarey AH (1994) Rubber products manufacturing technology. Marcel Dekker, New York
5. Billmeyer FW (1984) Textbook of polymer science. Wiley, New York
6. Coleman MM, Painter PC (2017) Fundamentals of polymer science: an introductory text.2nd edn. CRC Press, Taylor & Francis, New York
7. McCrum NG, Buckley CP, Bucknall CB, Bucknall CB (1997) Principles of polymer engineering.2nd edn. Oxford University Press, New York
8. Mark JE, Erman B, Eirich FR (2005) The science and technology of rubber.3rd edn. Elsevier, Cambridge
9. Hofmann W (1967) Vulcanization and vulcanizing agents. Maclaren and Sons Limited, London
10. Brydson JA (1988) Rubbery materials and their compounds. Elsevier, London, p 469
11. Hawkins WL (1984) Polymer degradation. In: Polymer degradation and stabilization. Springer, Berlin, pp 3–34
12. Van Amerongen G (1955) Oxidative and nonoxidative thermal degradation of rubber. J Ind Eng Chem 47(12):2565–2574
13. Kader MA, Bhowmick AK (2003) Thermal ageing, degradation and swelling of acrylate rubber, fluororubber and their blends containing polyfunctional acrylates. Polym Degrad Stab 79(2):283–295
14. Sarkar MD, Mukunda PG, De PP, Bhowmick AK (1997) Degradation of hydrogenated styrene-butadiene rubber at high temperature. Rubber Chem Technol 70(5):855–870
15. Bhowmick AK, Rampalli S, McIntyre D (1985) Effect of resin components on the degradation of guayule rubber. J Appl Polym Sci 30(6):2367–2388
16. Bhattacharjee S, Bhowmick AK, Avasthi BN (1991) Degradation of hydrogenated nitrile rubber. Polym Degrad Stab 31(1):71–87
17. Maiti M, Mitra S, Bhowmick AK (2008) Effect of nanoclays on high and low temperature degradation of fluoroelastomers. Polym Degrad Stab 93(1):188–200
18. Sadhu SD, Rajeev RS, Bhowmick AK (2008) Thermal degradation of elastomer based nanocomposites. Polymer Polymer Compos 16(5):283
19. Choudhury A, Bhowmick AK, Soddemann M (2010) Effect of organo-modified clay on accelerated aging resistance of hydrogenated nitrile rubber nanocomposites and their life time prediction. Polym Degrad Stab 95(12):2555–2562
20. Saha T, Bhowmick AK (2020) High-temperature degradation of butadiene-based model elastomers by reactive molecular dynamics simulation. J Appl Polym Sci 137(16):48592
21. Saha T, Bhowmick AK (2017) High-temperature pyrolysis simulation of acrylonitrile-butadiene model compound with experimental evidence. J Anal Appl Pyrolysis 125:243–257
22. Saha T, Bhowmick AK (2019) Influence of nanofiller on thermal degradation resistance of hydrogenated nitrile butadiene rubber. Rubber Chem Technol 92(2):263–285
23. Saha T, Bhowmick AK, Oda T, Miyauchi T, Fujii N (2016) Degradation of polyacrylic elastomers: theoretical and experimental studies. Polym Degrad Stab 134:60–75
24. Li C, Strachan A (2015) Molecular scale simulations on thermoset polymers: a review. J Polym Sci B 53(2):103–122
25. Eslami H, Müller-Plathe F (2010) Viscosity of nanoconfined polyamide-6, 6 oligomers: atomistic reverse nonequilibrium molecular dynamics simulation. J Phys Chem B 114(1):387–395
26. Saha S, Bhowmick AK (2019) An insight into molecular structure and properties of flexible amorphous polymers: a molecular dynamics simulation approach. J Appl Polym Sci 136(18):47457

27. Zhan Y, Mattice WL (1992) Simulation of the molecular dynamics of poly (1,4-trans-isoprene) and isoprene as inclusion complexes in crystalline perhydrotriphenylene. Macromolecules 25(13):3439–3442
28. Smith GD, Paul W (1998) United atom force field for molecular dynamics simulations of 1,4-polybutadiene based on quantum chemistry calculations on model molecules. J Phys Chem A 102(7):1200–1208
29. Sharma P, Roy S, Karimi-Varzaneh HA (2016) Validation of force fields of rubber through glass-transition temperature calculation by microsecond atomic-scale molecular dynamics simulation. J Phys Chem B 120(7):1367–1379
30. Gee RH, Boyd RH (1994) Conformational dynamics and relaxation in bulk polybutadienes: a molecular dynamics simulation study. J Chem Phys 101(9):8028–8038
31. Van Duin AC, Dasgupta S, Lorant F, Goddard WA (2001) ReaxFF: a reactive force field for hydrocarbons. J Phys Chem A 105(41):9396–9409
32. Chenoweth K, Van Duin AC, Goddard WA (2008) ReaxFF reactive force field for molecular dynamics simulations of hydrocarbon oxidation. J Phys Chem A 112(5):1040–1053
33. Chenoweth K, Cheung S, Van Duin AC, Goddard WA, Kober EM (2005) Simulations on the thermal decomposition of a poly (dimethylsiloxane) polymer using the ReaxFF reactive force field. J Am Chem Soc 127(19):7192–7202
34. Brown RP, Butler T, Hawley SW (2001) Ageing of rubber: accelerated heat ageing test results. iSmithers Rapra Publishing, Shropshire
35. Bueche F (1960) Mechanical degradation of high polymers. J Appl Polym Sci 4(10):101–106
36. Ginic-Markovic M, Choudhury NR, Dimopoulos M, Williams DR, Matisons J (1998) Characterization of elastomer compounds by thermal analysis. Thermochim Acta 316(1):87–95
37. Yang R, Zhao J, Liu Y (2013) Oxidative degradation products analysis of polymer materials by pyrolysis gas chromatography–mass spectrometry. Polym Degrad Stab 98(12):2466–2472
38. Shield SR, Ghebremeskel GN, Hendrix C (2001) Pyrolysis-GC/MS and TGA as tools for characterizing blends of SBR and NBR. Rubber Chem Technol 74(5):803–813
39. Hiltz JA (2000) Pyrolysis gas chromatography/mass spectrometry identification of poly (butadiene-acrylonitrile) rubbers. J Anal Appl Pyrolysis 55(2):135–150
40. Bhowmick AK, Rampalli S, Gallagher K, Seeger R, McIntyre D (1987) The degradation of guayule rubber and the effect of resin components on degradation at high temperature. J Appl Polym Sci 33(4):1125–1139
41. Deuri AS, Bhowmick AK (1986) Ageing of rocket insulator compound based on EPDM. Polym Degrad Stab 16(3):221–239
42. Garbarczyk M, Kuhn W, Klinowski J, Jurga S (2002) Characterization of aged nitrile rubber elastomers by NMR spectroscopy and microimaging. Polymer 43(11):3169–3172
43. Nyden MR, Coley TR, Mumby S (1997) Applications of molecular dynamics to the study of thermal degradation in aromatic polymers. I: polystyrene. Polym Eng Sci 37(9):1496–1500
44. Nyden MR, Stoliarov SI, Westmoreland PR, Guo ZX, Jee C (2004) Applications of reactive molecular dynamics to the study of the thermal decomposition of polymers and nanoscale structures. Mater Sci Eng A 365(1–2):114–121
45. Stoliarov SI, Westmoreland PR, Nyden MR, Forney GP (2003) A reactive molecular dynamics model of thermal decomposition in polymers: I. Poly (methyl methacrylate). Polymer 44(3):883–894
46. Stoliarov SI, Lyon RE, Nyden MR (2004) A reactive molecular dynamics model of thermal decomposition in polymers. II. Polyisobutylene. Polymer 45(25):8613–8621
47. Jee CSY, Guo ZX, Stoliarov SI, Nyden MR (2006) Experimental and molecular dynamics studies of the thermal decomposition of a polyisobutylene binder. Acta Mater 54(18):4803–4813
48. Diao Z, Zhao Y, Chen B, Duan C, Song S (2013) ReaxFF reactive force field for molecular dynamics simulations of epoxy resin thermal decomposition with model compound. J Anal Appl Pyrolysis 104:618–624

49. Liu X, Li X, Liu J, Wang Z, Kong B, Gong X, Yang X, Lin W, Guo L (2014) Study of high density polyethylene (HDPE) pyrolysis with reactive molecular dynamics. Polym Degrad Stab 104:62–70
50. Hu S, Sun W, Fu J, Zhang Z, Wu W, Tang Y (2018) Initiation mechanisms and kinetic analysis of the isothermal decomposition of poly (α-methylstyrene): a ReaxFF molecular dynamics study. RSC Adv 8(7):3423–3432
51. Zhao T, Li T, Xin Z, Zou L, Zhang L (2018) A ReaxFF-based molecular dynamics simulation of the pyrolysis mechanism for polycarbonate. Energy Fuel 32(2):2156–2162
52. Golub MA, Gargiulo RJ (1972) Thermal degradation of 1,4-polyisoprene and 1,4-polybutadiene. J Polym Sci B Polym Lett 10(1):41–49
53. Sanglar C, Quoc HN, Grenier-Loustalot MF (2010) Studies on thermal degradation of 1-4 and 1-2 polybutadienes in inert atmosphere. Polym Degrad Stab 95(9):1870–1876
54. Opfermann J, Kaisersberger E (1992) An advantageous variant of the Ozawa-Flynn-Wall analysis. Thermochim Acta 203:167–175
55. Starink MJ (2003) The determination of activation energy from linear heating rate experiments: a comparison of the accuracy of isoconversion methods. Thermochim Acta 404(1–2):163–176
56. Pazur RJ, Petrov I (2015) The thermo-oxidation of isoprene containing copolymers of isobutylene: activation energies and reactions from room temperature to 100°C. Polym Degrad Stab 113:55–65
57. Celina M, Wise J, Ottesen DK, Gillen KT, Clough RL (2000) Correlation of chemical and mechanical property changes during oxidative degradation of neoprene. Polym Degrad Stab 68(2):171–184
58. Saha T, Bhowmick AK, Oda T, Miyauchi T, Fujii N (2021) Understanding thermo-oxidative degradation of polyacrylic ester elastomer and its nanocomposites through molecular dynamics simulation and experiments. Polym Degrad Stab 183:109457
59. Saha T, Bhowmick AK, Oda T, Miyauchi T, Fujii N (2018) Reactive molecular dynamics simulation for analysis of thermal decomposition of oligomeric polyacrylicester model nanocomposite and its experimental verification. Polymer 137:38–53
60. Nakano T, Kawasaki T, Miyauchi T, Horiguchi T, Saha T, Bhowmick AK (2021) Indian Institute of Technology Kharagpur and Denka Co Ltd. Acrylic rubber composition. U.S. Patent 10982029
61. Saha T, Bhowmick AK, Oda T, Miyauchi T, Fujii N (2018) Influence of dianiline cross-linking systems on the structure, curing mechanism, and properties of polyacrylicester elastomer. Rubber Chem Technol 91(4):729–750
62. Pazur RJ, Mengistu T (2019) Activation energies of thermo-oxidized nitrile rubber compounds of varying acrylonitrile content. Rubber Chem Technol 92(1):129–151
63. Pazur RJ (2014) Activation energy of poly (isobutylene) under thermo-oxidative conditions from 40 to 100°C. Polym Degrad Stab 104:57–61
64. Pazur RJ, Petrov I (2015) The thermo-oxidation of chlorinated and brominated isobutylene-co-isoprene polymers: activation energies and reactions from room temperature to 100°C. Polym Degrad Stab 121:311–320
65. Gillen KT, Bernstein R, Celina M (2015) Challenges of accelerated aging techniques for elastomer lifetime predictions. Rubber Chem Technol 88(1):1–27

Adv Polym Sci (2023) 289: 239–272
https://doi.org/10.1007/12_2021_109

Published online: 25 November 2021

Experimental Analysis of Fast Crack Growth in Elastomers

Check for updates

Thanh-Tam Mai, Yoshihiro Morishita, Katsuhiko Tsunoda, and Kenji Urayama

Contents

Abstract Crack growth often leads to catastrophic failure of rubber products. Understanding the crack growth mechanism is important for toughening elastomers.

T.-T. Mai and K. Urayama (✉)
Department of Macromolecular Science and Engineering, Kyoto Institute of Technology, Kyoto, Japan
e-mail: urayama@kit.ac.jp

Y. Morishita and K. Tsunoda
Advanced Materials Division, Bridgestone Corporation, Tokyo, Japan

This article described in summarized form our recent experimental investigations on fast crack growth in elastomers. The crack-tip features, including the crack-tip profiles and the local crack-tip strain fields, are revealed for a quasi-stationary crack and fast-moving cracks ranging from subsonic to super-shear (intersonic) cracks. The velocity and crack-tip features of fast-moving cracks are discussed in relation to the nonlinear elasticity and viscoelasticity of bulk elastomers. The effects of anisotropic stress softening (anisotropic Mullins effect), which is pronounced in filler-reinforced elastomers, on the crack-tip properties are elucidated. We also describe the characterization of quasi-stationary cracks in elastomers subjected to various types of biaxial loading, providing a basis for the fracture mechanics of elastomers under multiaxial deformation.

Keywords Biaxial loading · Crack growth · Crack propagation · DIC · Elastomers · Filled rubbers · Intersonic crack · Mullins effect · Strain field · Super-shear crack · Viscoelasticity

1 Introduction

Fracture analysis of rubber materials has received considerable scientific and industrial interest because it is directly related to the safety of rubber products such as tires [1]. Crack propagation is often accompanied by the catastrophic failure of rubber materials [2–4]. Over the past few decades, significant progress has been made in the experimental and theoretical characterization of crack growth phenomena in rubber materials through crack growth velocity, strain energy release rate, crack-tip shape, and local strain/stress fields around the crack tip [5–7]. The crack growth velocity (V) of rubbers has been investigated primarily by two approaches, that is, quasi-static and cyclic crack propagation [2–4]. In cyclic crack tests, a sample is subjected to cyclic deformations, and V is measured via the increase in crack length (c) per cycle, dc/dn (n: number of cycles). In static crack tests, the steady-state crack growth (dc/dt) is evaluated after the crack is triggered on a strained sample. Various geometries, including uniaxial tension, plane strain (pure shear), angled, and trouser types, have been employed in crack growth tests [2–4].

Characteristically, V in static crack growth often exhibits a discontinuous change at a specific value of the strain energy release rate (input tearing energy) (Γ_c) [8–10]. At $\Gamma < \Gamma_c$, the crack growth is extremely slow (e.g., 10^{-5} to 10^{-3} m s^{-1}), while the crack moves fast (e.g., 10^{-1} to 10^{2} m s^{-1}) at $\Gamma > \Gamma_c$. This "velocity transition" phenomenon has been widely observed for various types of rubber since it was first recognized in the 1950s [11, 12]. Velocity transition phenomena include several important issues that remain to be completely understood. For example, crack-tip analysis for slow- and fast-moving cracks still remains incomplete, although it provides an important basis for understanding the growth mechanism. The fast crack dynamics at $\Gamma > \Gamma_c$ obey a simple power law ($V \sim \Gamma^{1/\beta}$), and the β value depends on the material. Persson et al. [5, 13] theoretically correlated β with the viscoelastic spectrum of the bulk material; however, their theory has not yet been fully assessed. The strategy for enhancing Γ_c is important for toughening rubber materials because Γ_c is a measure of resistance against catastrophic crack propagation [2–4]. The transition threshold Γ_c depends on the material; however, the type of material parameter that governs the magnitude of Γ_c remains unknown. Moreover, the physical origin of the velocity transition has been unclear since the discovery of this phenomenon in the 1950s. Our recent work [14–16] investigated each of these issues, the details of which are described in Sect. 3.

In general, filler reinforcement in rubber materials is required to provide sufficient toughness [17, 18]. Filler-reinforced elastomers exhibit pronounced stress softening (the so-called Mullins effect) [19–22] in which the second loading stress becomes significantly lower than the first loading stress. The stress softening in filled elastomers is attributed to the destruction of inherent physical structures, including filler networks and filler-rubber interfaces. Moreover, the stress-softening effect is considerably directional, resulting in anisotropic internal damage, in which the damage is maximized in the loading direction. Finite anisotropy in the stress-softening effect was revealed by the tensile behavior of subsamples cut out in various directions from pre-strained specimens [23–25]. Crack growth in filled elastomers simultaneously undergoes anisotropic stress softening. To elucidate the complicated effect of anisotropic stress softening on crack growth, the crack-tip properties of pristine and previously loaded elastomers should be compared. Section 4 describes our study [26].

In dynamic fracture mechanics [2, 3], fast crack propagation is classified into two types depending on the velocity regime, that is, "subsonic" cracks traveling with $V < C_s$ (shear wave speed) and "super-shear (or intersonic)" cracks, where $V > C_s$. Crack propagation where $V > C_s$ is beyond the framework of classical fracture theory [2, 3]. Several recent studies have shown that the velocity of a tensile crack in rubbers could exceed C_s because of the stability of the dynamic crack [27] and an increase in the local shear wave speed at the crack tip during crack growth [28]. They also reported that a super-shear crack in rubber exhibits a wedge-like crack-tip opening and accompanies oblique shock waves [27, 29, 30]. Some theories involving shockwave- [31, 32], lattice- [33], and hyperelasticity-based models [28] have been proposed to describe super-shear crack propagation. They suggest that the inertia effect plays a major role in the mechanism of super-shear crack growth [34, 35]. In Sect. 5, we provide insight into the transition from subsonic to super-shear cracks from several viewpoints, including the power-law behavior of $V \sim \Gamma^{1/\beta}$ and the crack-tip strain field [36].

Most existing studies have focused on the crack properties in rubbers under uniaxial or plane strain (pure shear stretching) normal to the crack growth direction [5–7]. However, uniaxial stretching is only one among the physically admissible deformations of rubbers [37]. In reality, rubber materials can be subjected to complex deformations in practical use. The classical theory assumes that the fracture phenomena are governed only by the stress or strain imposed normally to the crack growth direction [4]. Nevertheless, some studies on hard materials, such as metals under biaxial loading, have pointed out that the stress/strain in the direction parallel to the crack has pronounced effects on both stationary and fast-moving cracks [38–

40]. Several studies have investigated the fracture toughness, J-integral, stress intensity factor, or dissipative heating as a function of the biaxial strain ratio for various rubber compounds [41–45]. Recently, we experimentally characterized various aspects of quasi-stationary cracks in elastic hydrogels under several types of biaxial strains, including the strain energy release rate Γ, crack-tip opening displacement (CTOD), and local crack-tip strain fields [46]. We revealed that the Γ values depend on both the type and degree of deformation and that they are significantly influenced by the strain imposed parallel to the crack growth direction. In Sect. 6, we describe the corresponding results for a filler-reinforced rubber with low filler content.

2 Characterization of Fast Crack Growth

A custom-built experimental setup (Fig. 1a) was employed to investigate the Mode-I crack propagation in "plane strain (pure shear)" geometry using sheet samples with dimensions of $x \times y \times z = 175 \times 20 \times 1.0$ mm.

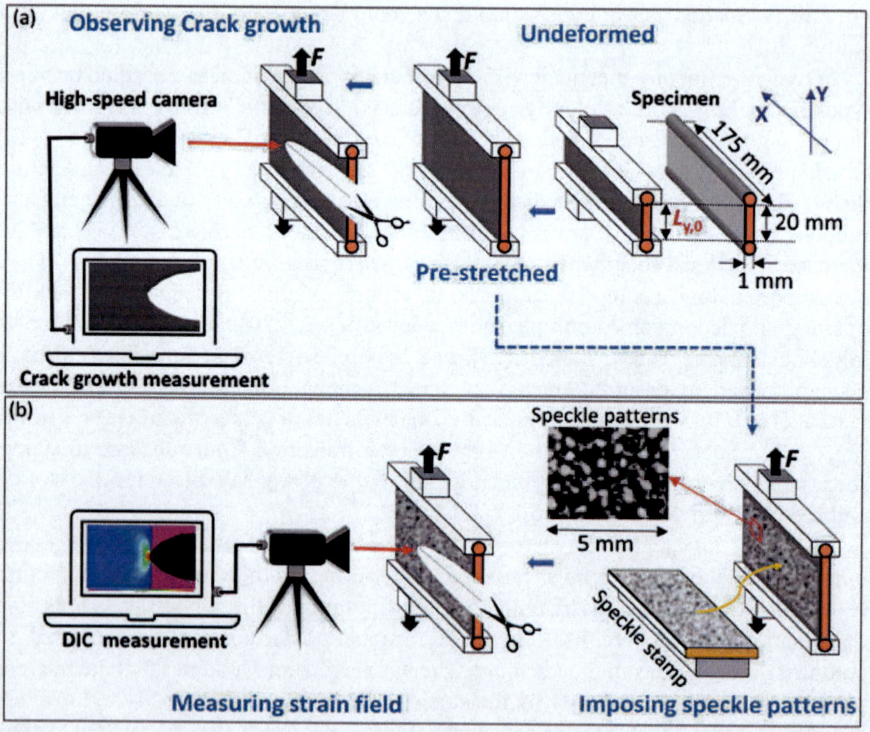

Fig. 1 Schematics of the measurements of (**a**) crack growth and (**b**) local strain field

In the crack growth tests, the specimens were first stretched to the desired elongation (λ_y^{pre}). The crack propagation in the x-direction was triggered by a cut at the edge of the sample midway between the y-boundaries. The crack growth behavior was observed using a high-speed digital camera (FASTCAM Mini AX100 with a maximum frame rate of 540,000 Hz and a maximum image resolution of 1,024 × 1,024 pixel) with the aid of two high-power LED light sources. The central region with a length of approximately 100 mm that was used for observation is sufficiently long to attain a steady state of crack propagation [14, 36].

The strain energy release rate (Γ) during crack propagation in plane strain geometry was evaluated using the relationship $\Gamma = WL_{y,0}$ [10, 47], where W and $L_{y,0}$ are the stored elastic energy density in the stretching of λ_y^{pre} ($W = \int \sigma d\lambda_y^{pre}$) and the initial length in the y-direction, respectively.

The two-dimensional (2D) strain field around the crack tip during crack growth was evaluated using digital image correlation (DIC) method (Fig. 1b). The speckle patterns for DIC analysis were imposed on the specimen surface in the pre-stretched state of λ_y^{pre} using a custom-made speckle stamp. The speckle patterns satisfied the requirements for the DIC analysis: isotropic and random distributions, approximately 50% of coverage, speckle size of 3–7 pixels (pixel size was approximately 80 μm), and high contrast (bright white dots on a dark black background). In addition, the first-standard deviation confidence in the match around the crack tip of the DIC measurement was less than 5%, as compared to the corresponding displacement magnitude [26, 36]. The 2D local displacement field $u(u_x, u_y)$ was measured using the image in the pre-stretched state of λ_y^{pre} (without crack) as the reference state with commercial software. The corresponding Lagrangian strain field was obtained from the partial derivatives of the displacement field. The measured strain corresponds to the incremental strain [$\Delta\varepsilon_{ij} = (i, j = x, y)$] induced by crack propagation, $\Delta\varepsilon_{ij} = \varepsilon_{ij} - \varepsilon_{ij}^{pre}$, where ε_{ij} is the total strain and ε_{ij}^{pre} is the pre-strain. The details of the experiments are described elsewhere [14, 22, 26, 36].

3 Velocity Transition in Crack Growth

3.1 *Crack Growth Velocity*

Figure 2a, b illustrates the relationships between the steady-state crack growth velocity (V) and strain energy release rate (Γ) for two types of elastomers, i.e., carbon black (CB)-filled acrylonitrile-butadiene copolymer (NBR/CB) and silica (SI)-filled styrene-butadiene copolymer (SBR/SI), with various volume fractions of filler (ϕ_f) and the same cross-linker concentration ($c_x = 1.4$ wt%) at 25°C. In the figures, V is reduced by the linear shear wave speed (C_s) of each specimen: $C_s = (\mu_{(1)}/\rho)^{1/2}$, where $\mu_{(1)}$ is the linear shear modulus and ρ is the mass density in the pristine state. The C_s value of each specimen is shown in the inset. In each elastomer, V exhibits a discontinuous-like change over several decades at a transition threshold

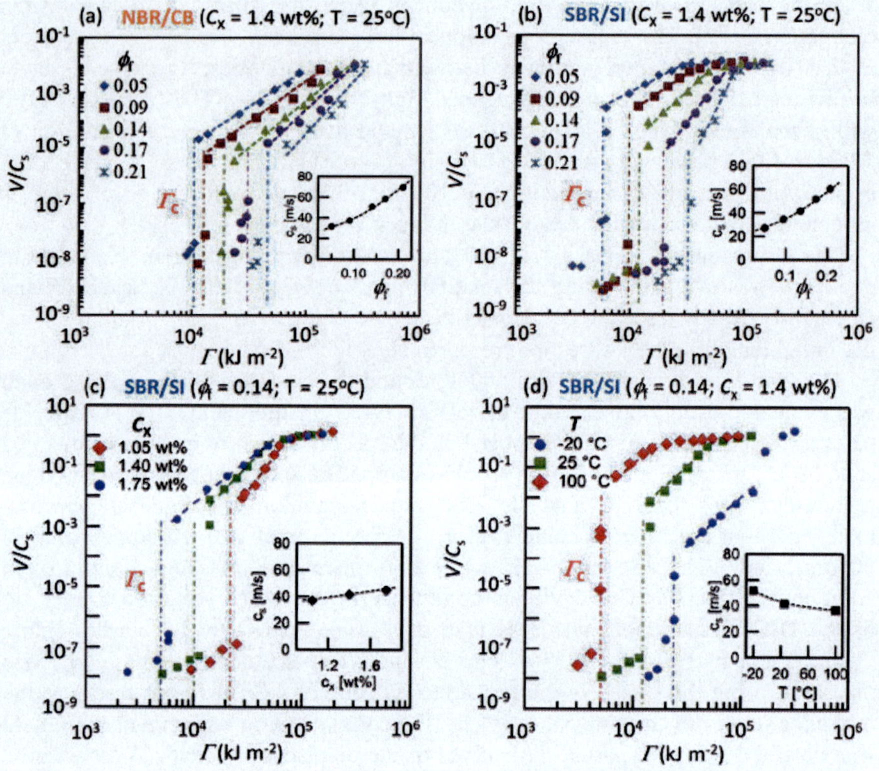

Fig. 2 Relationship between strain energy release rate (Γ) and steady-state crack growth velocity (V) reduced by the linear shear wave speed (C_s) for the filler-reinforced rubbers NBR/CB and SBR/SI. The effects of filler volume fraction (ϕ_f) on NBR/CB and SBR/SI are shown in (**a**) and (**b**), respectively. The effect of cross-linker concentration (c_x) and temperature (T) are shown in (**c**) and (**d**), respectively. The inset of each figure shows the C_s value of each specimen. The data in (**a–d**) are reproduced from Refs. [14, 15]

value of Γ (Γ_c), while V increases with increasing Γ. They are classified in the high V region of $V \approx 10^{-1}$ to 10^2 m s^{-1}, and the very slow V region of $V \approx 10^{-7}$–10^{-5} m s^{-1}, each of which is designated as fast or slow mode, respectively. In the fast mode regime, the V-Γ relationships are characterized by a power law of $V \sim \Gamma^{1/\beta}$, where β^{-1} corresponds to the slope in the figure. Figure 2c, d shows the corresponding plots for SBR/SI of $\phi_f = 0.14$ with various values of cross-linker concentration (c_x) and at various atmospheric temperatures (T), respectively. As shown in Fig. 2a–d, Γ_c depends on each of the rubber/filler combinations, ϕ_f, c_x, and T. The exponent β also depends on these material parameters, but is independent of T. The effects of the material parameters on β and Γ_c are discussed in Sects. 3.3 and 3.4, respectively. In the fast crack growth, the nonlinear elasticity effect appears as a non-parabolic crack-tip shape, as described in Sect. 3.2. The physical origin of the velocity transition phenomena is described in Sect. 3.5. The quasi-flattening of the V-

Γ relations at high Γ in which V is close to or higher than C_s results from the emergence of inertia effect of super-shear crack, which is mentioned in Sect. 5.

3.2 Crack-Tip Shape

Linear elastic fracture mechanics (LEFM) [2, 4] expect the crack-tip shape to be parabolic: $X = -a(Y)^2$, where X and Y are deformed configurations that move with the crack tip during crack growth. The non-parabolic crack-tip shapes reflect the finite deviation of the local strain/stress field from the expectation of LEFM. Slowly moving or stationary cracks have a (nearly) parabolic crack-tip shape (left figure in Fig. 3a). The deviation from the parabolic shape becomes significant when V increases, particularly in the fast mode region (center and right figures in Fig. 3a). A similar tendency has been observed in fast crack growth in an elastic hydrogel [48, 49] and elastomers, regardless of the rubber/filler combination, ϕ_f, c_x, or T [14, 15]. The non-parabolic crack-tip shapes are characterized by a modified parabolic function $X = -a(Y)^2 - \delta$, where the two parameters a and δ are the inverse of the CTOD and the deviation, respectively (Fig. 3b) [14, 15].

Figure 3c, d shows the δ and a values as a function of V/C_s, respectively, for SBR/SI with various ϕ_f values. In the slow-mode region of each elastomer, δ remains almost zero, reflecting the nearly parabolic crack-tip shapes. In contrast, δ in the fast crack increases steeply with increasing V. The δ value increases with increasing ϕ_f when compared at the same V/C_s. The a values in the slow crack are almost independent of V, whereas in the fast mode regime, a decreases (that is, CTOD increases) with increasing V.

The finite deviation from the parabolic shape indicates the emergence of a nonlinear elasticity effect. The weakly nonlinear theory (WNLT) in fracture mechanics has been applied to interpret the finite δ values in the crack growth of elastic hydrogels [50, 51]. The WNLT considers the contributions of the first-order (linear) shear modulus ($\mu_{(1)}$) and second-order nonlinear shear modulus ($\mu_{(2)}$) to the crack-tip strain field. The moduli $\mu_{(1)}$ and $\mu_{(2)}$ of each elastomer are evaluated by fitting a polynomial function to the stress (σ)-strain (ε) data in plane strain deformation: $\sigma(\varepsilon) = E_{(1)}\varepsilon - E_{(2)}\varepsilon^2 + E_{(3)}\varepsilon^3 - \ldots E_{(n)}\varepsilon^n$, where $E_{(i)}$ is the ith order modulus, and $E_{(1)}$ and $\mu_{(1)}$ are related to $\mu_{(1)} = E_{(1)}/4$ for plane strain extension of incompressible materials [52]. The $\mu_{(1)}$ and $\mu_{(2)}$ values are insensitive to the number of terms (n) in the polynomial function when n is sufficiently large [14, 15]. The WNLT has a characteristic length (l) below which the second-order nonlinear elasticity effect becomes pronounced and the behavior expected by LEFM, such as the $r^{1/2}$ strain singularity, breaks down. When δ is assumed to be proportional to l, the following scaling relationship is expected:

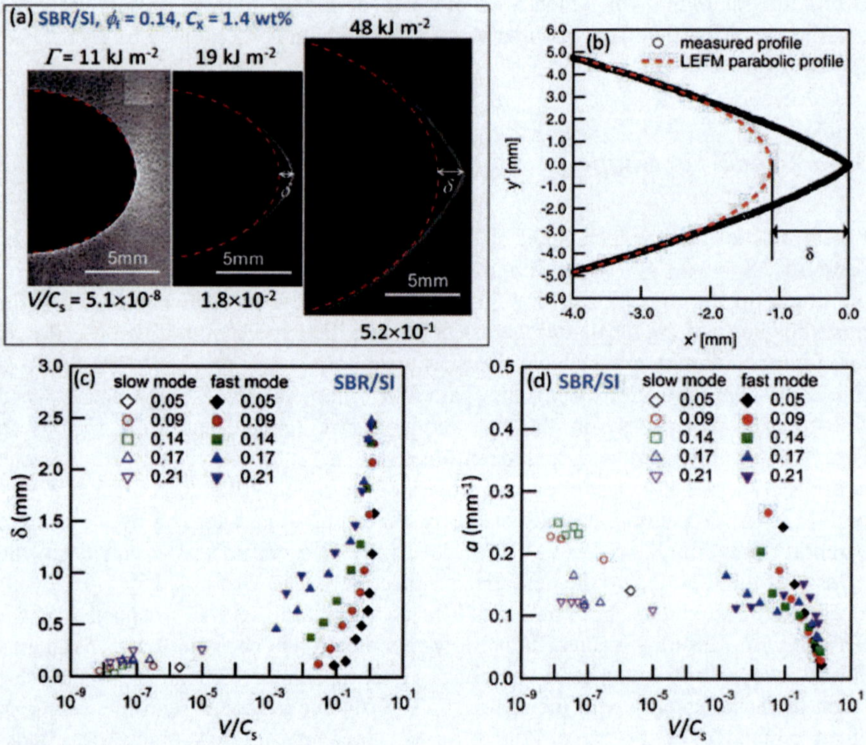

Fig. 3 (**a**) Images of crack-tip shape for SBR/SI with $\phi_f = 0.14$ and $c_x = 1.4$ wt.% at various V or Γ. (**b**) Characterization of the non-parabolic shape of the crack tip. (**c**) Dependence of the deviation δ from the parabolic shape, and (**d**) parabolic curvature a on V/C_s for SBR/SI with various ϕ_f. All figures are reproduced from Refs. [14, 15]

$$\delta \sim l \sim \frac{{\mu_{(2)}}^2}{{\mu_{(1)}}^3}\Gamma \tag{1}$$

Figure 4a assesses the validity of Eq. (1) for NBR/CB and SBR/SI with various ϕ_f and c_x values. The corresponding data at various T values for SBR/SI are included in the figure. The data are categorized into three groups: (i) the data in the slow-mode regime, (ii) the data at $\delta < \delta_c$ in the fast mode regime, which fall on a straight line, and (iii) the data at $\delta > \delta_c$ in the fast mode regime, which deviate downward from the linear relationship. The quantity δ_c is the onset δ value for the deviation, and δ_c depends on the rubber/filler combination: $\delta_c \approx 2$ mm for SBR/SI and $\delta_c \approx 1$ mm for NBR/CB. Group (i) includes the data at $\delta \approx 0$ in the low $\phi_f \leq 0.14$, the high $c_x \geq 1.4$ wt.%, or high temperature $T \geq 25$°C. The small but finite δ values ($<$ 0.3 mm) at $\phi_f > 0.14$ and $c_x < 1.4$ wt.%, or $T < 25$°C are also included in group (i), and they have a linear correlation with l (inset of Fig. 4a). Importantly, all data at $\delta < \delta_c$ in the fast mode (group (ii)) collapse into a single straight line, validating Eq. (1).

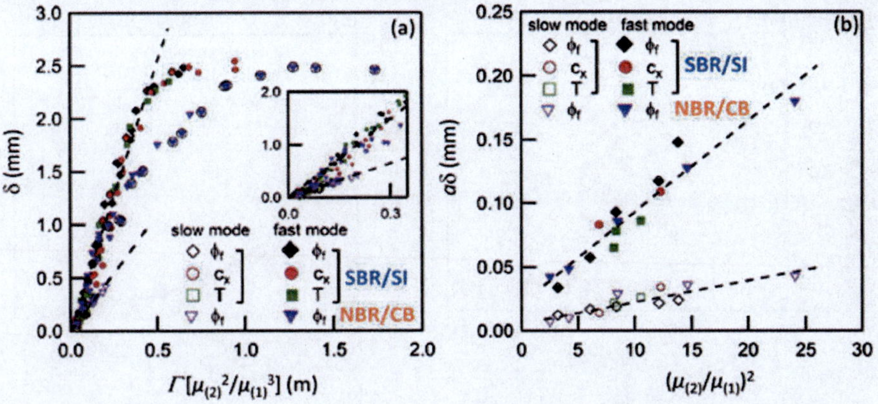

Fig. 4 (**a**) Relationship between δ and $\Gamma[\mu_{(2)}{}^2/\mu_{(1)}{}^3]$ for NBR/CB and SBR/SI with various values of ϕ_f, c_x, and T. (**b**) Relationship between $a\delta$ and $[\mu_{(2)}/\mu_{(1)}]^2$ in groups (i) and (ii) for each specimen. Unfilled and filled symbols illustrate the data in the slow and fast mode regions, respectively. The dashed lines indicate the linear fits to the data. The data in (**a**, **b**) are reproduced from Refs. [14, 15]

The WNLT also expects a simple scaling relation between the product $a\delta$ and the modulus ratio $\mu_{(2)}/\mu_{(1)}$, as

$$a\delta \sim \left[\frac{\mu_{(2)}}{\mu_{(1)}}\right]^2 \quad (2)$$

because a is proportional to $\mu_{(1)}/\Gamma$ [50, 51]. It can be seen in Fig. 4b that each of the data at $\delta < \delta_c$ in groups (i) and (ii) obeys Eq. (2) [15]. In contrast, no definite correlation was observed for the data in group (iii) at $\delta > \delta_c$ [15].

These results demonstrate that the non-parabolic crack-tip shapes are accounted for by the second-order nonlinear elasticity effect when the deviation is not large ($\delta < \delta_c$). The failure of the WNLT for group (iii) data at $\delta > \delta_c$ suggests the emergence of the effect of higher-order nonlinear elasticity. The NBR/CB specimens exhibited a smaller δ_c value ($\delta_c = 1$ mm) than did the SBR/SI ($\delta_c = 2$ mm). This is explained by the fact that the third-order nonlinear effect $\mu_{(3)}/\mu_{(1)}$ in NBR/CB is larger than that in SBR/SI when compared at the same ϕ_f [15].

3.3 *Correlation Between Fast Crack Growth and Bulk Viscoelasticity*

Fast crack propagation is characterized by the power law $V \sim \Gamma^{1/\beta}$. As shown in Fig. 2d, β is insensitive to T, allowing the superposition of the V–Γ data at various T values along the vertical axis (Fig. 5a). The vertical shift factors a_T in the

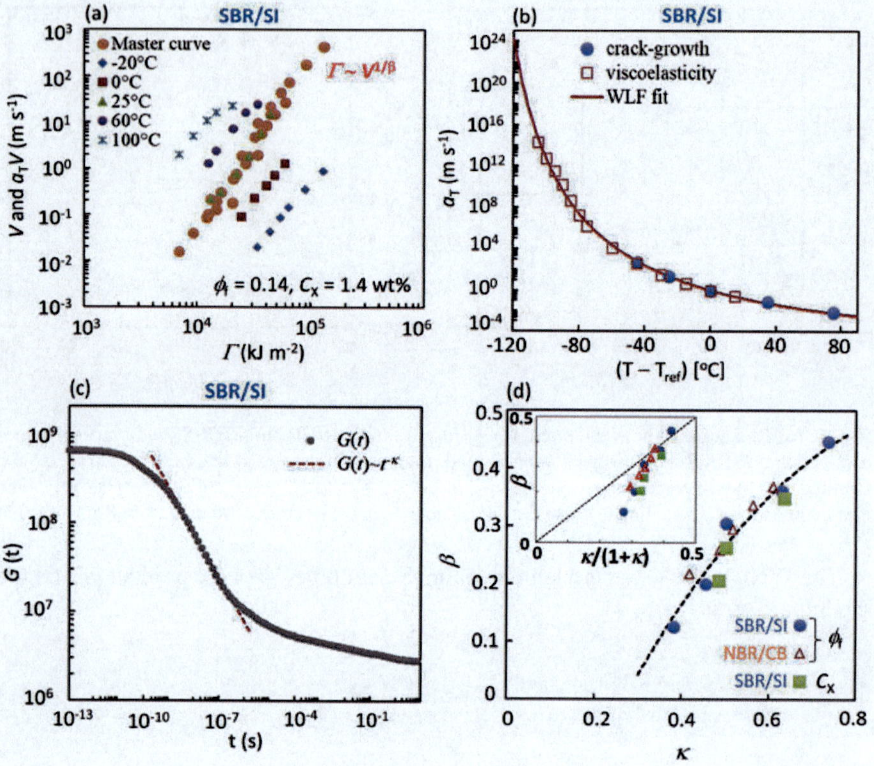

Fig. 5 (**a**) Γ dependence of V for SBR/SI with $\phi_f = 0.14$ and $c_x = 1.4$ wt.% at various temperatures from −20 to 100°C. A master curve is obtained by the vertical shifts at a reference temperature of 25°C. (**b**) Temperature dependence of the shift factors for the crack growth and the linear viscoelasticity. The solid line represents the fitting curve of the WLF equation using $C_1 = 6.83$ and $C_2 = 154$. (**c**) Relaxation modulus converted from the frequency dependence of dynamic shear moduli. The dashed line indicates the linear fit with $\kappa = 0.50$ to the data in the glass-rubber transition regime. (**d**) Correlation between β and κ for NBR/CB and SBR/SI with various values of ϕ_f and c_x. The inset shows the assessment of the Persson-Brener theory for the data. The data in (**a**–**d**) are reproduced from Ref. [16]

superposition agree with those for the linear dynamic viscoelasticity in the frequency-temperature superposition, and the T dependence of a_T obeys the Williams, Landel, and Ferry equation [53] (Fig. 5b). This result indicates that the fast crack growth dynamics are governed by the bulk viscoelasticity, which has also been reported in previous studies on vulcanized natural rubber (NR) or SBR [5, 47, 54].

The V–Γ relationship in correlation with bulk viscoelasticity has been theoretically discussed in previous studies [5, 13, 55, 56]. Because most viscoelastic dissipation occurs in the glass-rubber transition region, Persson and Brener [5, 13] correlated β with the exponent κ for the time (t) dependence of the shear relaxation modulus $G(t)$ in the transition regime, $G(t) \sim t^{-\kappa}$. Figure 5c shows the $G(t)$ obtained from the master curve of the linear dynamic moduli-frequency relationships [16]. The slope of the linear regression yields the value of κ.

Figure 5d displays plots of β vs. κ for various elastomers. All data fall on a single line, independent of the rubber/filler combination, ϕ_f, or c_x, indicating a strong correlation between the fast crack growth and viscoelasticity in the glass-rubber transition regime. As the corresponding viscoelastic spectrum becomes steeper, the V dependence of Γ becomes stronger. The inset of the figure assesses the prediction of Persson and Brener, $\beta = \kappa/(\kappa + 1)$ [5, 13]. The agreement is limited to $\beta > 0.3$. At $\beta < 0.3$, the deviation increases as β decreases. The specimens with $\beta < 0.3$ have high filler contents ($\phi_f > 0.09$) or low cross-linker concentrations ($c_x < 1.75$ wt.%). The deviation cannot be attributed to the nonlinear elasticity effect because the modulus ratio $\mu_{(2)}/\mu_{(1)}$, which is a measure of nonlinear elasticity, increases with increasing ϕ_f but decreases with decreasing c_x [14–16]. The universal β-κ correlation obtained experimentally (Fig. 5d) clearly indicates that the fast crack dynamics are governed by the bulk viscoelasticity, but understanding this relationship theoretically remains an open issue.

3.4 *Transition Threshold Γ_c*

The velocity transition between slow- and fast-moving cracks for each elastomer is characterized by the transition threshold Γ value (Γ_c) (Fig. 2). Understanding the material parameters governing Γ_c is important for practical applications of rubber products because Γ_c is a measure of the resistance against catastrophic crack propagation. Transition threshold Γ_c tends to increase with increasing ϕ_f and decreasing c_x or T. The bulk fracture toughness (W_c) of the elastomer is intuitively expected to govern Γ_c: W_c is the mechanical work per unit volume required for the fracture under extension, and $W_c = \int_1^{\lambda_c} \sigma d\lambda$, where λ_c is the elongation at break. As can be seen in Fig. 6a, Γ_c tends to increase with increasing W_c, but no definite correlation between them is observed for elastomers with high W_c. Most elastomers with a large W_c exhibit high nonlinear elasticity. The combined effects of the second-order nonlinear elasticity effect and W_c are considered here by multiplying W_c and the corresponding factor ${\mu_{(2)}}^2/{\mu_{(1)}}^3$ used in the WNLT analysis of the crack-tip shape in Eq. (2). Figure 6b shows a good correlation between Γ_c and $W_c[{\mu_{(2)}}^2/{\mu_{(1)}}^3]$, indicating that the increase in the degree of nonlinear elasticity and fracture toughness enhances Γ_c. This result provides an important basis not only for the theoretical interpretation of Γ_c, but also for the development of tough elastomers with high resistance against crack propagation.

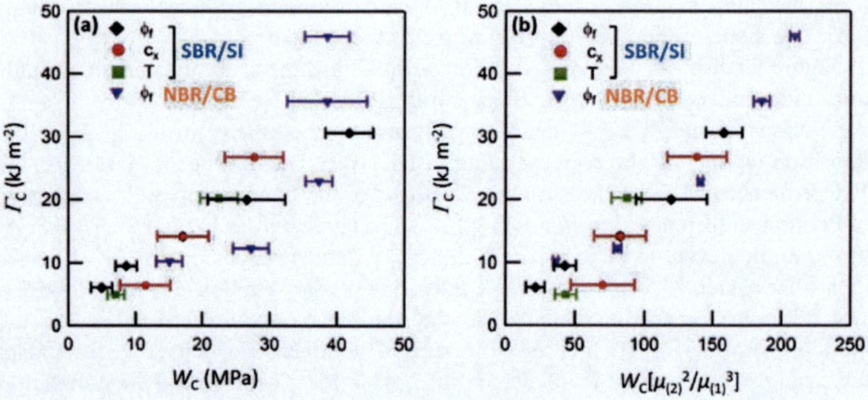

Fig. 6 Plots of the transition threshold value of strain energy release rate (Γ_c) for various specimens and at various T against (**a**) fracture toughness (W_c) and (**b**) the product of W_c and the nonlinear elasticity factor, $W_c[\mu_{(2)}{}^2/\mu_{(1)}{}^3]$. Error bars represent the standard deviations for the W_c values. The data in (**a, b**) are reproduced from Refs. [14, 15]

3.5 Origin of Velocity Transition

The velocity transition (or velocity jump phenomena) in the crack growth of elastomers has been recognized since the 1950s [5, 11, 12], but the origin of the abrupt acceleration of crack growth remains to be understood. A theoretical model considered the temperature effect at the crack tip [55, 57], but the corresponding temperature rise (~1,000 K) was unrealistically high, in contrast to the modest increase observed experimentally [58, 59].

Recently, Kubo and Sakumichi et al. [60] proposed a mechanism attributing the velocity jump to the dynamic glass transition in the vicinity of the propagating crack tip. Originally, each of them explained the velocity jump phenomena using an FEM simulation [61] and a minimum model for crack propagation [62] independently. The FEM simulation employed a phenomenological viscoelastic function that reflected the bulk viscoelasticity of the elastomers, together with a fracture criterion. They observed a nonmonotonic temporal variation in the stress near the crack tip, resulting in a velocity jump [61]. As a minimum model, they employed a 2D lattice model in which the viscoelastic response obeys a Zener element characterized by the viscosity and the two elastic moduli, each of which corresponds to the modulus in the rubber or glass regime, respectively [62]. The model showed a velocity jump that originated from the response of the glass-rubber transition in the Zener element. Thereafter, they showed that these two approaches are consistent and have the same origin for the transition [60]. This model expects a linear relationship between Γ_c and W_c, in agreement with the experimental results for unfilled elastomers with different glass transition temperatures and various c_x values, each of which has a modest degree of nonlinear elasticity and W_c (<30 MPa) [60]. As described in Sect. 3.4, the $\Gamma_c - W_c$ linearity fails at high W_c because of the pronounced nonlinear elasticity effect, which is not considered in this linear model. Direct observation of the occurrence of the local glass transition near the propagating crack tip will further validate the proposed mechanism, although it is not straightforward experimentally.

4 Stress-Softening Effects on Fast Crack Growth

4.1 *Anisotropic Stress-Softening*

The effects of stress softening and its anisotropy on crack growth behavior were investigated using three types of elastomers: (I) a low-filled elastomer with almost no stress softening, (II) a pristine highly filled elastomer with a pronounced stress-softening feature, and (III) a "softened" elastomer obtained by imposing previously uniaxial loading–unloading cycles on specimen (II). The SBR/CB specimens with $\phi_f = 0.05$ and 0.21 (designated as SBR5 and SBR21) correspond to specimens (I) and (II), respectively. The details of sample preparation were given in Ref. [26]. The softened elastomer (III) (designated as SBR21-SY) was obtained by imposing multiple plane strain loading–unloading cycles on SBR21 with a maximum stretch of $\lambda_{y,m} = 4.5$, which was larger than the maximum stretch used in the crack growth tests.

In Fig. 7a, the uniaxial loading–unloading curves obtained with increasing $\lambda_{y,m}$ confirm that the stress-softening feature is pronounced in SBR21, whereas it is almost absent in SBR5. The corresponding curves of the two subsamples of SBR21-SY in Fig. 7b show that SBR21-SY has a significant anisotropy related to the internal damage caused by the previous loading in the *y*-direction: The subsample SBR21-SY-x exhibits a considerable stress-softening feature, whereas the subsample SBR21-SY-y does not. Similar damage anisotropy caused by uniaxial stretching has been previously reported [23–25].

4.2 *Effect on Velocity Transition*

Figure 7c, d compares the velocity transitions of the three types of elastomers using λ_y^{pre} and Γ as variables, respectively. The strain energy release rate Γ ($\Gamma = L_{y,0}W$) was evaluated using the unloading stress-strain data, enabling a focus on the strain-induced softening occurring only in the crack growth [63–66] and to facilitate comparison between the pristine (SBR21) and softened specimens (SBR21-SY).

The *V* values in the fast mode regimes for SBR21 and SBR21-SY are identical at the same λ_y^{pre}, while the transition threshold elongation (λ_y^c) of the velocity transition for SBR21-SY is smaller than that for SBR21. This means that the pre-softening in the *y*-direction reduces λ_y^c, but it has no effect on *V* for fast crack growth in the *x*-direction. The pre-softening reduced Γ_c: Γ_c of SBR21-SY was lower than that of SBR21. This is explained by a reduction in fracture toughness (W_c) from 23 to

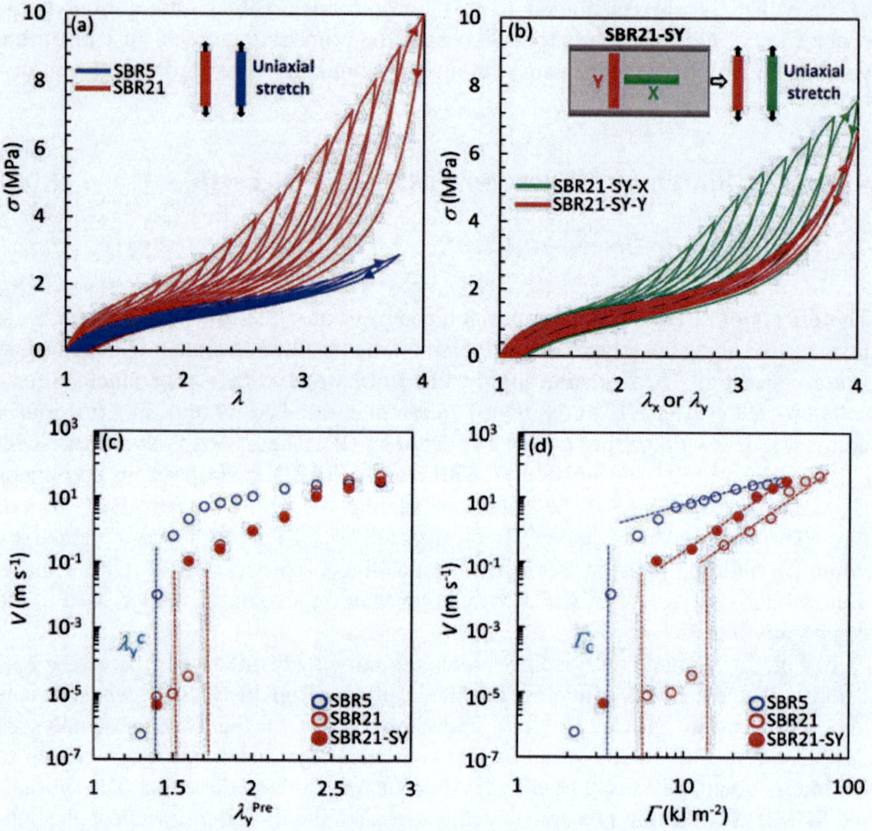

Fig. 7 Nominal stress (σ)–elongation (λ) relationships under cyclic uniaxial extension with various λ_m for (**a**) SBR5 and SBR21 and (**b**) the subsamples SBR21-SY-x and SBR21-SY-y, which were cut from the softened specimen SBR21-SY. The crack growth velocity (V) is shown as a function of (**c**) λ_y^{pre} and (**d**) Γ for SBR5, SBR21, and SBR21-SY. The data in (**a–d**) are reproduced from Ref. [26]

17 MJ m^{-3} caused by the pre-softening [26], according to the positive dependence of W_c on Γ_c in Fig. 6.

The β values ($\beta \approx 0.25$) in the power-law behavior of $V \sim \Gamma^{1/}\beta$ in the fast mode are similar for SBR21 and SBR-SY, indicating that the pre-softening in the y-direction has no appreciable influence on the fast crack growth in the x-direction. This is explained by the fact that the pre-softening does not affect the viscoelastic spectrum in the glass-rubber transition characterized by the exponent κ [67] and by the $\beta-\kappa$ correlation in Fig. 5d.

4.3 Effect on Crack-Tip Shape and Strain Fields

Figure 8 compares the 2D strain fields of $\Delta\varepsilon_{yy}$ and $\Delta\varepsilon_{xx}$ and the crack-tip shapes at each of the three Γ values among the three specimens. Figure 9a, b shows the extent of the strain field caused by the crack growth (A_{yy}) and CTOD as a function of Γ. The A_{yy} is evaluated from the area surrounded by the isostrain contour of $\Delta\varepsilon_{yy} = 0.02$ (inset in Fig. 9a). The CTOD value is obtained from the distance between two opposite crack faces and a distance, $X = X_b$, from behind the crack tip [68] (inset of Fig. 9b). In the slowly moving cracks with parabolic crack-tip shapes, no significant difference in the strain fields was observed among the specimens. In the fast crack growth, the pre-softening has significant effects in some aspects, whereas it has no appreciable effect in other aspects. SBR21SY and SBR21 are similar with respect to CTOD ($X_b = 1.0$ mm, Fig. 9b) and $\Delta\varepsilon_{xx}$ (Fig. 8b) [26]. In contrast, SBR21SY and SBR21 are considerably different regarding $\Delta\varepsilon_{yy}$ (or A_{yy}) in the region ahead of the crack tip (Figs. 8a and 9a), while $\Delta\varepsilon_{yy}$ (or A_{yy}) for SBR21-SY is comparable to that for SBR5 with no stress-softening feature.

SBR21-SY has inherent damage in the *y*-direction, but almost no damage in the *x*-direction (Fig. 7b). The maximum stretch in the softening pre-treatment ($\lambda_{y,m} = 4.5$) was larger than that in the crack growth test ($\lambda_y^{pre} = 3.0$). Therefore, crack growth caused no further damage in the *y*-direction for SBR21-SY. In contrast, the crack growth induces damage in the *x*-direction for SBR21-SY, which is as large as that in the corresponding pristine state (SBR21).

In Fig. 8, ahead of the crack tip, the magnitude of $\Delta\varepsilon_{yy}$ is considerably larger than that of $\Delta\varepsilon_{xx}$. In contrast, behind the crack tip, $\Delta\varepsilon_{yy}$ is contractive (negative) because of the unloading from the pre-stretched state, whereas $\Delta\varepsilon_{xx}$ is positive. These characteristics indicate that the energy dissipation ahead of or behind the crack tip is primarily caused by $\Delta\varepsilon_{yy}$ or $\Delta\varepsilon_{xx}$, respectively. This result also indicates that the CTOD is correlated with the strain fields of $\Delta\varepsilon_{xx}$ behind the crack tip. This correlation explains why SBR21 and SBR21-SY have similar CTOD values, although they have considerably different $\Delta\varepsilon_{yy}$ (or A_{yy}) values ahead of the crack tip, and why SBR5 and SBR21-SY are dissimilar with respect to CTOD, although they are similar regarding $\Delta\varepsilon_{yy}$ (or A_{yy}) ahead of the crack tip.

In the vicinity of the crack tip, $\Delta\varepsilon_{yy}$ sharply increases as the distance from the crack tip along the *x*-axis (r_x) approaches zero (Fig. 10a). The upturn behavior of $\Delta\varepsilon_{yy}$ near the crack tip, which is called strain singularity, is characterized by the relationship $\Delta\varepsilon_{yy} \sim r_x^{-\alpha}$, which has a singularity exponent α [2–4]. The power-law behavior is observed in a relatively long range of several millimeters with short and long cut-off lengths ($r_x{*}$ and r_x^0, respectively). The region $0 < r < r_x{*}$ very close to the crack tip is regarded as a damage or plastic zone where nanocavitation occurs predominantly [69], resulting in a deviation from the power-law behavior [4]. The $r_x{*}$ values for the three specimens were approximately 0.6 mm. It can be confirmed that this $r_x{*}$ value is not related to the resolution limit of DIC analysis, the details of which are described in the Supporting Information of Ref. [26]. The size of the damage zone [66, 70] could be much smaller than $r_x{*}$, but its effect on the strain

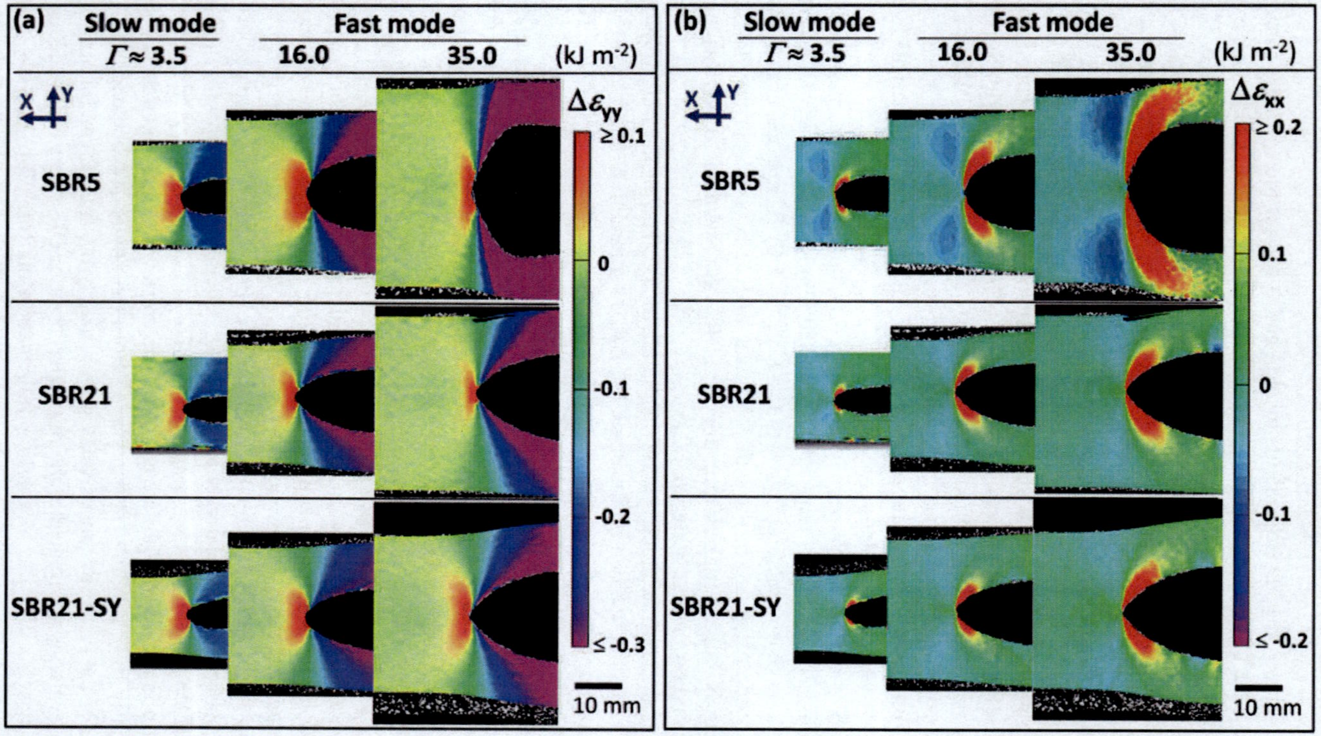

Fig. 8 2D crack-tip strain fields of (**a**) $\Delta\varepsilon_{yy}$ and (**b**) $\Delta\varepsilon_{xx}$ at various degrees of Γ for SBR5, SBR21, and SBR21-SY. The data in (**a**, **b**) are reproduced from Ref. [26]

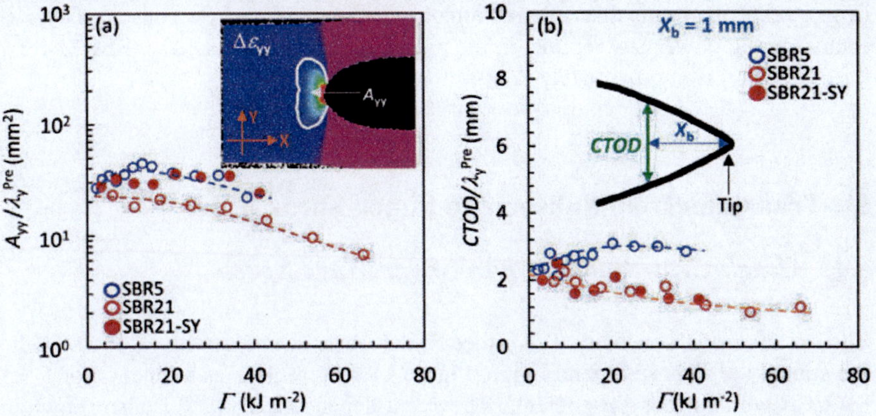

Fig. 9 (**a**) Extent of strain field of $\Delta\varepsilon_{yy}$ (A_{yy}) and (**b**) crack-tip opening degree (CTOD) as a function of Γ for SBR5, SBR21, and SBR21-SY. A_{yy} is estimated from the area enclosed by the isostrain contour of $\Delta\varepsilon_{yy} = 0.02$ [white line in the inset of (**a**)]. The evaluation of CTOD is illustrated in the inset of (**b**). The values of A_{yy} or CTOD are reduced by λ_y^{pre} to consider the difference in the degree of pre-stretching. The data in (**a, b**) are reproduced from Ref. [26]

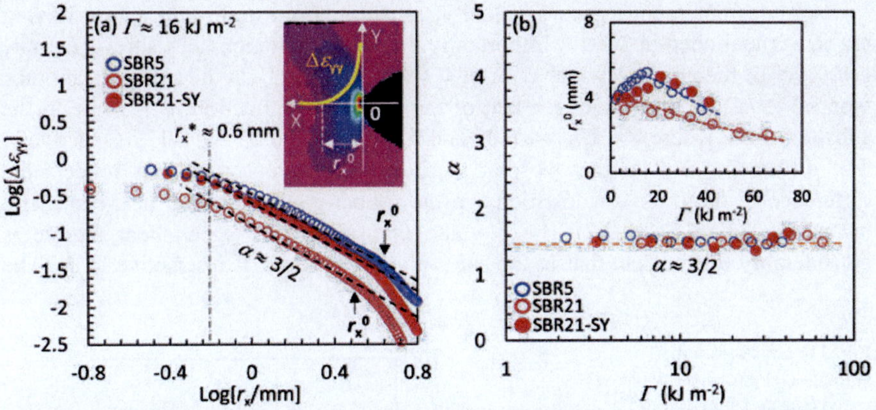

Fig. 10 (**a**) Double logarithmic plots of $\Delta\varepsilon_{yy}$ vs. r_x at $\Gamma \approx 16$ kJ/m^2 for SBR5, SBR21, and SBR21-SY. (**b**) Singularity exponent α ($\Delta\varepsilon_{yy} \sim r_x^{-\alpha}$) as a function of Γ. The inset of (**b**) shows r_x^0 as a function of Γ. The data in (**a, b**) are reproduced from Ref. [26]

singularity is expected to reach a size comparable to that of r_x*. The r_x^0 values are plotted against Γ in the inset of Fig. 10b.

The α values are constant ($\alpha \approx 1.5$) and independent of both Γ and specimen type. This value is considerably larger than $\alpha = 0.5$, predicted by LEFM [2, 4], and $\alpha \approx 1$, predicted by WNLT [50, 51]. The interpretation of the large α value requires consideration of the effect of higher-order nonlinear elasticity. Interestingly, the pre-softening had no effect on α, whereas it increased $\Delta\varepsilon_{yy}$ in the entire r_x regime

(Fig. 10a). This means that the pre-softening increases $\Delta\varepsilon_{yy}$ by a constant factor at each r_x in the power-law regime; in the present case, this factor [$\Delta\varepsilon_{yy}$(SBR21-SY)/$\Delta\varepsilon_{yy}$(SBR21)] is approximately 2.

5 Transition from Subsonic to Super-Shear Crack

5.1 *Crack Growth faster than Shear Wave Speed*

The crack growth velocity (V) exceeds the shear wave speed (C_s) when the pre-stretch (λ_y^{pre}) is sufficiently high. Figure 11 represents V as a function of Γ for SBR/CB with various CB contents. The strain-dependent $C_s(\lambda_y^{pre})$ is also shown in the figure, although the dependence is modest. The strain-dependent C_s was obtained from the nonlinear shear modulus (G_{non}) using the stress-strain data by the Mihai-Goriely method [71] and $C_s = (G_{non}/\rho)^{1/2}$. Prior to the crack growth tests, the specimens were subjected to several loading–unloading cycles with a maximum elongation close to the breaking point to exclude the complicated influences of the strain-induced softening during the crack growth, as described in Sect. 4.

Each elastomer undergoes a velocity transition from a slow to a fast-moving crack, as mentioned in Sect. 3. Importantly, in the fast V region, they show a definite crossover of the power-law behavior of $V \sim \Gamma^{1/\beta}$ at $V \approx C_s$. In the super-shear crack where $V > C_s$, $\beta \approx 1.8$ independent of the CB volume fraction (ϕ_f), while in the subsonic crack where $V < C_s$, $\beta \approx 0.40$ and 0.24 for $\phi_f = 0.05$ and 0.17, respectively. The difference in β in the subsonic regime for different ϕ_f values reflects the difference in the bulk viscoelasticity in the rubber-glass transition between them, as mentioned in Sect. 3.3. The β value ($\approx$ 1.8) in the super-shear regime is considerably higher than that in the subsonic crack, and it is insensitive to ϕ_f. The

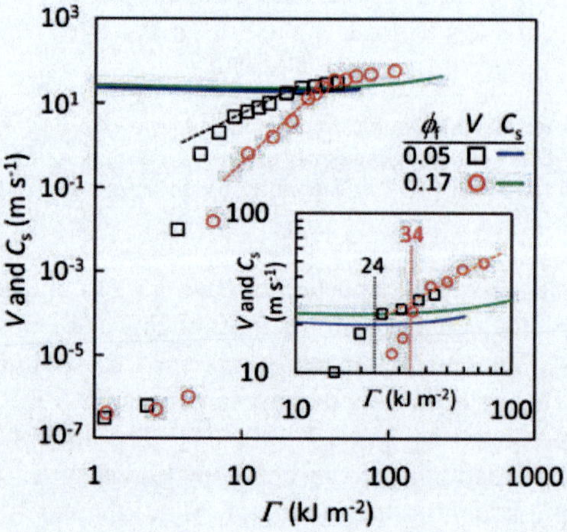

Fig. 11 Crack growth velocity (V) and shear wave speed (C_s) as a function of Γ for softened SBR/CB specimens with $\phi_f = 0.05$ and 0.17. The inset magnifies the plots in the large Γ region. The data are reproduced from Ref. [36]

origin of the considerably weak Γ dependence of V for the super-shear crack is described together with the features of the crack-tip strain field in the next section.

5.2 Crack-Tip Strain Fields in Subsonic and Super-Shear Cracks

The subsonic and super-shear cracks are significantly different in the 2D crack-tip strain field of $\Delta\varepsilon_{yy}$ (Fig. 12a). The quantity $\Delta\varepsilon_{yy}$ is the strain increment in the y-direction caused by crack propagation from the pre-stretched (reference) state. The extent of $\Delta\varepsilon_{yy}$ was estimated by the isostrain contour using $\Delta\varepsilon_{yy} = 0.02$ as a boundary (dashed white lines in Fig. 12a). Figure 12b illustrates the corresponding area (A_{yy}) and the distance from the crack tip along the x-axis (r_x^0) to the contour (inset of Fig. 12b) as a function of V [36]. In the subsonic crack, A_{yy} and r_x^0 tended to increase slightly with V, whereas they steeply decreased with increasing V in the super-shear crack. The results of the previously softened specimens are insensitive to ϕ_f in both subsonic and super-shear cracks because they have similar mechanical properties in the y-direction, as described in Sect. 4.

Figure 13a shows the upturn behavior of $\Delta\varepsilon_{yy}$ in the vicinity of the crack tip for the subsonic and super-shear cracks. Figure 13b displays the strain singularity exponent (α) at each V, which characterizes the strain upturn behavior according to the power law $\Delta\varepsilon_{yy} \sim r_x^{-\alpha}$. The V dependence of α is independent of ϕ_f, but is very different for subsonic and super-shear cracks. The exponent α is constant ($\approx$1.5) in the subsonic crack, but α increases steeply with V in the super-shear crack. Figure 13c compares the $\Delta\varepsilon_{yy}$ value at each V at the lower cut-off distance ($r_x^* = 0.6$ mm) ($\Delta\varepsilon_{yy}^*$) for the power-law behavior. The $\Delta\varepsilon_{yy}^*$ values at $V < C_s$ are almost insensitive to V, whereas those at $V > C_s$ decrease steeply with increasing V.

In the subsonic crack ($V < C_s$), the strain field characteristics ($\Delta\varepsilon_{yy}$ magnitude, A_{yy}, r_x^0, and α) are almost independent of V. This result indicates that rubber chains can rearrange themselves and transport energy around the crack tip from far away. In reality, the local energy required to break the network strands in the damage zone is considerably smaller than the entire input tearing energy. Most input energies are released through viscoelastic dissipation, and the crack growth velocity is primarily governed by the viscoelastic effect [5, 7, 13]. Therefore, the subsonic crack dynamics and the bulk viscoelasticity in the glass-rubber transition regime are strongly correlated, and β depends on ϕ_f, as described in Sect. 3.

In contrast, in the super-shear cracks, the strain field characteristics significantly depend on V, whereas the Γ dependence of V is considerably weak and independent of ϕ_f. These features reflect the dynamic behavior of the rubber chains under a swift crack, known as the inertia effect [34, 35, 72, 73]. The rubber networks do not have enough time to transport the total input energy to the region near the crack tip from

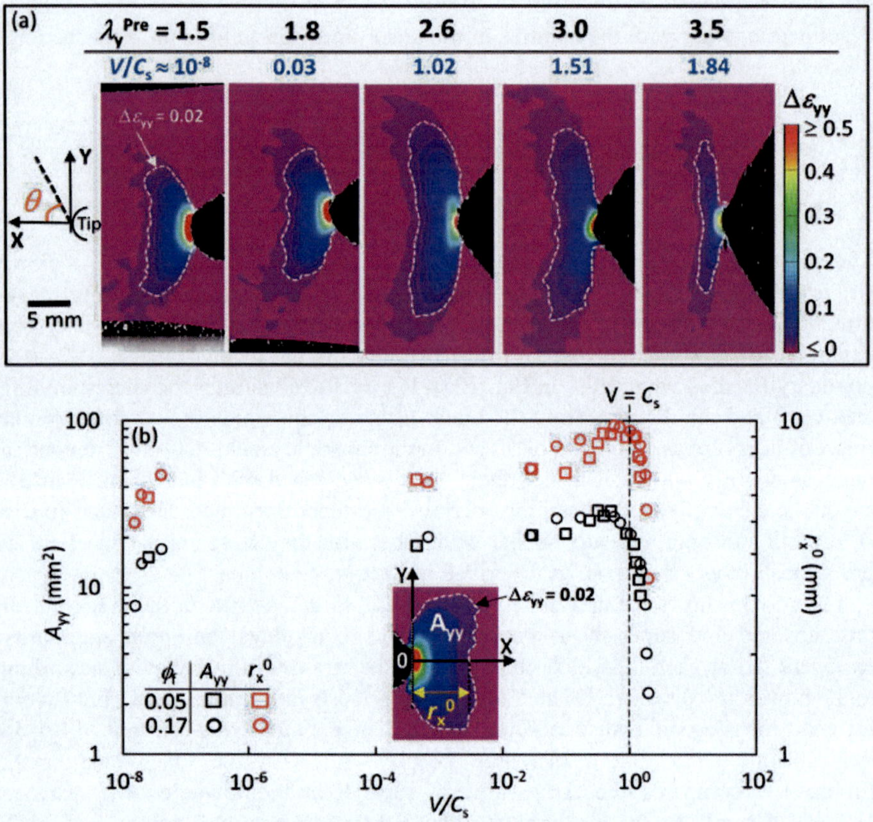

Fig. 12 (**a**) 2D crack-tip strain field of $\Delta\varepsilon_{yy}$ at various λ_y^{pre} or V for softened SBR/CB specimens with $\phi_f = 0.17$. The dashed white curve in each image is the isostrain contour of $\Delta\varepsilon_{yy} = 0.02$. (**b**) A_{yy} and r_x^0 as a function of V/C_s for the specimens of $\phi_f = 0.05$ and 0.17. The definitions of A_{yy} and r_x^0 are schematically illustrated in the inset of (**b**). The data are reproduced from Ref. [36]

far away during the super-shear crack propagation. The rubber networks in the damage zone ($r < r_x^*$) undergo rupture immediately (shock-like fracture), accompanied by a significantly small viscoelastic dissipation. The super-shear crack generates a shock wave in the form of a Mach cone [3, 27, 29–32], which appears as a wedge-like crack shape (Fig. 12a). The total input energies are mostly released through contraction, accompanied by energetic vibration of the material behind the crack tip. The movie in the Supporting Information in Ref. [36] shows super-shear crack propagation accompanied by vibration.

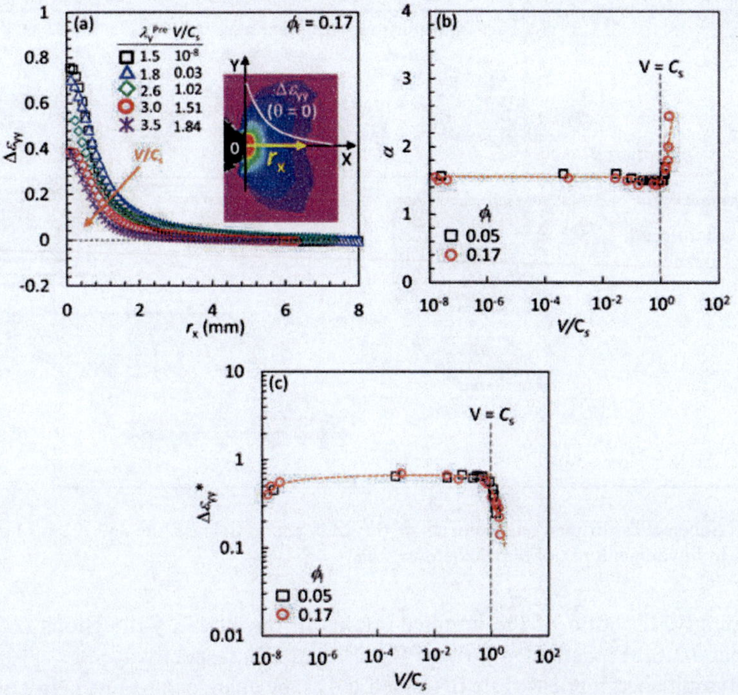

Fig. 13 (**a**) Plots of $\Delta\varepsilon_{yy}$ at Y = 0 as a function of r_x at various V/C_s for the specimen where $\phi_f = 0.17$. (**b**) Singularity exponent α ($\Delta\varepsilon_{yy} \sim r_x^{-\alpha}$) and the $\Delta\varepsilon_{yy}$ value at $r_x = r_x^* \approx 0.6$ mm ($\Delta\varepsilon_{yy}^*$) as a function of V/C_s for the specimens where $\phi = 0.05$ and 0.17. The orange dashed lines in (**b**) and (**c**) are guides for the eyes. The data in (**a**) and (**b**) are reproduced from Ref. [36]

6 Biaxial Loading Effect on Quasi-Stationary Crack

6.1 Strain Energy Release Rate Under Biaxial Loading

The strain energy release rate under biaxial loading was investigated using a single-edge notched SBR/CB ($\phi_f = 0.05$) sheet specimen with dimensions of 65 × 65 × 1.0 mm (Fig. 14). The stress-softening effect was very small because of the low filler content, which facilitated the measurements and data interpretation. A notch was made along the *x*-axis from one side. The notch length (X_c) represents the distance between the crack tip and center of the specimen. The notched specimens with various X_c values ($-20 < X_c < +15$ mm) were deformed by three types of biaxial stretching, i.e., planar extension (PE), unequal biaxial extension (UB), and equibiaxial extension (EB). The PE, UB, and EB were conducted at 25°C using a custom-built biaxial tensile tester [22] with a crosshead speed in the *x*-direction of $V_x = 0$, 0.15, or 0.3 mm s^{-1}, respectively, while $V_y = 0.3$ mm s^{-1} for each stretching.

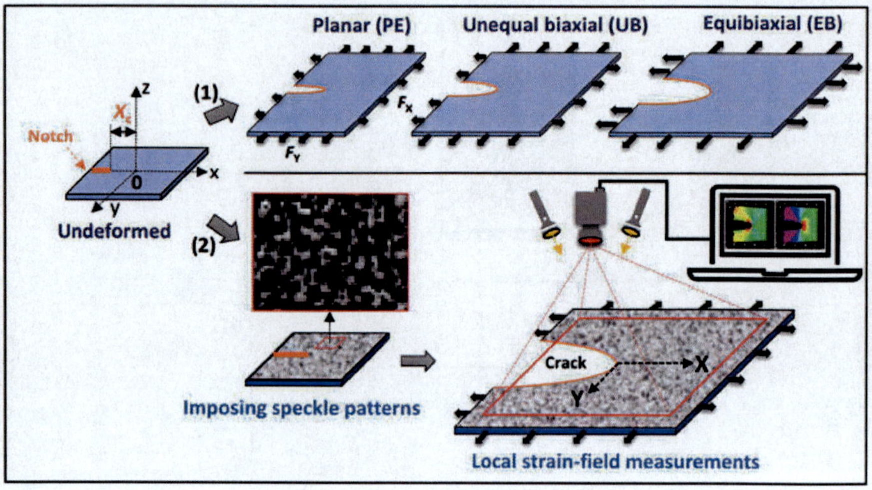

Fig. 14 Schematics for the measurements of (1) strain energy release rate and (2) local crack-tip strain field in various types of biaxial deformation

Accordingly, the ratio of the nominal strains in the x- and y-directions (E_x/E_y) is 0, 0.5, and 1.0, respectively, during PE, UB, and EB, respectively.

The strain energy release rate (Γ) is defined as the change in strain energy (w) with respect to X_c under constant displacement [11, 47, 74]. The expression of Γ for the biaxial deformation of the imposed nominal strains E_x and E_y is expressed by

$$\Gamma\left(E_x, E_y\right) = -\frac{1}{t}\left[\frac{\mathrm{d}w\left(X_c, U_x, U_y\right)}{\mathrm{d}X_c}\right]_{U_x, U_y} \tag{3}$$

where t is the thickness of the initial specimen. Figure 15a–c shows w as a function of displacement U_y for various values of X_c for each deformation. The w for each deformation was evaluated from the tensile force F_i–U_i (i = x, y) data using the relationship $w = w_x + w_y = \int F_x \mathrm{d}U_x + \int F_y \mathrm{d}U_y$, which is schematically illustrated in the inset of Fig. 15a. The cross symbols in the figure represent the onset of crack growth. Figure 15d–f illustrates the w-X_c relationship for each displacement obtained from the data in Fig. 15a–c. The relationship is well approximated by linear regression, allowing the evaluation of the dw/dX_c value at each displacement in Eq. (3) from the slope [11, 47, 74] The Γ value for each deformation was obtained using this procedure.

Figure 16a shows the double logarithmic plots of Γ vs. the imposed nominal strain E_y for each deformation. The corresponding data for an elastic poly N, N-dimethylacrylamide (PDMA) hydrogel obtained by the same method [46] are also shown in the figure. For each type of deformation, Γ increased with increasing E_y. When compared at the same E_y, Γ increases with increasing E_x, i.e., Γ(PE) < Γ (UB) < Γ (EB). A qualitatively similar dependence of Γ on the deformation type was

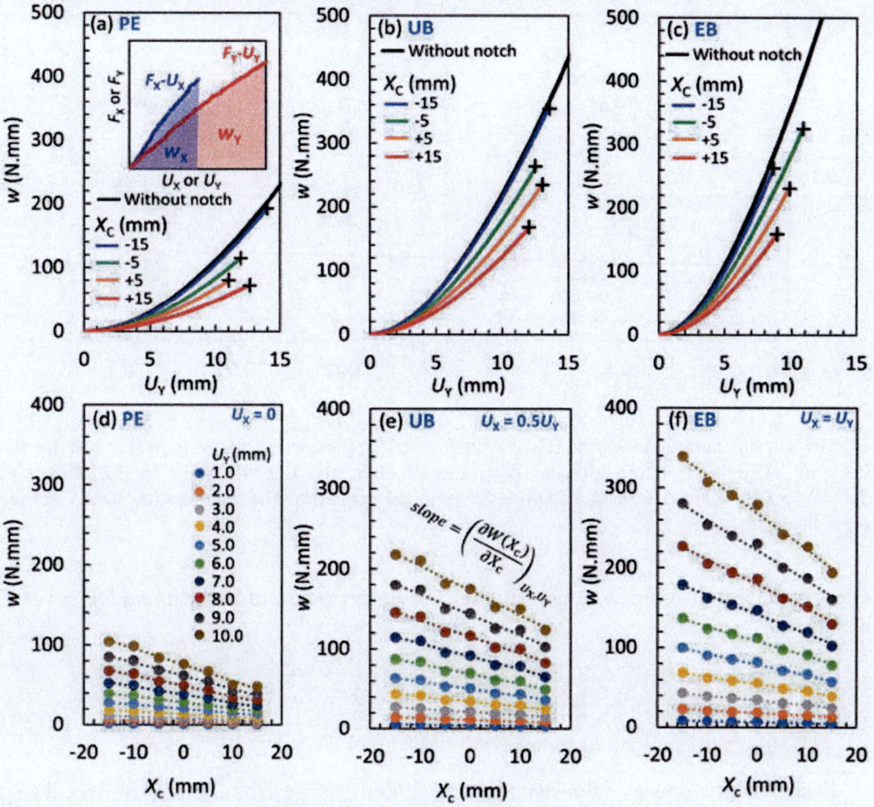

Fig. 15 **(a–c)** The U_y dependences of mechanical work (w) for SBR/CB with various X_c values in each extension type. **(d–f)** The X_c dependences of w for each deformation of (U_x, U_y) in each extension type. The cross symbol in **(a–c)** represents the onset of crack growth

observed in a simulation of hyperelastic materials [75]. The dependence of the Γ values on both E_x and E_y indicates the finite effect of strain biaxiality on Γ, which will be discussed later.

The quantity Γ in each biaxial deformation can be evaluated using another approach. Fracture mechanics originally defines Γ as the difference in the strain energy between the states with crack (far behind the crack tip) and without crack (far ahead of the crack-tip) under the same imposed deformation [2, 4, 11, 47]. Correspondingly, we can consider the strain energy difference between the following two types of deformation: one is the biaxial stretching of the specimens without a notch, corresponding to the state without a crack. The other is a type of constrained uniaxial extension in which one side is constrained by clamping, while the other side has no constraint (inset of Fig. 17d). This deformation, which is called residual extension (RE) for convenience, corresponds to the state after the crack passes. The strain

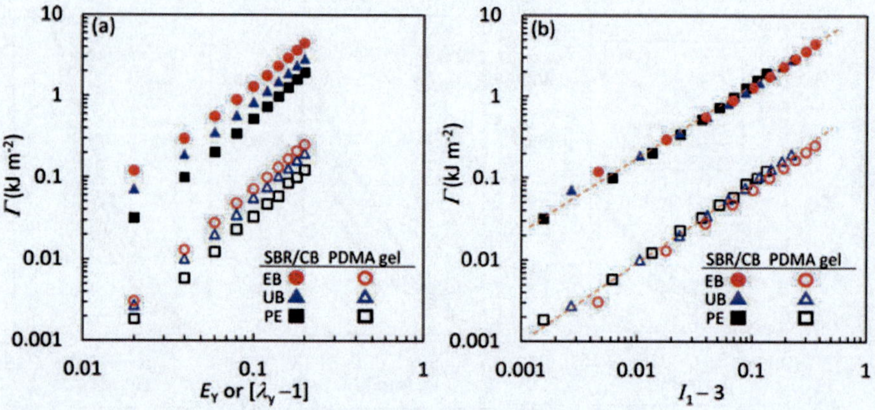

Fig. 16 Strain energy release rate (Γ) as a function of (**a**) imposed nominal strain (E_y) and (**b**) first invariant of gradient deformation tensor (I_1) in PE, UB, and EB extensions for SBR/CB with $\phi_f = 0.05$ (filled symbols) and a PDMA gel (unfilled symbols). The data for the PDMA gel are reproduced from Ref. [46]

energy density difference (ΔW) between the deformation of interest and RE gives Γ as

$$\frac{\Gamma\left(E_x, E_y\right)}{L_{y,0}} = \Delta W\left(E_x, E_y\right) \equiv W\left(E_x, E_y\right) - W_{\mathrm{RE}}(E_x) \tag{4}$$

Each strain energy density is calculated using the relationships $W(E_x, E_y) = W_x + W_y = \int \sigma_x \mathrm{d}E_x + \int \sigma_y \mathrm{d}E_y$ or $W_{\mathrm{RE}}(E_x) = \int \sigma_x \mathrm{d}E_x$, using the stress-strain relation of each deformation shown in Fig.17. In PE, $A_{\mathrm{RE}} = 0$ because $E_x = 0$, resulting in a familiar relationship of $\Gamma = WL_{y,0}$ [11, 47, 74]. Figure 18 shows the $\Gamma/L_{y,0}$ plots obtained from Eq. (3) vs. ΔW obtained from Eq. (4) for various deformations of the SBR/CB and PDMA gels. All data points in each specimen fall on a straight line with a slope of unity, ensuring the agreement of the Γ values obtained from the two independent methods. This agreement also signifies that the Γ values for general biaxial loading are straightforwardly obtained by the two types of biaxial stretching tests, without the laborious experiments varying X_c with Eq. (3).

Figure 16a shows the finite effect of strain biaxiality on Γ: Γ depending on both E_x and E_y. As a variable to express the total degree of deformation, we have the first and second invariants of the gradient deformation tensor (I_1 and I_2, respectively), and their combinations, $I_1 = \lambda_x^2 + \lambda_y^2 + \lambda_z^2$ and $I_2 = \lambda_x^2\lambda_y^2 + \lambda_y^2\lambda_z^2 + \lambda_z^2\lambda_x^2$, where λ_i ($i = x, y, z$) is the principal ratio in the i-direction, i.e., $\lambda_i = L_i/L_{i,0} \equiv E_i + 1$. Figure 16b illustrates the double logarithmic plots of Γ for various deformations against $I_1 - 3$. The λ_z value for each deformation was calculated from the volume conservation ($\lambda_x\lambda_y\lambda_z = 1$). For each specimen, the data for various deformations approximately fell on a straight line with a slope of unity. This result indicates a simple linear relationship between Γ and I_1, i.e., $\Gamma \sim (I_1 - 3)^1$. For each specimen, it was

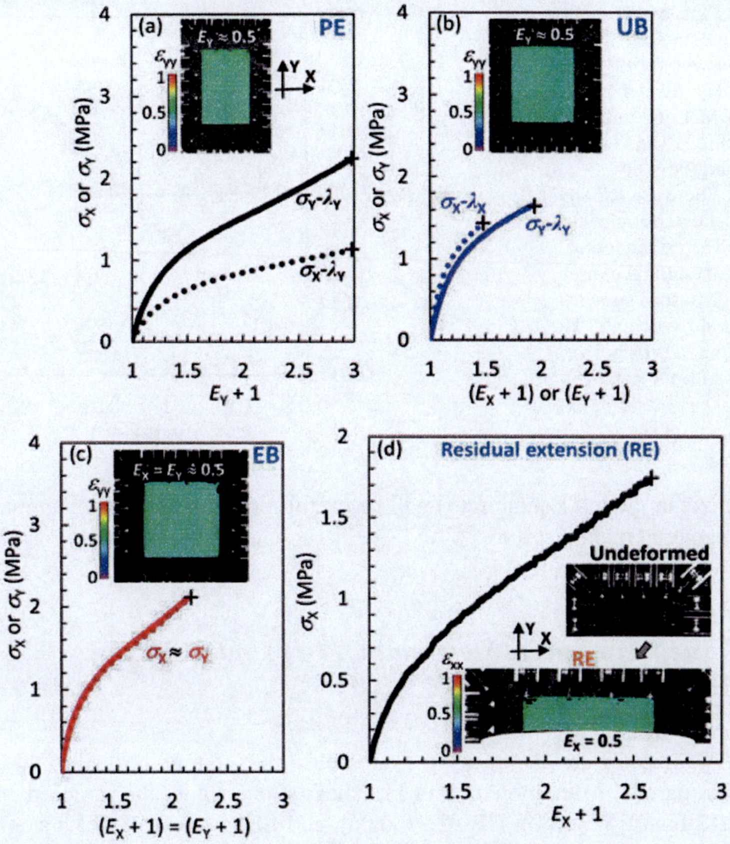

Fig. 17 Nominal stresses (σ_i; $i = x, y$) as a function of E_i for SBR/CB without notches in **(a)** PE, **(b)** UB, **(c)** EB extensions, and **(d)** residual extension (RE). The cross symbol indicates the fracture point. The insets indicate the uniformity of strain field of ε_{yy} for PE, UB, and EB at $E_y = 0.5$, and of ε_{xx} for RE at $E_x = 0.5$

confirmed that the proportional constant of this relation corresponds to $GL_{y,0}/2$, where G is the linear shear modulus. Thus, we obtain the following expression for Γ for the general biaxial strain:

$$\frac{\Gamma(E_x, E_y)}{L_{y,0}} = \frac{G}{2}(I_1 - 3) \tag{5}$$

Equation (5) agrees with the strain energy density function of a Neo-Hookean solid, which is the simplest model of rubber elasticity. The strain range examined here is modest ($I_1 < 3.4$) because it is inevitably limited to below the onset strain of crack propagation (Fig. 15a–c). In general, in the corresponding small strain range, the stress-strain relationships of most elastomers and gels approximately obey the

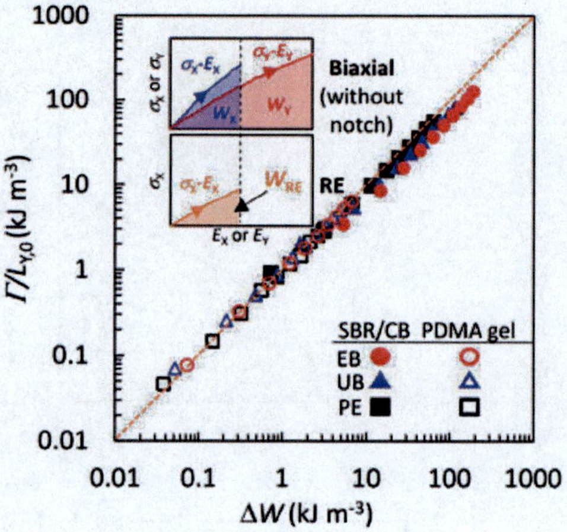

Fig. 18 Plots of $\Gamma/L_{y,0}$ vs. ΔW in each extension type for SBR/CB ($\phi_f = 0.05$) (filled symbols) and a PDMA gel (unfilled symbols). ΔW (Eq. 4) is the strain energy density difference between RE and the biaxial extension of interest. The estimation of W is illustrated in the inset. An orange dashed straight line has a slope of unity. The data for the PDMA gel are reproduced from Ref. [46]

Neo-Hookean model. Equation (5) will be useful for estimating the magnitude of Γ under a general biaxial strain.

6.2 Crack-Tip Strain Field and CTOD Under Biaxial Loading

Figure 19 indicates the 2D crack-tip strain fields of ε_{yy} (strain normal to the notch), ε_{xx} (strain parallel to the notch), and ε_{xy} (shear strain) under the imposed strain of $E_y = 0.12$ for PE ($E_x = 0$), UB ($E_x = 0.06$), and EB ($E_x = 0.12$) for a single-edge notched SBR/CB with $X_c = -5$ mm. The engineering strain ε_{ij} was measured using the undeformed state as the reference state. Each local strain increased with an increase in E_x. Three characteristic regions are observed: (I) far ahead of the crack tip, (II) near the crack tip, and (III) far behind the crack tip. Region (I), without the effect of crack opening, has a uniform strain field that is identical to the imposed strain field. In Region (III), even after the passage of the crack, ε_{xx} is positive for UB and EB, while $\varepsilon_{xx} \approx 0$ for PE. This explains the finite residual strain energy density (W_{RE}) in Eq. (4) for UB and EB after the passage of the crack. The contraction in the y-direction ($\varepsilon_{yy} < 0$) in Region (III) is caused by unloading. Region (II) has inhomogeneous strain fields because of the crack-tip opening. In Region (II) ahead of the crack tip, ε_{yy} is significantly larger than ε_{xx} and ε_{xy}. The ε_{yy} steeply increased near the crack tip.

The extent of strain (A_{yy}; area enclosed by the isostrain contour of $\Delta\varepsilon_{yy} = 0.02$), CTOD, and the singularity exponent (α; $\varepsilon_{yy} \sim r_x^{-\alpha}$) in each deformation were evaluated using the same procedures as in Sects. 4.3 and 5.2. Each of these quantities is plotted against E_y in Fig. 20a–c for SBR/CB. The corresponding data for a PDMA gel [46] are also shown in the figure. Every quantity depends not only on E_y, but also on E_x, indicating the pronounced effect of strain biaxiality. Importantly, when Γ is

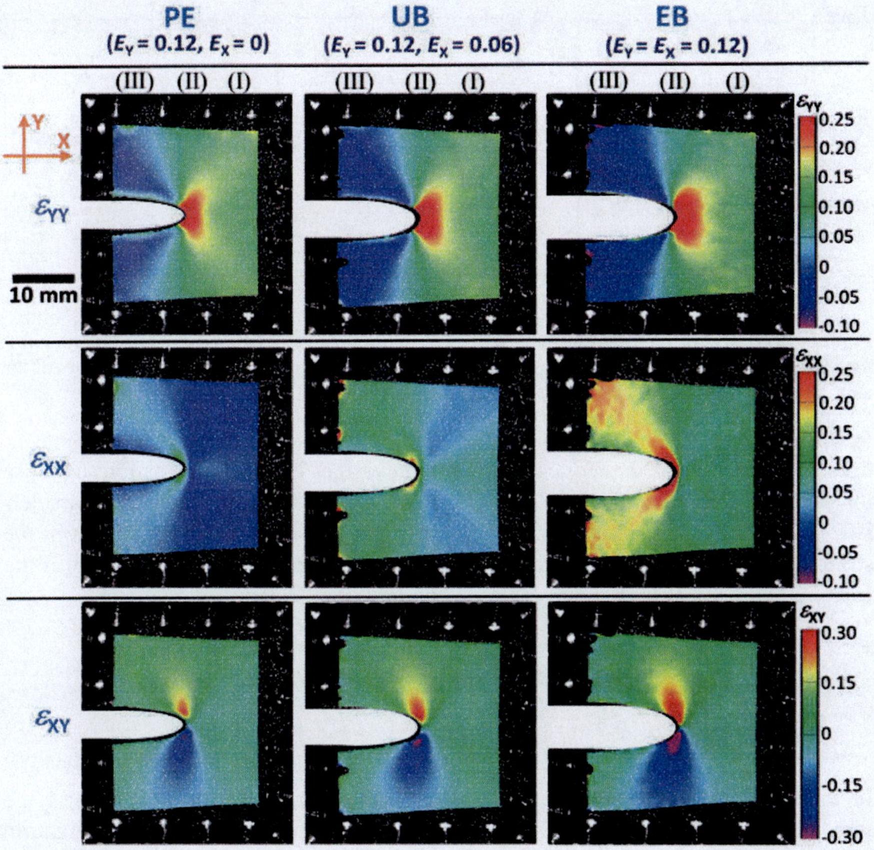

Fig. 19 2D local crack-tip strain fields of ε_{yy} (strain normal to the notch), ε_{xx} (strain parallel to the notch), and ε_{xy} (shear strain) at $E_y = 0.12$ for the stationary cracks in PE ($E_x = 0$), UB ($E_x = 0.06$), and EB ($E_x = 0.12$) extensions

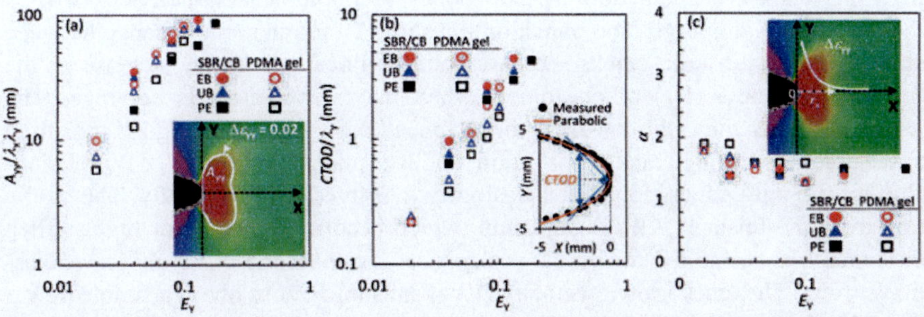

Fig. 20 **(a)** Extent of finite strain (A_{yy}), **(b)** CTOD, and **(c)** singular exponent α for ε_{yy} as a function of E_y for SBR/CB and PDMA gel. The data for the PDMA gel are reproduced from Ref. [46]

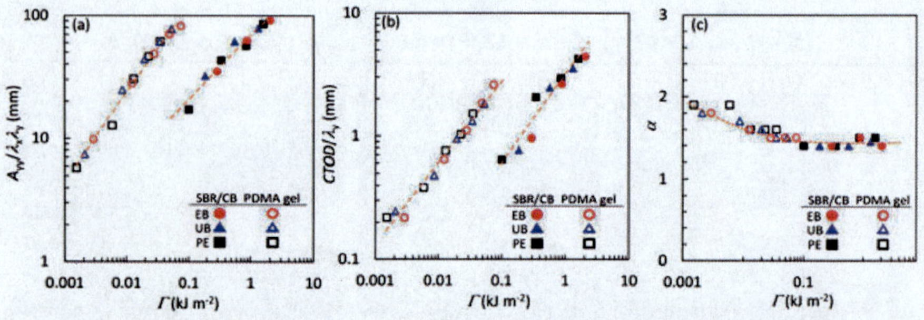

Fig. 21 (**a**) Extent of finite strain (A_{yy}), (**b**) CTOD, and (**c**) singular exponent α as a function of Γ for SBR/CB and PDMA gel. The orange dashed lines in each figure indicate the matter curves of the data. The data for a PDMA gel in (**b**) are reproduced from Ref. [46]

used as a variable, the data for each property in various biaxial deformations collapse into a single curve for each material (Fig. 21a–c). This signifies that the characteristics of the crack-tip strain field are governed exclusively by Γ, independent of the strain biaxiality. This simple result provides a comprehensive understanding of the fracture behavior of elastic soft materials under complicated deformations.

7 Conclusion

In this section, we summarize the conclusions of each section and discuss the remaining issues. We also highlight several issues that require further investigation with respect to the crack analysis of elastomers.

Section 3 describes the differences in the crack tip characteristics between slow- and fast-moving cracks. The non-parabolic feature of crack-tip profiles becomes pronounced as the strain energy release rate (Γ) increases. The deviation from the parabolic shape is accounted for by the second-order nonlinear elasticity effect when the deviation is not large. The transition threshold Γ_c for an abrupt change between slow and fast subsonic cracks increases almost linearly with an increase in the fracture toughness (W_c) of elastomers when the filler content is not high. The model attributes the origin of the velocity transition to the dynamic glass transition at the crack tip during crack propagation, and it explains the linear $\Gamma_c - W_c$ relationship for the unfilled and low-filled elastomers observed experimentally. The deviation from the linear $\Gamma_c - W_c$ relationship, which becomes obvious for highly filled elastomers, is accounted for by the emergence of the second-order nonlinear elasticity effect. The crack growth velocity (V) of subsonic cracks obeys a simple power law, $V \sim \Gamma^{1/\beta}$, and the β value depends on the filler and cross-linker concentrations and the type of rubber. The β value has a universal correlation with the characteristic exponent (κ) of the viscoelastic spectrum in the glass-rubber transition regime [$G(t) \sim t^{-\kappa}$)], which was confirmed for elastomers with various filler and cross-linker concentrations and those composed of chemically dissimilar polymers. However, no existing theory accounts for the universal β-κ relationship obtained experimentally.

The effects of anisotropic stress softening (Mullins effect), which is pronounced in highly filled elastomers, on various aspects of crack propagation are described in Sect. 4. The softening by the previous loading, which was imposed normally to the crack growth (x-) direction, results in an increase in the magnitude and extent of local strain in the previous loading direction ($\Delta\varepsilon_{yy}$), as well as a reduction in Γ_c. In contrast, the previous softening has no appreciable effect on $\Delta\varepsilon_{xx}$ and the crack-tip shape, suggesting that the strain field of $\Delta\varepsilon_{xx}$ behind the crack tip governs the crack-tip shape. The exponent β ($V \sim \Gamma^{1/\beta}$) is also insensitive to the previous loading, which is explained by the fact that softening has no effect on the viscoelastic spectra in the glass-rubber transition regime. Moreover, the previous loading has no significant influence on the exponent α ($\Delta\varepsilon_{yy} \sim {r_x}^{-\alpha}$), indicating that the same degree of softening in the y-direction occurs everywhere near the crack tip.

According to the transition from subsonic ($V < C_s$) to super-shear ($V > C_s$) cracks, the dependence of various quantities on V exhibits a definite crossover at $V \approx C_s$ (Sect. 5). The exponent β (≈ 1.8) at $V > C_s$ is independent of the filler content, whereas the β values at $V < C_s$ are dependent on it. The magnitude and extent of $\Delta\varepsilon_{yy}$ are not significantly affected by V in the subsonic regime, whereas they markedly decrease with an increase in V in the super-shear regime. As V increases, the exponent α remains unchanged in the subsonic crack, but α steeply increases in the super-shear crack. The crossover behavior in these quantities reflects the difference in the propagation mechanism between subsonic and super-shear cracks. While viscoelastic dissipation plays an important role in the subsonic crack, the inertia effect comes into play in the super-shear crack. The input energies in the super-shear cracks are dissipated through the contraction of the material behind the crack tip accompanied by a vibration, with negligibly small viscoelastic dissipation. Further theoretical development explicitly considering the inertia effects is required to quantitatively understand the propagation mechanism of super-shear cracks.

Section 6 describes the biaxial loading effect on quasi-stationary cracks in elastomers. The strain energy release rate (Γ) in general biaxial loading was experimentally quantified using two independent methods. The Γ values at various degrees and types of biaxial strain linearly depend on the first invariant (I_1) of the deformation gradient tensor, indicating that the strain dependence of Γ obeys the Neo-Hookean model, which is commonly observed for low-filled elastomers and elastic hydrogels. This finding provides a useful estimate of the magnitude of Γ under a general deformation. The crack-tip shape and local crack-tip strain fields depend on the strain biaxiality, but they are exclusively governed by Γ. This simple feature provides an important basis for the nonlinear fracture mechanics of elastomers under general deformation.

Finally, we raise several issues that need to be addressed. Nonlinear elasticity and viscoelasticity specific to rubber materials provide intriguing and complicated aspects of crack growth phenomena, which require the development of theories for interpretation. The weakly nonlinear fracture theory (WNLT) considering the second-order nonlinear effect is successful for explaining the results when the deviation from the linear elastic fracture mechanics (LEFM) is small (Sect. 3.2). However, the development of a theory that considers the higher-order nonlinear effect is required to interpret the crack-tip features of a subsonic crack with high velocity. The bulk viscoelasticity is closely related to the velocity of fast crack growth (Sect. 3.3), but the effect of viscoelasticity on the crack-tip features, such as the crack-tip profile and strain field, remains unclear. The elucidation requires a systematic experimental survey using elastomer specimens with various and known degrees of viscous components.

The biaxial loading effect on quasi-stationary cracks was characterized (Sect. 6), but the crack growth dynamics under general biaxial loading is still a challenging issue. For instance, an interesting open question relates to how strain/loading biaxiality affects the crack growth direction and crack-tip properties. This investigation will provide an important basis for crack growth dynamics in elastomers and for the practical design of rubber materials.

In reality, elastomers exhibiting strain-induced crystallization (SIC), such as natural rubber (cis1,4-polyisoprene rubber), have often been employed for practical use, requiring high reliability and safety. The occurrence of SIC near a crack tip subjected to large deformation has been experimentally detected for quasi-stationary cracks or dynamic cracks under imposed cyclic strain [76–85]. It has been considered that the SIC near the crack tip can reinforce the resistance against crack growth. In fast-moving cracks, the competition between the time required for SIC and the crack growth rate is expected to govern the crack dynamics, but the crack growth mechanism of elastomers undergoing SIC is still a matter of investigation. Further investigations of these remaining issues will deepen the understanding of crack growth phenomena in elastomers.

Acknowledgments This work was supported by JST, CREST grant number JPMJCR2091, Japan, and the ImPACT Program of the Council for Science, Technology and Innovation (Cabinet Office, Government of Japan).

References

1. Rodgers B, Waddell W (2013) Tire engineering. In: Mark JE, Erman B, Roland M (eds) The science and technology of rubber4th edn. Elsevier, Waltham, pp 653–695
2. Freund LB (1990) Dynamic fracture mechanics. Cambridge University Press, Cambridge
3. Ravi-Chandar K (2004) Dynamic fracture. Elsevier
4. Anderson TL (2017) Fracture mechanics. CRC Press
5. Persson BNJ, Albohr O, Heinrich G, Ueba H (2005) Crack propagation in rubber-like materials. J Phys Condens Matter 17. https://doi.org/10.1088/0953-8984/17/44/R01
6. Long R, Hui C-Y (2015) Crack tip fields in soft elastic solids subjected to large quasi-static deformation – a review. Extrem Mech Lett 4:131–155. https://doi.org/10.1016/j.eml.2015.06.002
7. Creton C, Ciccotti M (2016) Fracture and adhesion of soft materials: a review. Rep Prog Phys 79:046601. https://doi.org/10.1088/0034-4885/79/4/046601

8. Kadir A, Thomas AG (1981) Tear behavior of rubbers over a wide range of rates. Rubber Chem Technol 54:15–23. https://doi.org/10.5254/1.3535791
9. Horst T, Heinrich G (2008) Crack propagation behavior in rubber materials. Polym Sci Ser A 50:583–590. https://doi.org/10.1134/S0965545X08050131
10. Tsunoda K, Busfield JJC, Davies CKL, Thomas AG (2000) Effect of materials variables on the tear behaviour of a non-crystallizing elastomer. J Mater Sci 35:5187–5198. https://doi.org/10.1023/A:1004860522186
11. Greensmith HW, Thomas AG (1955) Rupture of rubber. III. Determination of tear properties. J Polym Sci 18:189–200. https://doi.org/10.1002/pol.1955.120188803
12. Greensmith HW (1956) Rupture of rubber. IV. Tear properties of vulcanizates containing carbon black. J Polym Sci 21:175–187. https://doi.org/10.1002/pol.1956.120219802
13. Persson BNJ, Brener EA (2005) Crack propagation in viscoelastic solids. Phys Rev E Stat Nonlinear Soft Matter Phys 71:1–8. https://doi.org/10.1103/PhysRevE.71.036123
14. Morishita Y, Tsunoda K, Urayama K (2016) Velocity transition in the crack growth dynamics of filled elastomers: contributions of nonlinear viscoelasticity. Phys Rev E 93:043001. https://doi.org/10.1103/PhysRevE.93.043001
15. Morishita Y, Tsunoda K, Urayama K (2017) Crack-tip shape in the crack-growth rate transition of filled elastomers. Polymer 108:230–241. https://doi.org/10.1016/j.polymer.2016.11.041
16. Morishita Y, Tsunoda K, Urayama K (2019) Universal relation between crack-growth dynamics and viscoelasticity in glass-rubber transition for filled elastomers. Polymer 179:121651. https://doi.org/10.1016/j.polymer.2019.121651
17. Heinrich G, Klüppel M, Vilgis TA (2002) Reinforcement of elastomers. Curr Opin Solid State Mater Sci 6:195–203. https://doi.org/10.1016/S1359-0286(02)00030-X
18. Donnet J-B, Custodero E (2013) Reinforcement of elastomers by particulate fillers. In: Mark JE, Erman B, Roland M (eds) The science and technology of rubber4th edn. Elsevier, Waltham, pp 367–400
19. Mullins L, Tobin NR (1957) Theoretical model for the elastic behavior of filler-reinforced vulcanized rubbers. Rubber Chem Technol 30:555–571. https://doi.org/10.5254/1.3542705
20. Mullins L (1969) Softening of rubber by deformation. Rubber Chem Technol 42:339–362. https://doi.org/10.5254/1.3539210
21. Diani J, Fayolle B, Gilormini P (2009) A review on the Mullins effect. Eur Polym J 45:601–612. https://doi.org/10.1016/j.eurpolymj.2008.11.017
22. Mai T-T, Morishita Y, Urayama K (2017) Novel features of the Mullins effect in filled elastomers revealed by stretching measurements in various geometries. Soft Matter 13:1966–1977. https://doi.org/10.1039/C6SM02833K
23. MacHado G, Chagnon G, Favier D (2012) Induced anisotropy by the Mullins effect in filled silicone rubber. Mech Mater 50:70–80
24. Mai T-T, Morishita Y, Urayama K (2017) Induced anisotropy by Mullins effect in filled elastomers subjected to stretching with various geometries. Polymer 126:29–39. https://doi.org/10.1016/j.polymer.2017.08.012
25. Mai T-T, Matsuda T, Nakajima T et al (2019) Damage cross-effect and anisotropy in tough double network hydrogels revealed by biaxial stretching. Soft Matter 15:3719–3732. https://doi.org/10.1039/C9SM00409B
26. Mai T-T, Okuno K, Tsunoda K, Urayama K (2021) Anisotropic stress-softening effect on fast dynamic crack in filler-reinforced elastomers. Mech Mater 155:103786. https://doi.org/10.1016/j.mechmat.2021.103786
27. Marder M (2006) Supersonic rupture of rubber. J Mech Phys Solids 54:491–532. https://doi.org/10.1016/j.jmps.2005.10.002
28. Buehler MJ, Abraham FF, Gao H (2003) Hyperelasticity governs dynamic fracture at a critical length scale. Nature 426:141–146. https://doi.org/10.1038/nature02096
29. Petersan P, Deegan R, Marder M, Swinney H (2004) Cracks in rubber under tension exceed the shear wave speed. Phys Rev Lett 93:015504. https://doi.org/10.1103/PhysRevLett.93.015504

30. Chen CH, Zhang HP, Niemczura J et al (2011) Scaling of crack propagation in rubber sheets. EPL 96. https://doi.org/10.1209/0295-5075/96/36009
31. Rosakis AJ (1999) Cracks faster than the shear wave speed. Science 284:1337–1340. https://doi.org/10.1126/science.284.5418.1337
32. Marder M (2005) Shock-wave theory for rupture of rubber. Phys Rev Lett 94:1–4. https://doi.org/10.1103/PhysRevLett.94.048001
33. Guozden TM, Jagla EA, Marder M (2010) Supersonic cracks in lattice models. Int J Fract 162: 107–125. https://doi.org/10.1007/s10704-009-9426-4
34. Kroon M (2011) Steady-state crack growth in rubber-like solids. Int J Fract 169:49–60. https://doi.org/10.1007/s10704-010-9583-5
35. Kroon M (2014) Energy release rates in rubber during dynamic crack propagation. Int J Solids Struct 51:4419–4426. https://doi.org/10.1016/j.ijsolstr.2014.09.010
36. Mai T, Okuno K, Tsunoda K, Urayama K (2020) Crack-tip strain field in Supershear crack of elastomers. ACS Macro Lett 9:762–768. https://doi.org/10.1021/acsmacrolett.0c00213
37. Urayama K (2006) An experimentalist's view of the physics of rubber elasticity. J Polym Sci B Polym Phys 44:3440–3444. https://doi.org/10.1002/polb.21010
38. Adams NJI (1973) Some comments on the effect of biaxial stress on fatigue crack growth and fracture. Eng Fract Mech 5:983–991. https://doi.org/10.1016/0013-7944(73)90063-5
39. Liebowitz H, Lee JD, Eftis J (1978) Biaxial load effects in fracture mechanics. Eng Fract Mech 10:315–335. https://doi.org/10.1016/0013-7944(78)90015-2
40. Eftis J, Jones DL, Liebowitz H (1990) Load biaxiality and fracture: synthesis and summary. Eng Fract Mech 36:537–574. https://doi.org/10.1016/0013-7944(90)90112-T
41. Marano C, Calabrò R, Rink M (2010) Effect of molecular orientation on the fracture behavior of carbon black-filled natural rubber compounds. J Polym Sci B 48:1509–1515. https://doi.org/10.1002/polb.22054
42. Caimmi F, Calabrò R, Briatico-Vangosa F et al (2015) Toughness of natural rubber compounds under biaxial loading. Eng Fract Mech 149:250–261. https://doi.org/10.1016/j.engfracmech.2015.08.003
43. Schneider K, Calabrò R, Lombardi R et al (2017) Grellmann W, Langer B (eds) Characterisation of the deformation and fracture behaviour of elastomers under biaxial deformation. Springer, Cham, pp 335–349
44. Ahmad D, Sahu SK, Patra K (2019) Fracture toughness, hysteresis and stretchability of dielectric elastomers under equibiaxial and biaxial loading. Polym Test 79:106038. https://doi.org/10.1016/j.polymertesting.2019.106038
45. Dedova S, Schneider K, Stommel M, Heinrich G (2021) Dissipative heating, fatigue and fracture behaviour of rubber under multiaxial loading. Adv Polym Sci 286:421–443. https://doi.org/10.1007/12_2020_75
46. Mai T, Urayama K (2021) Biaxial loading effects on strain energy release rate and crack-tip strain field in elastic hydrogels. Macromolecules. https://doi.org/10.1021/acs.macromol.1c00445
47. Rivlin RS, Thomas AG (1953) Rupture of rubber. I. Characteristic energy for tearing. J Polym Sci 10:291–318. https://doi.org/10.1002/pol.1953.120100303
48. Livne A, Bouchbinder E, Svetlizky I, Fineberg J (2010) The near-tip fields of fast cracks. Science 327:1359–1363. https://doi.org/10.1126/science.1180476
49. Livne A, Bouchbinder E, Fineberg J (2008) Breakdown of linear elastic fracture mechanics near the tip of a rapid crack. Phys Rev Lett 101:1–4. https://doi.org/10.1103/PhysRevLett.101.264301
50. Bouchbinder E, Livne A, Fineberg J (2010) Weakly nonlinear fracture mechanics: experiments and theory. Int J Fract 162:3–20. https://doi.org/10.1007/s10704-009-9427-3
51. Bouchbinder E, Livne A, Fineberg J (2008) Weakly nonlinear theory of dynamic fracture. Phys Rev Lett 101:2–5. https://doi.org/10.1103/PhysRevLett.101.264302
52. Treloar LRG (1975) The physics of rubber elasticity.3rd edn. Oxford University Press Inc., New York

53. Williams ML, Landel RF, Ferry JD (1955) The temperature dependence of relaxation mechanisms in amorphous polymers and other glass-forming liquids. J Am Chem Soc 77:3701–3707. https://doi.org/10.1021/ja01619a008
54. Gent AN, Lai S-M, Nah C, Wang C (1994) Viscoelastic effects in cutting and tearing rubber. Rubber Chem Technol 67:610–618. https://doi.org/10.5254/1.3538696
55. Carbone G, Persson BNJ (2005) Crack motion in viscoelastic solids: the role of the flash temperature. Eur Phys J E 17:261–281. https://doi.org/10.1140/epje/i2005-10013-y
56. Carbone G, Persson BNJ (2005) Hot cracks in rubber: origin of the giant toughness of rubberlike materials. Phys Rev Lett 95:9–12. https://doi.org/10.1103/PhysRevLett.95.114301
57. Carbone G, Persson BNJ (2005) Hot cracks in rubber: origin of the Giant toughness of rubberlike materials. Phys Rev Lett 95:114301. https://doi.org/10.1103/PhysRevLett.95.114301
58. D'Amico F, Carbone G, Foglia MM, Galietti U (2013) Moving cracks in viscoelastic materials: temperature and energy-release-rate measurements. Eng Fract Mech 98:315–325. https://doi.org/10.1016/j.engfracmech.2012.10.026
59. Knauss WG (2015) A review of fracture in viscoelastic materials. Int J Fract 196:99–146. https://doi.org/10.1007/s10704-015-0058-6
60. Kubo A, Sakumichi N, Morishita Y et al (2021) Dynamic glass transition dramatically accelerates crack propagation in rubberlike solids. Phys Rev Mater 5:073608. https://doi.org/10.1103/PhysRevMaterials.5.073608
61. Kubo A, Umeno Y (2017) Velocity mode transition of dynamic crack propagation in hyperviscoelastic materials: a continuum model study. Sci Rep 7:4–6. https://doi.org/10.1038/srep42305
62. Sakumichi N, Okumura K (2017) Exactly solvable model for a velocity jump observed in crack propagation in viscoelastic solids. Sci Rep 7:1–11. https://doi.org/10.1038/s41598-017-07214-8
63. Zhang T, Lin S, Yuk H, Zhao X (2015) Predicting fracture energies and crack-tip fields of soft tough materials. Extrem Mech Lett 4:1–8. https://doi.org/10.1016/j.eml.2015.07.007
64. Diani J, Brieu M, Batzler K, Zerlauth P (2015) Effect of the Mullins softening on mode I fracture of carbon-black filled rubbers. Int J Fract 194:11–18. https://doi.org/10.1007/s10704-015-0030-5
65. El Yaagoubi M, Juhre D, Meier J et al (2018) Tearing energy and path-dependent J-integral evaluation considering stress softening for carbon black reinforced elastomers. Eng Fract Mech 190:259–272. https://doi.org/10.1016/j.engfracmech.2017.12.029
66. Mzabi S, Berghezan D, Roux S et al (2011) A critical local energy release rate criterion for fatigue fracture of elastomers. J Polym Sci B 49:1518–1524. https://doi.org/10.1002/polb.22338
67. Diaz R, Diani J, Gilormini P (2014) Physical interpretation of the Mullins softening in a carbon-black filled SBR. Polymer 55:4942–4947. https://doi.org/10.1016/j.polymer.2014.08.020
68. Schreier H, Orteu J-J, Sutton MA (2009) Image correlation for shape, motion and deformation measurements. Springer, Boston
69. Zhang H, Scholz AK, De Crevoisier J et al (2015) Nanocavitation around a crack tip in a soft nanocomposite: a scanning microbeam small angle X-ray scattering study. J Polym Sci B 53: 422–429. https://doi.org/10.1002/polb.23651
70. Slootman J, Waltz V, Yeh CJ et al (2020) Quantifying rate- and temperature-dependent molecular damage in elastomer fracture. Phys Rev X 10:041045. https://doi.org/10.1103/PhysRevX.10.041045
71. Mihai LA, Goriely A (2017) How to characterize a nonlinear elastic material? A review on nonlinear constitutive parameters in isotropic finite elasticity. Proc R Soc A Math Phys Eng Sci 473:20170607. https://doi.org/10.1098/rspa.2017.0607
72. Huang Y, Gao H (2001) Intersonic crack propagation-part I: the fundamental solution. J Appl Mech Trans ASME 68:169–175. https://doi.org/10.1115/1.1357871

73. Bouchbinder E, Fineberg J, Marder M (2010) Dynamics of simple cracks. Annu Rev Condens Matter Phys 1:371–395. https://doi.org/10.1146/annurev-conmatphys-070909-104019
74. Thomas AG (1960) Rupture of rubber. VI. Further experiments on the tear criterion. J Appl Polym Sci 3:168–174. https://doi.org/10.1002/app.1960.070030805
75. Long R, Hui C-Y (2010) Effects of triaxiality on the growth of crack-like cavities in soft incompressible elastic solids. Soft Matter 6:1238. https://doi.org/10.1039/b917148g
76. Trabelsi S, Albouy PA, Rault J (2002) Stress-induced crystallization around a crack tip in natural rubber. Macromolecules 35:10054–10061. https://doi.org/10.1021/ma021106c
77. Saintier N, Cailletaud G, Piques R (2011) Cyclic loadings and crystallization of natural rubber: an explanation of fatigue crack propagation reinforcement under a positive loading ratio. Mater Sci Eng A 528:1078–1086. https://doi.org/10.1016/j.msea.2010.09.079
78. Brüning K, Schneider K, Roth SV, Heinrich G (2012) Kinetics of strain-induced crystallization in natural rubber studied by WAXD: dynamic and impact tensile experiments. Macromolecules 45:7914–7919. https://doi.org/10.1021/ma3011476
79. Brüning K, Schneider K, Heinrich G (2012) Deformation and orientation in filled rubbers on the nano- and microscale studied by X-ray scattering. J Polym Sci B 50:1728–1732. https://doi.org/10.1002/polb.23148
80. Brüning K, Schneider K, Heinrich G (2013) Grellmann W, Heinrich G, Kaliske M et al (eds) In-situ structural characterization of rubber during deformation and fracture. Springer, Berlin, pp 43–80
81. Brüning K, Schneider K, Roth SV, Heinrich G (2013) Strain-induced crystallization around a crack tip in natural rubber under dynamic load. Polymer 54:6200–6205. https://doi.org/10.1016/j.polymer.2013.08.045
82. Brüning K, Schneider K, Roth SV, Heinrich G (2015) Kinetics of strain-induced crystallization in natural rubber: a diffusion-controlled rate law. Polymer 72:52–58. https://doi.org/10.1016/j.polymer.2015.07.011
83. Samaca Martinez JR, Balandraud X, Toussaint E et al (2014) Thermomechanical analysis of the crack tip zone in stretched crystallizable natural rubber by using infrared thermography and digital image correlation. Polymer 55:6345–6353. https://doi.org/10.1016/j.polymer.2014.10.010
84. Rublon P, Huneau B, Verron E et al (2014) Multiaxial deformation and strain-induced crystallization around a fatigue crack in natural rubber. Eng Fract Mech 123:59–69. https://doi.org/10.1016/j.engfracmech.2014.04.003
85. Xiang F, Schneider K, Heinrich G (2020) New observations regarding fatigue crack paths and their fracture surfaces in natural rubber: influences of R-ratio and pre-load. Int J Fatigue 135:105508. https://doi.org/10.1016/j.ijfatigue.2020.105508

Adv Polym Sci (2023) 289: 273–292
https://doi.org/10.1007/12_2022_129

Published online: 25 July 2022

The Effect of Apparent Cross-Link Density on Cut and Chip Wear in Natural Rubber

M. Pöschl, R. Stoček, and P. Zádrapa

Contents

Abstract Natural rubber is a polymer that, by inducing crystallization at a certain level of stress, contributes significantly to reducing cut and chip (CC) damage to rubber articles when exposed to harsh conditions. This unique property is dependent on several factors, including the processing conditions, the cross-linking system and the type of additives used, resulting in varying apparent cross-link density (CLD) of the cross-linked CB filled rubber. Therefore, this work focuses on the systematic investigation of CC phenomena as a function of CLDs represented by conventional (CV), semi-efficient (SEV) and efficient (EV) cross-linking systems. Rubber samples based on different cross-linking systems were prepared by varying the concentration of the accelerator N-tert-butylbenzothiazolesulfonamide (TBBS) at a constant concentration of 2.5 phr sulfur as a cross-linking agent. The different CLDs were achieved by different concentration ratios (A/S) between accelerator (A) and sulfur (S), using A/S = 0.1, 0.3, 0.6 for the CV system, A/S = 0.7, 1.0, 1.5, 2.0, 2.5 for the SEV system and A/S = 3.0 for the EV system. First, the basic mechanical behaviour was presented as a function of CLD, with the optimal behaviour found in the range of 181–241 μmol × cm^{-3}. The CC resistance is independent of the CLD when the

M. Pöschl and P. Zádrapa
Tomas Bata University in Zlín, Centre of Polymer Systems, Zlín, Czech Republic

R. Stoček (✉)
PRL Polymer Research Lab s.r.o., Zlín, Czech Republic
e-mail: stocek@utb.cz

rubber specimens are loaded with a normal force of 100 N. However, at higher load, the optimal range of CLD decreases rapidly from 136 to 241 μmol × cm^{-3}. Furthermore, a significant influence of SIC on CC resistance was confirmed in the range of CLD from 181 to 241 μmol × cm^{-3}. Moreover, in the range of CLD from 181 to 241 μmol × cm^{-3} the predominant effect of NR on CC resistance was observed. Finally, an effect of degradation of cross-link network on CC properties due to rubber curing in the reversion has been discussed.

Keywords Accelerators · Cross-link density · Cut and chip wear · Laboratory testing · Natural rubber · Rubber · Strain induced crystallization · Sulfur

1 Introduction

When operating rubber products, such as tyres, conveyor belts, rubber tracks, are exposed to harsh severity conditions one observes Cut and Chip (CC) wear phenomena. Sometimes, these wear phenomena are also termed as CCC which stands for cut, chip and chunk wear field conditions [1]. In the case of tyres, CC wear refers to the detachment or breakage of rubber material from their treads when riding on a rough road surface (e.g. gravel roads, roots, stalks). This is most commonly associated with off-the-road tyres (OTR), agricultural, industrial, light truck (LTT), SUV as well as moto-cross/enduro (MX) tyres. Moreover, this phenomenon is also observed with highway commercial truck as well as all-season passenger car, touring moto-bike tyres and even in the multi-purpose tyres (MPT). In Fig. 1, photographs representing CC damage occurring on a worn tyre tread of highway commercial truck tyre, MX tyre as well as on MPT tyre which are accompanied by a photograph of a typical CC effect visible on the surface of a conveyor belt in a mine.

Beside the elastomer type and fillers, the cross-link density and distribution are important parameters which affect the physical, mechanical and viscoelastic

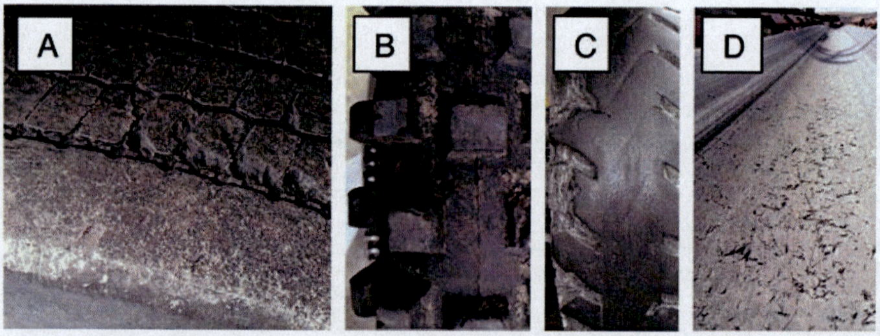

Fig. 1 The Cut and Chip wear phenomena of a truck tyre tread (**a**), MTX tyre (**b**), MPTR tyre (**c**) and a conveyor belt (**d**)

properties of a cured rubber [2–5]. In Refs. [6, 7], it has been shown that the introduction of carbon black filler increased the apparent cross-link density of conventional vulcanized rubber due to filler–filler or polymer–filler interaction during the cross-linking process. And since carbon black filled rubbers are investigated in this study, the effect of apparent cross-link density (CLD) is studied. Sulfur and organic peroxides are in general the two most commonly used vulcanizing agents, whereas elastomers used for tyres, conveyor as well as rubber tracks are exclusively cured with accelerated sulfur curing systems [8–10]. Sulfur curing made the dimensions of elastomer products more stable in a wide range of temperatures and conditions, and improved numerous physical and chemical properties. Further development in the elastomer technology was stimulated by the use of curing activators and accelerators [11]. These organic components served to shorten the curing time and additionally improve the properties. During the curing process, one of the critical steps is the formation of Monosulfidic (C-S-C), Disulfidic ($C\text{-}S_2\text{-}C$) and Polysulfidic ($C\text{-}S_x\text{-}C$) accelerator terminated groups, which are bound to polymer chains [10]. In further reactions, they recombine with other pendant groups or directly with the polymer chains [12, 13]. In this way, intermolecular cross-links are formed, which connect the chains. The number of cross-links per unit volume in a polymer network is called cross-link density (CLD) [14]. Type of accelerator and then the ratio of the amount of accelerator to sulfur (A/S) gives the CLD and the resulting designation for the curing system. High sulfur levels and low levels of accelerator, resulting in A/S ranging from 0.1 to 0.6 is generally described as conventional vulcanizing (CV) systems, which generally produce a CLD with 95% poly- and disulfidic cross-links and 5% monosulfidic with good mechanical properties [15–17]. The range of A/S from 0.7 to 2.5 based on intermediate sulfur and accelerator loadings is characteristic for semiefficient vulcanization (SEV) system, which is mainly formed by 50% poly- and disulfidic cross-links and the remaining bonds are monosulfidic [18, 19]. SEV was introduced to eliminate poor cut growth of the compounds based on natural rubber [20]. Low sulfur levels with high accelerator levels given A/S ratio over 2.5 stands for efficient vulcanization (EV) system, which is characterized by 80% monosulfidic and remaining part consists of poly- and disulfidic cross-links. Rubber with a high content of monosulfidic cross-link's is less elastic, but the energy dissipation is also lower, which makes this type of cross-links preferable for low rolling resistance tyre tread material. Furthermore, EV leads to low stress relaxation and good resistance to ageing. It is also important to mention that the bonding energy of sulfidic cross-links decreases with higher the number of sulfur atoms [21].

Unfortunately, higher CLD increases the crack propagation and worsens cut growth, which results in decrease in resistance against CC [22, 23]. Rubbers achieving elevated tensile strength values are in general resistant against cutting and because tensile strength is related with the CLD, whereas tensile strength passes through a maximum value with increasing CLD [24], it is expected the identical behaviour for CC.

The usual difficulties in predicting CC resistance of tyre treads from the tests of rubber compounds in the lab condition are often appropriate to resort to costly research programmes involving the iterative tyre design and in field testing [25]. Therefore, a reliable direct laboratory prediction of the CC behaviour, before the final fabrication of the tyre, is of great importance. Traditional lab CC test methods with small rotating rubber wheels, presented, e.g., in Refs [1, 26–29], employ simple devices during which loading conditions cannot be applied sufficiently and reproducibly due to the very low applied impact energy, which is not comparable with the real loading conditions of a tyre. Such simple CC test methods show for certain specimen geometries and loading conditions how the rate of abrasion can be predicted only in a qualitatively approximate manner. In other words, standard characterization of CC wear takes into consideration just only the wear loss of a sample after a single fixed test time. Standard CC tests provide only a wear loss, which is a qualitative parameter, and any physical parameter directly describing the CC process is missing. Additionally, the wear rate is not considered to be time-dependent which, in principle, is a misleading proposal, leading to adverse correlations with field tests. E.g., in Ref. [28], it was reported that the rubber compound rating obtained by means of simple CC test method showed a completely opposite trend to that from the field test. Therefore, an advanced testing method and fully instrumented equipment labelled Instrumented Chip&Cut Analyser (ICCA, Coesfeld GmbH & Co. KG, Germany) has been introduced by Stoček et al. [30, 31]. The ICCA method is now successfully integrated as a standard lab method in the rubber respective tyre industry. In Refs. [30–34], it is shown that the type of rubber polymers and the rubber blend composition significantly influence the CC behaviour over the varied range of applied normal forces. A study comparing the CC behaviour of tyres in the field with laboratory tests has been carried out [32], and laboratory investigations studying the CC behaviour of studied rubber close to tyre application have also been carried out, but without direct comparison to field tests [35, 36]. CLD was only studied on SBR based tread compounds with respect to CC properties [36], but the previous original equipment measuring the weight loss was used for laboratory analysis. An almost linearly increasing weight loss as a function of CLD was measured over the entire CLD range, and this trend is to be expected compared to natural rubber-based rubbers, which are subject to SIC at higher loads. For NR it has been found that CC damage and temperature go through a maximum at critical values of the impacting normal load and over this value gain significantly decreases for higher load. This effect has been associated with the appearance of strain induced crystallization (SIC) in the NR during cyclic impacts above a critical level [33, 35, 37]. The results impressively explain the empirical preference for NR or NR-blends in practice when it comes to minimizing CC wear. SIC for NR rubber depends on the cross-link density [38], whereas SIC becomes more significant for low CLD, which is characteristic for CV systems. With increasing CLD, the SIC effect decreases, which is especially critical for EV systems [39].

Therefore, this work is fully focussed on the investigation of CC behaviour of NR material in terms of CLD over the broad range of normal force to cover the entire loading range of natural rubber without as under the SIC formation, so that the effect of CLD on CC properties with respect to SIC action is clearly established.

Because of the poor thermal stability of the polysulfidic cross-links, the NR exhibits reversion when exposed to a temperature-time treatment. The breakage of the polysulfidic cross-links formed during the initial stages of the vulcanization is considered as one of the major reasons behind the reversion process [40–43]. Therefore, an additional part of this study is to investigate the degradation of CC behaviour for rubber cured over the optimal curing time t_{90} in the range of reversion and relate it to CLD.

2 Experiment and Material

In this study, NR SVR – CV 60 (standard Vietnamese rubber with a Mooney viscosity ML(1 + 4) at 100°C: 60 ± 5) obtained from Binh Phuoc has been investigated. Rubber was filled with 58 phr of carbon black N220 (Makrochem, Poland), whereas this CB concentration was chosen to achieve optimum tensile strength values [44]. The other raw materials contained in the rubber recipe are as follows: ZnO from SlovZink a.s. (Košeca, Slovakia), stearic acid from Setuza a.s. (Ústí nad Labem, Czech Republic), N-(1,3-Dimethylbutyl)-N´-phenyl-p-phenylenediamine (Vulkanox 4020) antiozonant from Lanxess (Köln, Germany), Varazon 5998 Antiozonant, (Lanxess, Sandton South Africa), antidegradant 2,2,4-trimethyl-1,2-dihydroquinoline (TMQ) from Hardwick (California, USA), Tudalen 11 aromatic Oil from Fraunhofer (München, Germany), accelerators N-tert-butyl-benzothiazole sulfonamide (TBBS) from Duslo a.s. (Šal'a, Slovakia) and Sulfur OT 33 from Eastman Chemical Company (Kingsport, Tennessee, USA). To ensure the production of rubbers with different CLD values, varied accelerator to sulfur (A/S) ratios were used, with a constant sulfur content of 2.5 phr for all compounds and varying the accelerator (TBBS) content. The following different A/S ratios with respect to varied curing systems has been applied:

CV: A/S ratio = 0.1, 0.3 and 0.6,
SEV: A/S ratio = 0.7, 1.0, 1.5, 2.0 and 2.5,
EV: A/S ratio = 3.0.

Table 1 lists the complete formulations of the compounds used.

A two-step mixing procedure was employed to prepare all rubber compounds. Both of the steps were carried out in an internal mixer SYD-2L (Everplast, Taiwan) with a fill factor of 0.7. First, the masterbatch was prepared after masticating the virgin rubber for 1 min. To this, carbon black together with Tudalen oil 11 were added and then mixed for another 3.5 min. Additionally, ZnO + Stearic acid has been incorporated to the mixture and mixed for another 2.5 min. The rotor speed was 50 rpm to reaching the temperature of the mixture maximum of 150°C, whereas the initial temperature was 70°C. The masterbatch thus prepared was milled over a period of 3 min using a double-roll mill and sheeted out at a rolls temperature of 70°C.

The final batch was then prepared by mixing the masterbatch together with sulfur and accelerator for 1 min at a rotor speed of 35 rpm and at an initial temperature of

Table 1 Used rubber formulas

	Sample indication								
	$CV_{0.1}$	$CV_{0.3}$	$CV_{0.6}$	$SEV_{0.7}$	$SEV_{1.0}$	$SEV_{1.5}$	$SEV_{2.0}$	$SEV_{2.5}$	$EV_{3.0}$
Ingredients	Loading in [phr]								
NR SVR-CV60	100								
TMQ	2								
N220	58								
6PPD	2.5								
Varazon 5998	2								
Tudalen oil 11	6								
ZnO	5								
Stearic acid	3								
TBBS	0.25	0.75	1.5	1.75	2.5	3.75	5.0	6.25	7.5
Sulfur OT33	2.5								

[phr] – parts per hundred rubber

70°C followed by adding the complete curing system and mixed until the temperature of the mixture reached 91°C. The final batch was again milled using a double-roll mill at 60°C of rolls for 3 min. After storing for 24 h, the curing properties were determined by Premier MDR moving die rheometer from Alpha Technology (Hudson, Ohio, USA) according to ASTM D5289 at a temperature of 160°C. This was done to determine the scorch time, t_{s2}, minimum torque, M_L, maximum torque, M_H, and t_{90} of the all compounds.

The specimens having different geometries were cured in a heat press LaBEcon 300 (Fontijne Presses, The Netherland) at a curing temperature of $T_t = 160$°C under constant pressure 200 kPa and under the estimated optimum curing time t_{90} + 0.5 min per 1 mm of the thickness. The cylindrical samples for CC analyses of the geometry 55 mm in diameter and 13 mm in thickness (see, e.g., Refs. [32, 33]), standard sheets (150 mm × 150 mm × 2 mm) for the determination of tensile properties as well as plates having 6 mm in thickness used for hardness test were cured. The samples, which were designed to analyse the effect of reversion when a time over t_{90} was reached, were additionally examined by additional curing time t_{90} + 10 min.

First, the tensile properties, Shore-A hardness, rebound resilience, CLD and DIN abrasion test were measured. The tensile test was performed using the T10D tensile testing machine from Alpha Technology (Hudson, Ohio, USA) according to ISO 37 standard at the extension rate of 500 mm/min for standard test specimens of the Type S2 with 2 mm. The tensile properties reported in this paper are averages for 5 specimens. The measurement of the Shore-A hardness was performed in accordance with ISO 7619-1 and measured on 5 samples for each compound. Rebound resilience was determined according to ISO 4662 standard using an identical sample as for Shore-A hardness. Average values of the rebound resilience and their standard deviation were subsequently determined from 6 replicates. Measurements of rebound elasticity and Shore-A hardness were performed at room temperature of 23°C. The CLD calculation is based on the Kraus modified Flory–Rehner equation for filled rubber vulcanizate from equilibrium swelling in toluene [45]. Whereas, firstly the mass of solvent in the rubber is determined as

$$m_s = m_q - m_0, \text{[g]} \tag{1}$$

where m_q is the mass of the rubber sample in the swollen equilibrium state and m_0 is the initial mass of the rubber sample before swelling. As next it is required to determine the volume of solvent V_q in 1 g of rubber sample using the equation

$$V_q = \frac{m_s}{\rho_s m_0}, \quad \left[\text{cm}^3\,\text{g}^{-1}\right] \tag{2}$$

where ρ_s is the density of the solvent. The volume of solvent in a given rubber sample can then be read as

$$V_s = m_0 V_q = \frac{m_0 m_s}{\rho_s m_0} = \frac{m_s}{\rho_s}, \left[\text{cm}^3\right] \tag{3}$$

Volume of rubber in the rubber compounds inclusive additives without fillers will be calculated as follows:

$$V_{k0} = \frac{m_r m_0}{\rho_k (m_r + m_{\text{add}})}, \left[\text{cm}^3\right] \tag{4}$$

where ρ_k is the density of the rubber, m_r is the mass of rubber and m_{add} is the mass of additives both in [phr]. The volume of rubber in the complete rubber compound inclusive additives and filler can then be estimated using the following equation:

$$V_k = \frac{m_r m_0}{\rho_k \left(m_r + m_{\text{add}} + m_f\right)}, \left[\text{cm}^3\right] \tag{5}$$

where m_f is the mass of fillers in [phr]. In the next step the volume fraction of the gel in an equilibrium swollen vulcanizate without fillers needs to be calculated using the equation:

$$V_{r0} = \frac{V_{k0}}{V_{k0} + V_s}, [-] \tag{6}$$

whereas the volume fraction of gel in the equilibrium swollen vulcanizate with fillers reads

$$V_r = \frac{V_k}{V_k + V_s}, [-] \tag{7}$$

Finally, the Flory–Rehner equation for the apparent cross-link density is given as follows:

$$\nu = \frac{-10^6 V_{ro}\left[\ln(1 - V_r) + V_r + \chi V_r^2\right]}{V_M\left(\sqrt[3]{V_{ro}^2 V_r} - 0.5V_r\right)}, \ \left[\mu\text{mol} \ \times \text{cm}^{-3}\right] \tag{8}$$

where V_M is the molar volume of toluene (106.27 cm^3 mol^{-1}). χ is Flory–Huggins polymer–solvent interaction parameter, whereas in our case for NR and Toluene $\chi = 0.414$ [46]. Flory–Huggins polymer–solvent interaction parameter is tabular value. For natural rubbers and isoprene rubbers in toluene is near 0.4. Samples for swelling test were rectangular in shape of a dimension 20 × 30 × 2 mm. The experiment was performed at room temperature at 23°C. Three samples were tested for each vulcanized rubber compound. The rubber samples were weighed at the beginning of the test. The initial weight was obtained. The samples were then immersed in toluene to swell to equilibrium, with equilibrium being reached in 10 days. The samples were then weighed again and the equilibrium weight was determined. The average CLD value was calculated from the values obtained.

The determination of rubber resistance against DIN abrasion for 3 replicate samples per compound was performed in accordance with ISO 4649 B, where this method is based on the determination of the mass loss, Δm_{rel} as relative number in %. Δm_{rel} has been determined due to the abrasive action of sliding a test piece over 50 m length of a specified grade of abrasive sheet under the acting force of 10 N and has been determined as follows:

$$\Delta m_{\text{rel}} = \frac{100(m_1 - m_2)}{m_1}, \ [\%] \tag{9}$$

where m_1 is the weight of sample in [g] prior and m_2 is the weight in [g] after the abrasion. The measurement was performed on samples that had the shape of a cylinder 16 mm of diameter and 6–12 mm in height.

Finally, the CC behaviour represented with the physical parameter CC damage, P, of all tested compounds was analysed at room temperature with the Instrumented Chip&Cut Analyser (ICCA, Coesfeld GmbH & Co. KG, Germany) using the following test conditions:

- Rotational speed: 150 rpm,
- Normal (impact) forces: 100, 150 and 200 N,
- Sliding time: 50 ms,
- Impact frequency: 5 Hz,
- Total no. of impact cycles: 3,000.

The complete ICCA analyses have been performed in accordance to the testing protocol described in detail previously in Refs. [17–19].

3 Results

In Fig. 2, left, the curing characteristics for all compounds are plotted, whereas Table 2 lists all evaluated parameters of the individual curing characteristics. In general, the increasing A/S ratio results in the increase of M, however it was not possible to trace a similar simple dependence for the remaining rheological parameters M_L, t_{s2} and t_{90}. For the CV system, the value of M_L increases with increasing A/S ratio. For all analysed materials, regardless of the A/S ratio, reversion is clearly visible, and except for $CV_{0.1}$ it can be said that the trend of reversion is decreasing in intensity with increasing A/S ratio. It can also be determined that the reversion starts in the shortest time for material $SEV_{0.7}$ compared to the other materials. $EV_{3.0}$ shows clearly the lowest reversion value due to the highest content of monosulfide bonds [19]. Moreover, the graphical representation of the rheological curing parameters t_{s1} and t_{90} in dependence on A/S ratio are plotted in Fig. 2, right. It can be seen that the

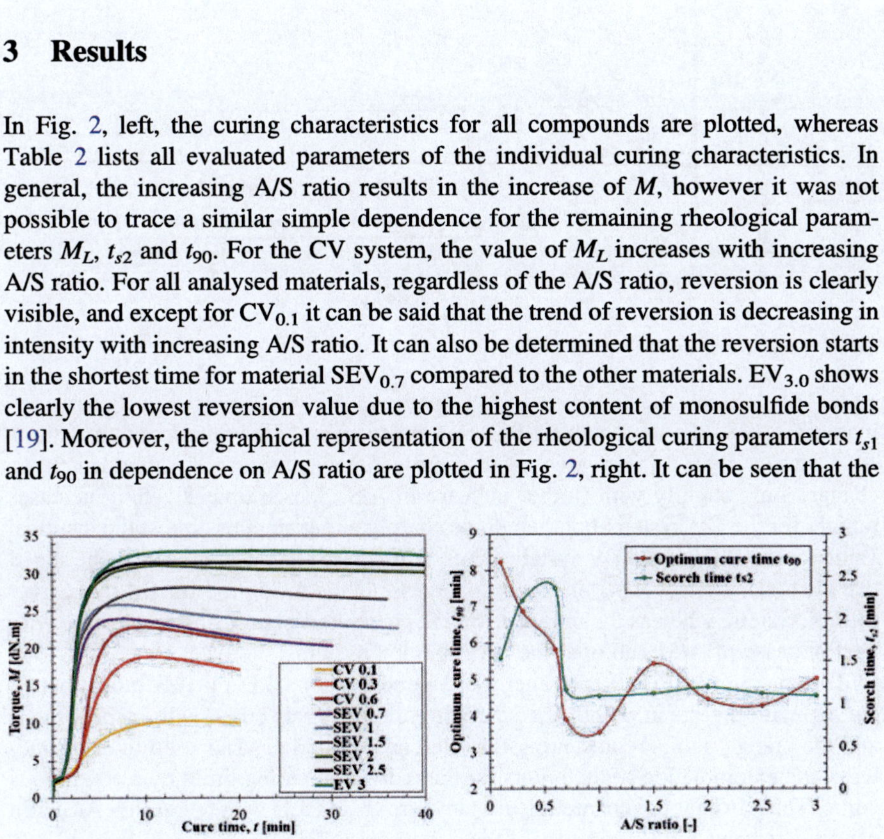

Fig. 2 Curing characteristic of rubber compounds (left) and dependence of curing parameters t_{90}, t_{s2} on A/S ratio (right)

Table 2 List of parameters of individual curing characteristics

Rubber type	M_H [dN m]	M_L [dN·m]	t_{s2} [min]	t_{90} [min]
$CV_{0.1}$	10.30	1.78	1.53	8.22
$CV_{0.3}$	18.81	2.34	2.22	6.87
$CV_{0.6}$	22.83	2.48	2.36	5.81
$SEV_{0.7}$	24.00	2.40	1.17	3.90
$SEV_{1.0}$	25.88	2.45	1.10	3.56
$SEV_{1.5}$	28.20	2.14	1.13	5.42
$SEV_{2.0}$	31.08	2.30	1.2	4.46
$SEV_{2.5}$	31.51	2.18	1.13	4.30
$EV_{3.0}$	33.04	2.17	1.10	5.02

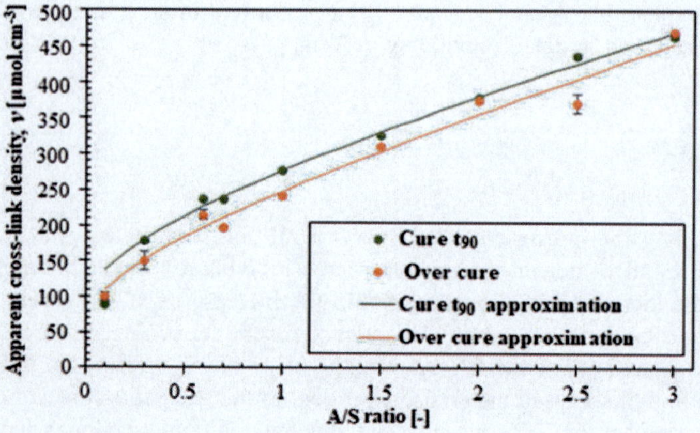

Fig. 3 Dependence of CLD on A/S ratio

increasing A/S ratio up to value 1.0 causes the decrease in the optimum curing time, t_{90}, however then t_{90} increases again rapidly for a value of 1.5, after which t_{90} changes only slightly with further increase in A/S. Scorch time, t_{s2}, first increases rapidly for the CV system, but then drops sharply and maintains almost the identical value for the SEV as the EV system. Thus, from these characteristics it can be stated that the highest effect on t_{90} and t_{s2} is an increase in the A/S ratio from 0.1 to 0.6 over the CV system, whereas the influence on the parameters significantly decreases with next increase of A/S ratio over the SEV and EV systems.

The diagram in Fig. 3. presents the dependence of CLD on A/S ratio for two curves, with the green points characterizing the materials cured with respect to t_{90} and the orange points representing the overcured materials. The solid lines characterize the extrapolation of the points for the individual curing times by a continuous curve. Thus, from these characteristics, an increase in CLD with increasing A/S ratio can be determined independently of the curing time. If values for the individual A/S ratios representing by the points will be observed, the higher CLD values for all A/S ratios except lowest A/S = 0.1 and highest A/S = 3.0 have been determined just for materials cured on t_{90}.This finding clearly correlates with the observation of reversion, where for these two A/S ratios in particular, as seen in Fig. 2, left, the reversion is minimal compared to the other A/S ratios, and therefore there is no degradation of the cross-link network, on the contrary, in the added time 10 min, there is still more network formation. In order to describe the relationship between the A/S ratio and the CLD with a continuous analytical function, moreover, with the highest possible accuracy, the specific function was required to be found. In order to find such a function with a coefficient of determination R2 as high as possible, a least squares method was chosen to determine the function, as described in detail in an earlier publication by the authors [47].The most appropriate function valid for both search dependencies meeting the assumption of the highest R^2 was found to be

Table 3 Coefficients *a*, *b*

	Curing conditions	
Coefficient	t_{90}	t_{90} + 10
a	0.053134	0.059648
b	0.202220	0.191122

Table 4 CLD related to A/S ratio

	CLD, ν [μmol cm^{-3}]		
A/S ratio	t_{90}	t_{90} + 10	Difference, $\Delta\nu$ [μmol cm^{-3}]
0.1	136	110	26
0.3	181	152	29
0.6	227	198	29
0.7	241	211	30
1.0	277	249	28
1.5	331	305	26
2.0	380	356	24
2.5	426	406	20
3.0	470	453	17

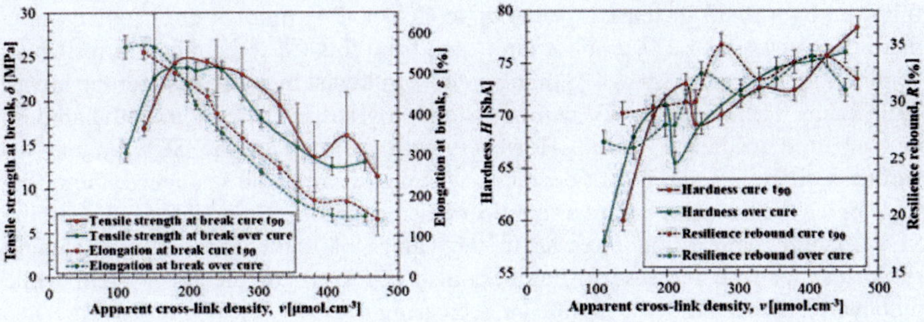

Fig. 4 Tensile properties (left) and hardness as well as rebound resilience (right) in dependence on CLD

$$\mathrm{CLD} = \left[\left(a \cdot \sqrt[3]{A/S} + b\right)^2\right]^3, \tag{10}$$

where the numerically determined coefficients *a*, *b* are listed for the individual curing condition in Table 3.

Table 4 summarizes the CLD values corresponding to each A/S ratio, with the numerical CLD values determined from the approximation function Eq. 10. The approximation is further used to relate the mechanical properties to the CLD values studied within this work. Table 4 lists additionally values $\Delta\nu$, which is the difference between t_{90} and t_{90} + 10. If all these values $\Delta\nu$ will be averaged, the single value $\Delta\nu_{aver}$ = 25 μmol × cm^{-3} will be given.

The tensile strength properties are shown in Fig. 4 on the left, and the hardness and rebound elasticity in Fig. 4 on the right, with the overall behaviour related to the CLD values. In general, it is clear that one the individual dependencies for both t_{90} and t_{90} + 10 cured materials have the identical character, only they are shifted to lower CLD values for the overcured state by an average value $\Delta\nu_{aver} = 25$ μmol × cm^{-3}. Therefore, the results for these properties are discussed below with respect to the CLD values of the material cured at t_{90}.

While the values for stress at break for CLD up to 181 μmol × cm^{-3} increase sharply so that they settle to a constant value in the range from 181 to 227 μmol. cm^{-3} and then decrease for higher CLD values, the stress at break values has an almost linear decreasing character over the entire range of analysed CLD. Thus, it can be said that while for the CV system stress at break increases, for the SEV and EV system its values decrease and for strain at break the values decrease for all curing systems in the order CV > SEV > EV.

For hardness, there is a significant increase in its value for CLD in the range from 136 to 227 μmol × cm^{-3}, after which its value drops sharply up to CLD = 241 μmol × cm^{-3} followed by an increase of hardness in the remaining CLD range. A similar character can be observed for the rebound resilience, where the initial increase in values for the CLD range up to 227 μmol × cm^{-3} is only slight followed by a sharp increase in value up to CLD = 241 μmol × cm^{-3}, after which there is a decrease to 277 μmol × cm^{-3} and from this CLD the rebound resilience continuously increases up to 426 μmol × cm^{-3} followed by a decrease for the higher CLD value. Therefore, the CV curing system provides an increase in both hardness and rebound resilience values. However, with the start of the SEV system for hardness, there is a significant decrease, whereas for rebound resilience, the initial CLD of the SEV system causes a significant increase. For the remaining region of the SEV system, there is an increase in the values of both hardness and rebound resilience, with an increase in hardness also observed for the EV system while rebound resilience shows a significant decreasing trend.

From the nature of the weight loss characteristics describing the resistance of NR material to DIN abrasion, it is evident that for both curing times they have a convex shape, while also for these data the characteristic for the overcured rubbers is shifted towards lower CLD values. It can also be seen that there is a significant decrease in the mass loss values in the range of CV system, whereas, in the CLD range for SEV and EV, the increase in mass loss is characterized by a lower slope compared to the slope of decrease in the CV system. Comparing this trend with the almost constant increase in hardness over the entire CLD range, the increase in mass loss over CLD region characteristic for the SEV and EV systems is surprising. This trend is unexpected given the assumption that the coefficient of friction decreases with increasing hardness and hence the shear forces acting on the rubber during abrasion decrease. The only explanation is directed to the fracture properties of the cross-links, where the proportion of monosulfide bonds increases with increasing CLD and the no. of polysulfide bonds decreases and hence the bonding energy increases. Cross-links with higher bonding energy are more brittle and therefore less resistant to fracture processes at the micro level and therefore also to fatigue abrasion.

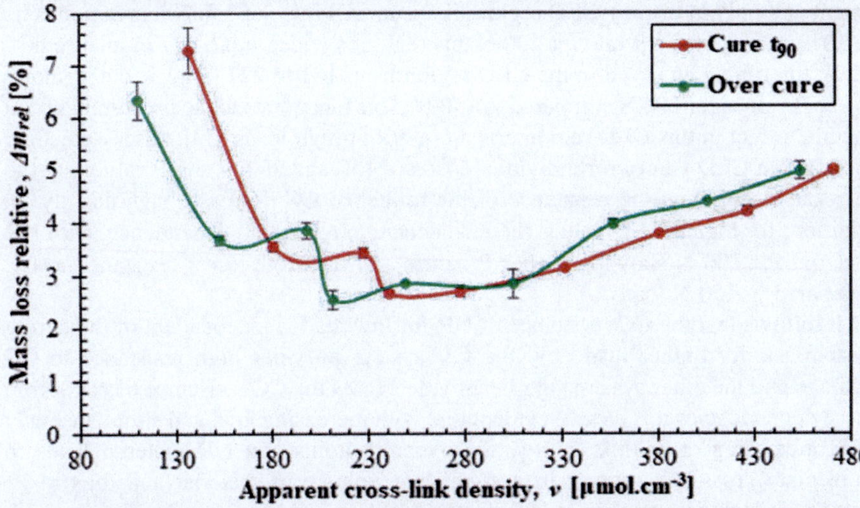

Fig. 5 Mass loss for DIN abrasion in dependence on CLD

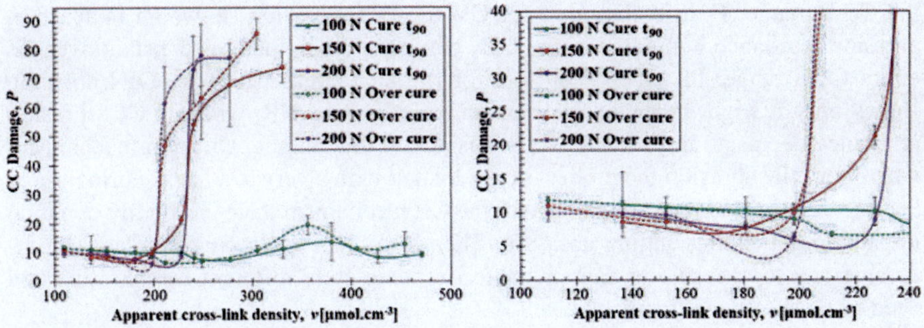

Fig. 6 Dependence of CC damage, *P* on cross-link density over the complete range of CLD (left) and detailed view up to 240 μmol × cm^{-3}

From the above characteristics presented in diagrams Figs.4 and 5 it is possible to determine the CLD region with respect to the optimum mechanical properties, which is the range from 181 to 241 μmol × cm^{-3}.

Finally, Fig. 6. presents the dependence of CC damage, P on CLD over the complete range of CLD (left) and detailed view in to the range of CLD up to 240 μmol × cm^{-3}. The diagrams show the dependencies for both the t_{90} and t_{90} + 10 curing systems, where the similar character for all presented curves is also quite evident, shifted to lower CLD values for the overcured materials, and therefore the CC behaviour is discussed again only for materials cured at t_{90}. It can be seen that the CC properties are dependent on the applied loading respective normal force. While for a normal force of 100 N the CC properties are completely independent of the CLD value, for higher normal forces this independence is only valid to a limited extent and only in the low CLD region. For a normal force of 150 N this region is up to 181 μmol × cm^{-3}, while for 200 N this region is wider, up to 227 μmol × cm^{-3}. Thus, the rubber analysed in the CLD region from 181 to 227 μmol × cm^{-3} shows lower *P* values for 200 N compared to 150 N. This fact points to the predominance of the SIC effect in this CLD region arising in NR at high loads [39]. After exceeding these given CLD values for individual loads of 150 and 200 N, the P values start to increase steeply and the resistance of the rubber to CC decreases significantly. In addition, for higher CLD values there is a change in the load dependence, with NR material for 200 N showing higher P values and therefore lower resistance to CC compared to 150 N loads.

It follows that the CC resistance of NR for low load is independent of the curing system but for higher load only the CV system provides high resistance to CC damage and the other system significantly decreases the CC resistance of NR. From these dependencies it is clearly evident that with increasing load and simultaneously with increasing cross-link energy, the dynamic impact on NR material leads to immediate cross-link rupture in a significant volume of material and thus to its cutting and chipping. This is due to the brittleness of the cross-link with high bonding energy.

Finally, if we compare the weight loss determined by DIN abrasion method with the CC damage, *P*, it is clear that for CV the DIN abrasion shows an increase in abrasion resistance with increasing CLD, while the P parameter is constant over the entire CLD range. It is then seen that for further increase in CLD, DIN abrasion shows only a slight decrease in abrasion resistance of NR, whereas CC abrasion resistance decreases tremendously. This is due to the inherent abrasion mechanism, where for DIN abrasion the rubber is only loaded with a very low static normal force leading to fatigue abrasion, whereas for the CC mechanism there is a highly dynamic impaction generating cutting abrasion. Therefore, the two mechanisms are incomparable and their resulting data characterize completely different mechanisms and need to be strictly distinguished.

It is evident that the ranges of CLD where optimal mechanical as well as CC properties occur more or less overlap, while the range of CLD with optimal CC properties is slightly larger, ranging from 136 to 241 μmol × cm^{-3}. For samples cured at t_{90} + 10, an increase in CC damage, *P* can already be seen at lower CLD values due to degradation of NR over t_{90}. This effect generally is called reversion. During reversion, changes occur in the structure of the resulting cross-linking. The polysulfide bonds formed during curing as well as rubber chains are cleaved. This process additionally reduces the crystallization in NR [44].

The nature of the CC damage can be clearly seen in the photographs of the surface of each individual sample for the applied load in relation to CLD value, which are shown in Table 5. A direct correlation between the worn surface topology as well as the CC Damage, *P* is evident for all loads over the entire CLD region analysed. For CLD values up to 227 μmol × cm^{-3}, a low abrasion topology is evident regardless of load, while the same topology is visible for the lowest applied load of 100 N over the entire CLD region. Additionally, significant abrasion is evident for the surface of samples with CLD in the range >227 μmol × cm^{-3} for applied loads of 150 and 200 N.

Table 5 Photograph of worn rubber surface after CC test for samples cured at t_{90}

Curing system	CLD [ν] [μmol cm^{-3}]	Normal force F_N [N]		
		100	150	200
CV	136			
	181			
	227			
SEV	241			
	277			
	331			
	380			
	426			
EV	470			

4 Conclusion

The study was aimed at determining the effect of cross-linking density on the CC properties of the NR blend when the accelerator was varied and the sulfur content was constant. The intent of the experimental research was to determine the region of apparent cross-link density (CLD) with enhanced resistance of NR rubber to CC abrasion and to compare this region with the range of CLD in which enhanced mechanical properties are achieved. Standard methods of analysis in tensile, hardness, rebound elasticity as well as DIN abrasion were chosen to characterize the mechanical properties, while an advanced method of determining the physical parameter describing the CC abrasion topology on the surface of the test body implemented in a commercially available and industrially validated Instrumented Chip&Cut Analyser (ICCA) produced by Coesfeld GmbH & Co. KG, Germany was used to determine the CC resistance. It was clearly determined that the CC properties of the NR for the analysed CLD range from 136 to 470 μmol × cm^{-3} were dependent on the applied load, with CC resistance being constant over the entire CLD range for low loads, but for higher loads, an enhanced CC resistance was achieved only in the low CLD range from 136 to 241 μmol × cm^{-3}. This region of enhanced CC resistance corresponds more or less to the region where enhanced mechanical properties were achieved, which is however slightly narrower from 181 to 241 μmol × cm^{-3}. Finally, a significant effect of strain induced crystallization (SIC) on the CC resistance was evident in the CLD region from 181 to 227 μmol × cm^{-3}, where for higher loads of 200 N, the CC resistance increases compared to 150 N. The intersection of the CLD range in which the optimum mechanical as well as CC properties are achieved shows that the effective CLD range is from 181 to 227 μmol × cm^{-3}, which corresponds to the conventional (CV) curing system. The study was carried out for rubber cured at the optimum curing time t_{90} as well as for the time t_{90} + 10 min, when thermal degradation of the network already occurs. It has been verified that the nature of all the properties analysed is not affected in any way during the overcuring, but the overall properties are shifted towards lower absolute CLD values, which is due to the degradation of the network.

This study is the first research work aimed at describing the effect of CLD on the CC properties of NR rubber, which is a basic elastomer used in the manufacture of truck tyres as well as rubberized tracks and conveyor belts, where all these products are subjected to constant cutting and chipping abrasion and therefore optimization of rubber compounds with respect to achieving optimum CC resistance is highly desirable. So that the effect of CLD on CC properties is fully described, a future intention is to analyse NR with respect to the formation of network density due to sulfur variation.

Acknowledgements This work was supported by the Ministry of Education, Youth and Sports of the Czech Republic – DKRVO (RP/CPS/2022/006).

References

1. Scherbakov M, Gurvich MR (2003) A method of wear characterization under cut, chip and chunk conditions. J Elastom Plast 35:73–84. https://doi.org/10.1177/009524403031097
2. Aleman JV, Chadwick AV, He J, Hess M, Horie K, Jones RG, Kratochvil P, Meisel I, Mita I, Moad G, Penczek S, Stepto RFT (2007) Definitions of terms relating to the structure and processing of sols, gels, networks, and inorganic-organic hybrid materials. Pure Appl Chem:1801–1829. https://doi.org/10.1351/pac200779101801
3. Lal J (1970) Effect of crosslink structure on properties of natural rubber. Rubber Chem Techol 43:664–686
4. Moore CG, Trego BR (1961) Structural characterization of vulcanizates. Part II. Use of triphenylphosphine to determine the structures of sulfur linkages in unaccelerated natural rubber–sulfur vulcanizate networks. J Appl Polym Sci 5:299–302. https://doi.org/10.1002/app.1961.070051508
5. Kok CM, Yee VH (1986) The effects of crosslink density and crosslink type on the tensile and tear strengths of NR, SBR and EPDM gum vulcanizates. Eur Polym J 22:341–345. https://doi.org/10.1016/0014-3057(86)90203-X
6. Porter M (1967) Structural characterization of filled vulcanizates part 1. Determination of the concentration of chemical crosslinks in natural rubber vulcanizates. Rubber Chem Technol 40(3):866–882. https://doi.org/10.5254/1.3539101
7. Fei Z, Long C, Qingyan P, Shugao Z (2012) Influence of carbon black on crosslink density of natural rubber. J Macromol Sci B 51(6):1208–1217. https://doi.org/10.1080/00222348.2012.664494
8. Dogadkin BA (1976) Vulcanization of elastomers, chemistry of elastomers.1st edn. WNT, Warsaw, p 201
9. Datta RN (2001) Rubber curing systems. Smithers Rapra Publishing
10. Mark JE, Erman B, Roland M (2013) The science and technology of rubber. Academic Press
11. Heideman G, Datta RN, Noordermeer JW, van Baarle B (2004) Activators in accelerated sulfur vulcanization. Rubber Chem Technol 77:512–541. https://doi.org/10.5254/1.3547834
12. Morrison NJ, Porter M (1984) Temperature effects on the stability of intermediates and crosslinks in sulfur vulcanization. Rubber Chem Technol 57:63–85. https://doi.org/10.5254/1.3536002
13. Koenig JL (2000) Spectroscopic characterization of the molecular structure of elastomeric networks. Rubber Chem Technol 73:385–404. https://doi.org/10.5254/1.3547598
14. Porter M (1968) The chemistry of the sulfur vulcanization of natural rubber. Chem Sulphides
15. Bateman L, Cunneen JI, Moore CG, Mullins L (1963) The chemistry and physics of rubber-like substances. Wiley, New York
16. Dogadkin BA, Tarasova ZN (1954) Vulcanization structures and their influence on the heat resistance and fatigue of rubber. Rubber Chem Techol 27:883–898. https://doi.org/10.5254/1.3543540
17. Saunders KJ (1988) Organic polymer chemistry.2nd edn. Chapman & Hall, London
18. Sathy SG, Stoček R, Kratina O (2020) Reversion free high-temperature vulcanization of cis-polybutadiene rubber with the accelerated-sulfur system. eXPRESS Polym Lett:823–837. https://doi.org/10.3144/expresspolymlett.2020.68
19. Bornstein D, Pazur RJ (2020) The sulfur reversion process in natural rubber in terms of crosslink density and crosslink density distribution. Polym Test:88. https://doi.org/10.1016/j.polymertesting.2020.106524
20. Cichomski E, Dierkes WK, Noordermeer JW, Schultz SM, Tolpekina TV, Reuvekamp LAEM, Blume A (2015) Effect of the crosslink density and sulfur-length on wet-traction and rolling resistance performance indicators for passenger car tire tread materials. In: Fall 188th technical meeting of Rubber Division. ACS Cleveland, OH October 13–15

21. Mensah B, Agyei-Tuffour B, Nyankson E, Bensah YD, Dodoo-Arhin D, Bediako JK, Yaya A (2018) Preparation and characterization of rubber blends for industrial tire tread fabrication. Int J Polym Sci. https://doi.org/10.1155/2018/2473286
22. Hamed GR, Rattanasom N (2002) Effect of crosslink density on cut growth in gum natural rubber vulcanizates. Rubber Chem Techol 75:323–332. https://doi.org/10.5254/1.3544981
23. Hamed GR, Rattanasom N (2002) Effect of crosslink density on cut growth in black-filled natural rubber vulcanizates. Rubber Chem Techol 75:935–942. https://doi.org/10.5254/1.3547693
24. Greensmith HW, Mullins L, Thomas AG (1963) The chemistry and physics of rubber like substances. Maclaren and Sons, London, p 262
25. Haws JR, WILDER C, Cooper WT (1969) Effect of carbon black structure on abrasion resistance. Rubber Chem Techol 101:69
26. Beatty JR (1979) Testing apparatus and method for measuring cutting, chipping and abrasion resistance. U.S. Pat. 740
27. Beatty JR, Miksch B (1982) A laboratory cutting and chipping tester for evaluation off-the-road and heavy-duty tire treads. Rubber Chem Techol 55:1531–1546. https://doi.org/10.5254/1.3535947
28. Nah C, Jo BW, Kaang S (1998) Cut and chip resistance of NR–BR blend compounds. J Appl Polym Sci 68:1537–1541. https://doi.org/10.1002/(SICI)1097-4628(19980531)68:9<1537::AID-APP17>3.0.CO;2-W
29. Ma JH, Wang YX, Zhang LQ, Wu YP (2012) Improvement of cutting and chipping resistance of carbon black-filled styrene butadiene rubber by addition of nanodispersed clay. J Appl Polym Sci 125:3484–3489. https://doi.org/10.1002/app.36710
30. Stoček R, Mars WV, Robertson CG, Kipscholl R (2018) Characterizing rubber's resistance against chip and cut behavior. Rubber World 257:38–40
31. Stoček R, Mars WV, Kipscholl R, Robertson CG (2019) Characterisation of cut and chip behaviour for NR, SBR and BR compounds with an instrumented laboratory device. Plast Rubber Compos 48:14–23. https://doi.org/10.1080/14658011.2018.1468161
32. Stoček R, Heinrich G, Schulze A, Wunde M, Klüppel M, Vatterott C, Tschimmel J, Lacayo-Pineda J, Kipscholl R (2020) Chip & cut wear of truck tire treads: comparison between laboratory and real tire testing. Kautschuk Gummi Kunststoffe 73:51–55
33. Kipscholl R, Stoček R (2019) Quantification of chip and cut behaviour of basic rubber (NR, SBR). RFP Rubber Fibre Plast 2:88–91
34. Lake GJ, Lindley PB (1965) The mechanical fatigue limit for rubber. J Appl Polym Sci 9:1233–1251. https://doi.org/10.1002/app.1965.070090405
35. Stoček R, Ghosh P, Machů A, Chanda J, Mukhopadhyay R (2020) Fatigue crack growth vs. chip and cut wear of NR and NR/SBR blend-based rubber compounds. In: Heinrich G, Kipscholl R, Stoček R (eds) Fatigue crack growth in rubber materials. Advances in polymer science, vol 286. Springer, Cham. https://doi.org/10.1007/12_2020_67
36. Elangovan K, Josephraj FX, Murugesan AK, Pandian B (2021) Effect of crosslink density on cut and chip resistance of 100% SBR based tire tread compound. Mater Plast 58:34–46. https://doi.org/10.37358/Mat.Plast.1964
37. Stoček R, Heinrich G, Kipschooll R, Kratina O (2021) Cut & chip wear of rubbers in a range from low up to high severity conditions. Appl Surf Sci Adv 6:100152. https://doi.org/10.1016/j.apsadv.2021.100152
38. Boyard N (2016) Heat transfer in polymer composite materials – forming processes.1st edn. Wiley, Hoboken
39. Chenal J-M, Gauthier C, Chazeau L, Guy L, Bomal Y (2007) Parameters governing strain induced crystallization in filled natural rubber. Polymer 48:6893–6901. https://doi.org/10.1016/j.polymer.2007.09.023
40. Bhowmick AK, Mukhopadhyay R, De SK (1979) High temperature vulcanization of elastomers. Rubber Chem Techol 52:725–734. https://doi.org/10.5254/1.3535236

41. Loo CT (1974) High temperature vulcanization of elastomers: 2. Network structures in conventional sulphenamide-sulphur natural rubber vulcanizates. Polymer 15:357–365. https://doi.org/10.1016/0032-3861(74)90177-3
42. Sathi SG, Harea E, Machů A, Stoček R (2021) Facilitating high-temperature curing of natural rubber with a conventional accelerated-sulfur system using a synergistic combination of bismaleimides. eXPRESS Polym Lett 15:16–27. https://doi.org/10.3144/expresspolymlett.2021.3
43. Pöschl M, Sathi SG, Stoček R (2021) Identifying the co-curing effect of an accelerated-sulfur/bismaleimide combination on natural rubber/halogenated rubber blends using a rubber process analyzer. Polymers 13:4329. https://doi.org/10.3390/polym13244329
44. Demassieux Q, Berghezan D, Cantournet S, Proudhon H, Creton C (2019) Temperature and aging dependence of strain-induced crystallization and cavitation in highly crosslinked and filled natural rubber. J Polym Sci Phys B Polym Phys 57:780–793. https://doi.org/10.1002/polb.24832
45. Kruželák J, Sýkora R, Hudec I (2015) Influence of mixed sulfur/peroxide curing system and thermo-oxidative ageing on the properties of rubber magnetic composites. J Polym Res 22:1–9. https://doi.org/10.1007/s10965-014-0636-8
46. Kim DY, Park JW, Lee DY, Seo KH (2020) Correlation between the crosslink characteristics and mechanical properties of natural rubber compound via accelerators and reinforcement. Polymer 12. https://doi.org/10.3390/polym12092020
47. Stoček R (2020) Some revisions of fatigue crack growth characteristics of rubber. In: Heinrich G, Kipscholl R, Stoček R (eds) Fatigue crack growth in rubber materials. Advances in polymer science, vol 286. Springer, Cham. https://doi.org/10.1007/12_2020_72

Adv Polym Sci (2023) 289: 293–316
https://doi.org/10.1007/12_2022_115

Published online: 29 April 2022

Parameters Influencing Fatigue Characteristics of Tyre Tread Rubber Compounds

P. Ghosh, J. Chanda, and R. Mukhopadhyay

Contents

Abstract Tyre acts as a linkage between a vehicle and road. Out of many functions a tyre must perform in its service life, durability is one of them. In service, a tyre is exposed to all kinds of road hazards (sharp asperities, pot holes, nails, etc.) and this may cause cuts on different parts of a tyre, viz. tread, sidewall, etc. These cuts propagate during subsequent loading and tyre may fail on reaching a critical crack length. To ensure tyre durability, knowledge of fatigue crack propagation is essential. In a tyre, majorly Natural rubber (NR), Butadiene rubber (BR) and Styrene butadiene rubber (SBR) are used. In this work, fatigue crack growth (FCG) characteristics of these rubbers and their blends using a Tear & Fatigue Analyser (TFA, Coesfeld GmbH & Co. KG, Germany) are studied in detail. This study emphasized on capturing the influence of material composition (rubber, rubber blend and filler) and operational conditions (temperature, waveform and R ratio) on FCG resistance of tyre rubber compounds. It has been observed that rubber blends have shown superior FCG resistance over single rubber compounds up to a certain tearing energy

P. Ghosh (✉), J. Chanda, and R. Mukhopadhyay
Hari Shankar Singhania Elastomer and Tyre Research Institute, 437, Hebbal Industrial Area, Mysore, India
e-mail: pghosh@jkmail.com

level. The influence of filler on FCG resistance has varied effect on crystallized and non-crystallized rubbers. In addition to filler, much higher property enhancement is observed in non-crystallized rubber compared to crystallized one. FCG resistance found to be increased with decrease in carbon black particle size. Temperature has a deleterious effect on FCG resistance and this effect is more prominent in blend compounds. Gaussian pulse load form exhibited higher FCG rate compared to sine loading. Positive R ratio exhibited superior FCG resistance, especially in NR compound. It has also been observed that advantage of positive R ratio effect in terms of superior FCG resistance diminished due to increase in measurement temperature. Fractured surface micrographs of NR compounds are distinctly different from BR and SBR compounds and reflected as higher root mean square roughness parameter.

Keyword Frequency · Fatigue crack growth · R ratio · Rubber · Rubber blend · Tear and fatigue analyser · Tearing energy · Temperature

1 Introduction

By definition, fatigue is the progressive weakening of the material under cyclic or repetitive loading. This weakening is due to the cumulative damages, which occurs over a period of time during loading and unloading cycles. These damages initiate at the intrinsic defect site in micro levels and subsequently evolved as a catastrophic failure because of loss of strength in the material. On application of forces, the stress concentration occurs at the flaw sites which cause increase in the dimension of the flaw size or micro-crack length. This is a cumulative process and materials fail at certain stress which is much lower than the ultimate strength of the material. In any solid material including rubber vulcanisates, these intrinsic defect sites are inevitable [1] and vary in terms of quantities, topology and length scales and mostly depending on the manufacturing process. It is a difficult task to quantify these defect sites and hence caution must be exercised while designing the product by ensuring adequate safety factor.

Rubber has been recognized as an engineering material for more than two and a half centuries by now. The unique ability of simultaneously energy storage and dissipation makes rubber as preferred choice for many sophisticated engineering applications over metals. Due to these distinctive characteristics, rubber has become an indispensable material and found its usage in all areas of engineering applications, ranging from household to space science. Rubber also widely used in various other areas, such as biomedical applications, soft robotics, sensor applications, etc. The application environment of these rubber products also varies widely, static, dynamic, as well as cyclic in nature. Hence, rubber products which are used in dynamic and cyclic applications are more prone to failure compared to static applications and require adequate understanding of failure modes and causes of failures. Tyre is one such product that operates under highly dynamic condition and undergoes millions of fatigue cycles in its service life. Application is much more demanding in electric vehicle tyres due to higher load and torque requirement.

In a tyre, majorly Natural Rubber (NR), Polybutadiene Rubber (BR) and Styrene Butadiene Rubber (SBR) are used in various tyre components (tread, sidewall, skim, etc.). It is important to note that some properties are unique to a particular type of rubber due to their varied chemical structures. For instance, NR provides higher strength due to its ability to crystallize under strain, BR exhibits better abrasion resistance because of lower glass transition temperature, SBR offers superior traction property due to possessing higher friction coefficient. Therefore, the application of these rubbers in a tyre is selected based on the vehicle type, e.g., passenger car, commercial vehicle, or farm. For example, in commercial vehicle tyre, NR is more suitable due to higher load carrying requirement, whereas in passenger car tyre, SBR and BR are majorly used to meet high speed performance and safety requirement. It is also to be noted that in a tyre various components (tread, sidewall, etc.) have some specific functions to perform and therefore it requires selection of appropriate rubber that helps in achieving the desired functionality of that component. To elucidate further, in tread either NR or blends of NR with BR or SBR are used, whereas blend of NR & BR is prevalent in sidewall. Furthermore, with the advancement in automobile technology, improved road infrastructure and enhanced customer expectations, meeting tyre performance property targets are also becoming more and more challenging. Blending of rubbers is one of the remedies to these challenges and target properties are realized by harnessing unique properties of individual rubbers. However, blending of rubbers poses many challenges, mainly in processing due to thermodynamic immiscibility of these rubbers and that needs great attention. As reported by Mangaraj [2], in polymer–polymer blends, thermodynamic immiscibility arises due to low entropy gain during mixing because of higher molecular weight that prevents free movement of polymer molecules. At the same time, enthalpy gain during mixing process is high and resulting in net positive free energy change. However, achieving miscibility is possible as these rubber blends have upper critical solution temperature (UCCT). This implies that above this critical temperature, components are miscible in all proportions. Apart from this, rubber compound is a mixture of filler and multiple chemicals. The interaction of these constituents with rubber matrix is a complex phenomena. It is important to mention here that in a rubber compound, both physical (rubber-filler or filler-filler) and chemical interactions (crosslinking) are in existence. Due to this fact, properties of rubber composites are highly influenced by these interactions and always remained as an important research topic.

Mars and Fatemi [3] reported a comprehensive review on important influencing factors that affect fatigue properties of rubber. These factors were categorized under four major groups: mechanical loading history, environmental effects, material formulation and constitutive behaviour. These factors have influence on both short-term and long-term fatigue properties of rubber compounds and for further details, this publication may be referred. One of the important inferences drawn in this review is that successful integration of results of all these individual factors in accurate fatigue life prediction remains as a challenge for the future research. Tee et al. [4] discussed various fatigue life prediction approaches used by various researchers in their review article on advances on rubber fatigue research carried out post 2004.

It is very much evident from both the above review articles that fatigue properties can be greatly influenced by using various rubber types and its composition. According to Young [5], strain crystallizable rubbers such as NR, choloroprene rubber (CR) have distinct advantage over other rubbers with respect to FCG resistance. Similarly, other researchers [6–11] have clearly brought out the benefit of rubber blending to accomplish improvement in FCG resistance.

Use of fillers, mainly carbon black in tyre is well known and its effect on fatigue properties of rubber was investigated since 1945 [4]. However, mechanism involved in reinforcement with respect to strength is complex and depends on type of rubber as well as carbon black. Silica as a filler in tyre also gained importance in the last few decades due its better wet grip and low hysteresis characteristics. Mars and Fatemi [3] reported that fillers alter fatigue properties through various mechanisms which include enhancement of stiffness characteristics, changes in dissipation characteristics, crack tip blunting, crack path deviation, and increase in effective initial flaw size due to agglomeration of filler particles. It is worth mentioning here that there is an optimal filler volume fraction up to which fatigue property improves. Beyond this point, fatigue property deteriorates due to increase in effective initial flaw size. Kim and Jeong [12] studied the influence of N330, N650, and N990 carbon black on fatigue properties of NR compound. They have observed the fatigue life as follows: the ranking of N330 is greater than N990 or N650. Nie et al. [13] showed that NR with N330 carbon black exhibited superior FCG characteristics due to higher viscoelastic dissipation compared to N770 carbon black. Reincke et al. [14] showed that at similar filler loading, silica provides better FCG resistance compared to carbon black.

The deleterious effect of temperature on fatigue properties was reported by many researchers [15, 16] for different NR based compounds. According to Lake and Lindley [17], effect of temperature on fatigue properties is much pronounced in amorphous rubbers (SBR, BR, etc.) compared to crystalline rubber, NR According to their study, unfilled SBR exhibited 10^4 times higher crack growth rate measured at 100°C compared to the FCG measurement at 0°C. Whereas in the case of NR compound, only four fold increase in FCG rate was observed in the same temperature range. Influence of aging on FCG was studied by Stoček et al. [18] using NR, SBR and NR/SBR blends.

Loading frequency and frequency waveform also attracted attention to many researchers in the past. Tyre in service subjected to cyclic loading and hence type of loading and loading frequency plays an important role in fatigue performance. Harbour et al. [19] reported two times reduction in fatigue life and 25°C rise in sample surface temperature occurred due to the test frequency increase from 0.2 to 1 Hz. With respect to frequency waveform, several researchers [8, 19–22] have investigated in detail and they have found that FCG rate is considerably higher in pulse than sine waveform.

Table 1 Summary of research work on effect of various parameters on rubber fatigue

Parameter	Research work carried out by
Rubber & rubber blends	Young [5], Lee [6], Chung and Chang [7], Ghosh et al. [8, 9], Stoček et al. [10, 11]
Filler	Mars and Fatemi [3], Kim and Jeong [12], Reincke et al. [14], Nie et al. [13]
Temperature	Lake and Lindley [17], Young and Danik [15], Legorju-jago and Bathias [16], Qazvini [28]
Waveform	Harbour et al. [19, 20], Andreini et al. [21, 22], Ghosh et al. [8]
R ratio	Lindley [23], Mars and Fatemi [24], Abraham et al. [25], Saintier et al. [26], Stadlbauer et al. [27], Ghosh et al. [8]

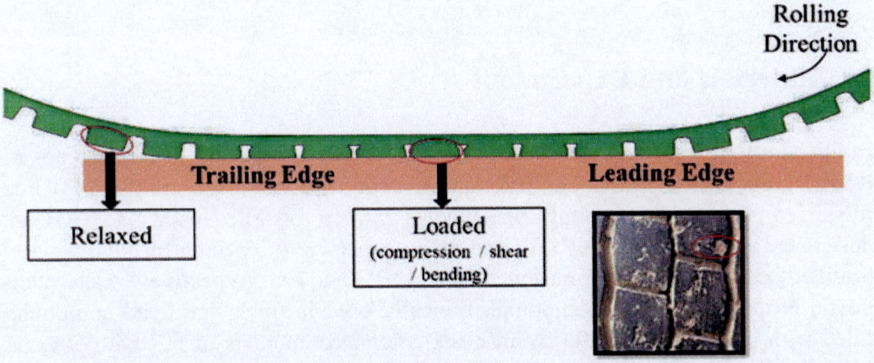

Fig. 1 Schematic of rolling process and image of damage on tyre tread surface

Researchers have studied the influence of R ratio on FCG resistance of rubber compounds. In majority of these works [8, 23–26] FCG tests were conducted under positive R ratio. However, Stadlbauer et al. [27] carried out FCG test under both positive and negative R ratio. From the above works it is evident that FCG rate decreases under positive R ratio, whereas under negative R ratio, FCG rate shows a reverse trend.

A brief summary of previous research work discussed so far is presented in Table 1. However, most of the previously reported research works portrayed a broad perspective and are not specific to any product application, like tyre. This is one of the major reasons that motivates the authors to undertake the present study, which is focused on understanding the specifics pertaining to tyre application, mainly tread as a component A schematic of tyre rolling process and a representative image of damage occurs in tread during its lifecycle are provided in Fig. 1. To illustrate further, tyre components at the contact patch experience loading for a very short duration and after leaving the contact patch these components get relaxed for a longer duration. The typical ratio of short to long duration is 1:4. This type of loading resembles Gaussian pulse waveform rather than sinusoidal and will be taken into consideration during experimentation. In this study a tyre tread compound is taken as a reference compound as tyre tread directly comes in contact with various road surfaces. Further, due importance is also given to understand the fundamental aspects of fatigue properties which is necessary for developing a successful product. In this work, major influencing parameters that may affect fatigue properties are considered here. These are types of rubber, blends of rubber, filler types, temperature, frequency and R ratio. A special emphasis has been given towards understanding of interaction of rubber blends and selected filler. In this work, fatigue properties of NR, BR, SBR and their blends will be discussed in detail. Three types of carbon blacks (N134, N234, & N330), mainly used in tyre tread applications have considered to understand its influence on fatigue properties. A compound containing 100 parts of NR filled with 50 parts of N339 is used as a reference tread compound.

2 Materials and Experiment

In this study, three kinds of rubbers (NR, BR & SBR) and four different types of ASTM grade (N134, N234, N330 & N339) carbon blacks (CB) were used. Pristine rubber compounds having only one type of rubber and filled with N339 CB are designated as N, B & S for NR, BR and SBR, respectively. Their corresponding gum (unfilled) compounds are represented as NG, BG and SG, respectively. Here compound N is the reference compound, typically used in truck tyre tread compound filled with 50 parts of N339 carbon black. Blend compounds of NR with BR and SBR in 60–40 proportion and filled with N339 CB are represented as NB & NS, respectively. Pristine NR compounds filled with N134, N234 and N330 are designated as N1, N2 and N3, respectively. Recipe of all these compounds is detailed in Table 2 containing information about types and quantity of rubber, CB, vulcanizing agent (sulphur & accelerator) and other additives constitute zinc oxide, stearic acid, anti-degradants, etc. It is important to mention here that the reference compound N is a proprietary compound and therefore exact quantities under other additives were not disclosed. Rubber compounds were prepared using a laboratory Banbury mixture

Table 2 Formulation of rubber compounds

Ingredients	N	NB	B	NS	S	NG	BG	SG	N1	N2	N3
NR	100	60	–	60	–	100	–	–	100	100	100
BR	–	40	100	–	–	–	100	–	–	–	–
SBR	–	–	–	40	100	–	–	100	–	–	–
CB1[a]	50	50	50	50	50	0	0	0	–	–	–
CB2[a]	–	–	–	–	–	–	–	–	50	–	–
CB3[a]	–	–	–	–	–	–	–	–	–	50	–
CB4[a]	–	–	–	–	–	–	–	–	–	–	50
Sulphur	2	1.5	1.3	1.8	1.6	2	1.3	1.6	2	2	2
Accelerator	0.5	0.8	1.15	1.0	1.8	0.5	1.15	1.8	0.5	0.5	0.5
Other	9.25	9.25	9.25	9.25	9.25	9.25	9.25	9.25	9.25	9.25	9.25

[a] CB1 – N339, CB2 - N134, CB3 – N234, CB4 – N330

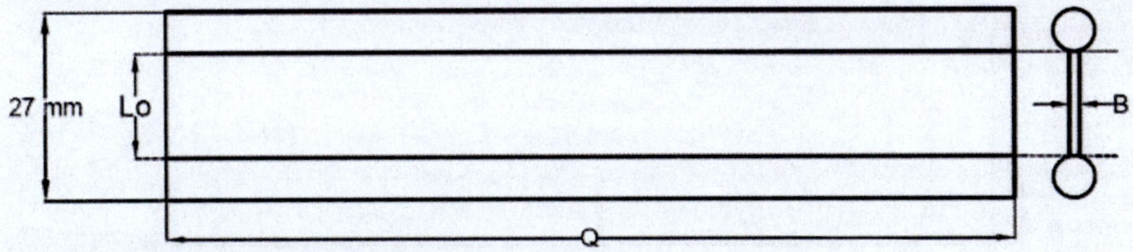

Fig. 2 Schematic of plane strain specimen

Table 3 Fatigue crack growth test condition

Parameters	Temperature effect	R ratio effect	Waveform effect	
			Sine	Gaussian pulse
Strain (%)	13–25	13–25	13–25	13–25
Temperature (°C)	30 & 70	30	30	30
R ratio	0	0 & 0.14	0	0
Waveform	Sine	Sine	Sine	Gaussian pulse
Frequency (Hz)	10	10	10 Hz	10 Hz repeat and 50 Hz pulse

(Make: Stewart Boling, UK) of 1.5 l capacity in three stages. In the first stage, rubbers and all chemicals except curatives were mixed at 60 rotations per minute (rpm) for 4 min time at about 140°C temperature. Then the mixed compound is re-milled at 30 rpm for 3 min time at about 120°C. The final batch containing curatives is mixed at 30 rpm for 3 min at 100°C. Plane strain specimens for FCG testing were prepared using a compression mould at 140°C temperature, 14.7 MPa pressure and cured for 1 h time. Mechanical properties of these compounds were measured using UTM (Zwick Z010, Germany).

FCG properties of all the compounds were measured using a Tear & Fatigue Analyser (TFA, Coesfeld GmbH & Co. KG, Germany). The detailed description of the equipment is described elsewhere [29, 30] and the testing methodology is detailed in the reference [10, 11]. Plane strain specimen of length $L_0 = 15$ mm, width $Q = 120$ mm and thickness $B = 1.5$ mm was used for FCG measurement as shown in Fig. 2 (schematic). Initial cuts of 21 mm provided on both edges of the specimens using a notch provider supplied by Coesfeld GmbH & Co. KG, Germany. The test conditions followed for conducting the FCG experiments are given below in Table 3. Figure 3 shows the schematic of sine and Gaussian pulse waveforms used for FCG measurement.

As a general test procedure, the fatigue loading cycle was applied as per the condition mentioned against each test tabulated in Table 3 and the crack length was monitored continuously for calculating the crack growth rate (*da/dn*) (Fig. 3). Un-notched test specimens were loaded up to the desired strain level to obtain elastic strain energy density, *W* and the tearing energy *T* is then calculated using the following equation:

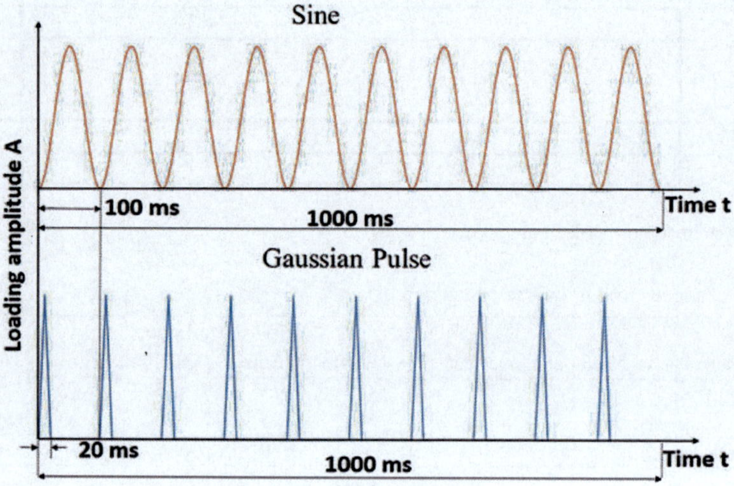

Fig. 3 Schematic of sine and Gaussian pulse waveforms

$$T = WL_0 \tag{1}$$

3 Results and Discussion

3.1 Mechanical Properties

Mechanical properties of all the rubber compounds are provided in Table 4. This table portrays the influence of rubber blending and filler addition on mechanical properties of these rubber compounds.

From this table it can be observed that the blending of BR and SBR with NR has significant influence on tensile strength (TS) and elongation at break (EB). TS and EB improved in both NB and NS compared to B and S compounds, with higher degree of influence observed in NB compound. Whereas, in the case of hardness and 100% modulus, change is very minimal. The influence of carbon black on mechanical properties of rubber compounds is noteworthy, mainly in non-crystallizing rubbers (BR, SBR, etc.). Figure 4 shows the influence of carbon black addition on various mechanical properties that includes hardness, modulus at 100% elongation, TS and EB.

Table 4 also contains property enhancement (ratio of property of pristine compound to gum compound) factor with a purpose of highlighting the reinforcement ability of CB in rubber, especially in synthetic rubbers. From this table it can be seen that property enhancement factor lies in the range of 1.4–1.6 for hardness and 2.2–2.7 for 100% modulus. In the case of elongation at break, property enhancement

Table 4 Mechanical properties of rubber compounds

Compound	Hardness (ShoreA)	100% modulus (MPa)	Tensile strength (MPa)	Elongation at break (%)
N	62	1.9	25.8	596
NB	60	2.0	25.0	585
B	67	2.6	15.3	398
NS	61	1.9	23.3	571
S	66	2.2	17.4	429
NG	38	0.7	21.1	742
BG	44	0.9	1.60	288
SG	47	0.9	1.70	271
N1	59	1.8	27.3	620
N2	57	1.7	26.2	632
N3	55	1.5	24.2	614
N/NG[a]	1.6	2.7	1.20	0.8
B/BG[a]	1.5	2.9	9.50	1.4
S/SG[a]	1.4	2.4	10.2	1.5

[a] Property enhancement factor – property of pristine to gum compound

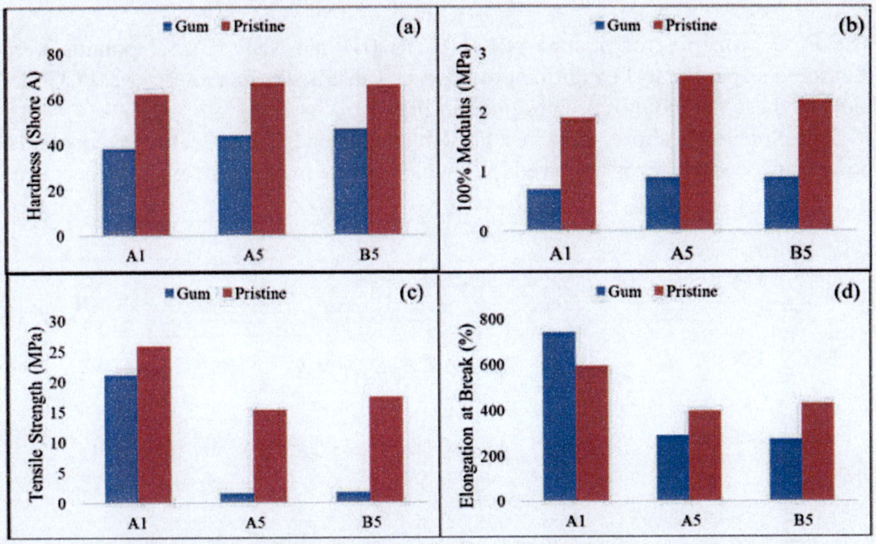

Fig. 4 Mechanical property comparison: gum (NG, BG, SG) vs. filled compounds (N, B & S): (**a**) Hardness, (**b**)100% Modulus, (**c**)Tensile Strength, (**d**) Elongation at break

follows the similar trend, with an exception in compound N. The reduction in elongation at break for N compound happens due to the restriction of chain movement in presence of filler particles. Furthermore, it is interesting to note that the enhancement factors for tensile strength are much higher (~10 times) in non-strain crystallizing rubber (BR and SBR) compounds B & S than that of strain crystallizing rubber (NR). In non-crystallizable rubbers, the strength enhancement happens due to minimization of stress concentration in the rubber network [31]. In presence of active fillers, rubber chains in the spatial network get absorbed on the filler surface. In the event of high stress condition, these rubber-filler bonds get broken in the stressed portion and transfer the load to the low stressed region and thereby prevent the chain rupture. However, in unfilled strain crystallizing rubbers, upon stretching crystallites formation takes place and plays the same role as that of filler. So crystallization phenomena dominates over filler reinforcement effect in strength enhancement resulting in lower property enhancement factor. Though small improvement in TS of compound N (property enhancement factor – 1.2) could be observed. This may be attributed to the fact that in presence of active filler, during stretching, active fillers acts as a nucleation point for crystal formation.

3.2 Fatigue Crack Growth Properties

3.2.1 Rubber and Rubber Blends

The FCG properties of pristine NR (N), BR (B) and SBR (S) compounds were measured as per the test condition provided in Table 3. Tearing energy vs. FCG rate plots of these compounds are provided in Fig. 5.

This plot clearly shows that crack growth rate depends on the tearing energy and a power law dependency is observed [32], which can be mathematically represented as

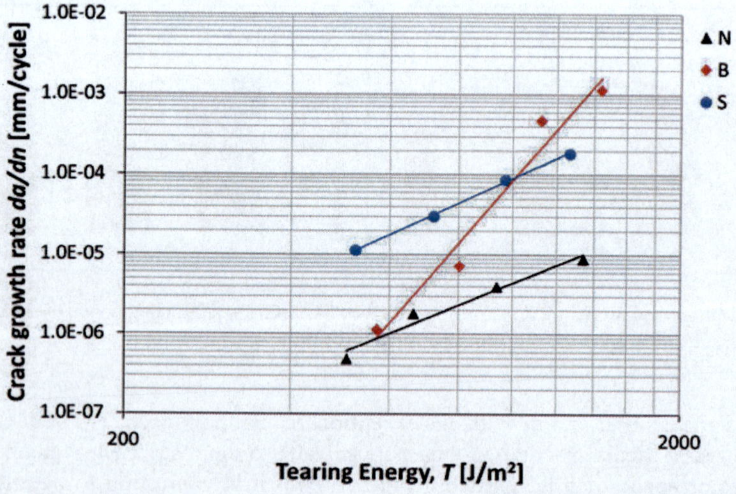

Fig. 5 FCG curves of N, B and S compound

Table 5 Fatigue crack growth parameters of N, B,S, NB & NS compounds

Compound	*b*	*m*
N	1.03E-14	2.87
B	5.03E-29	8.07
S	2.56E-14	3.18
NB	1.34E-16	3.14
NS	7.26E-29	7.44

$$\frac{da}{dn} = bT^m \tag{2}$$

where T is the tearing energy, b and m are the material constants. Fatigue crack growth parameters (b and m) are computed from these plots listed in Table 5.

It has been observed that S compound showed the highest FCG rate in low to moderate tearing energy level compared to the other two compounds. Between N and B, it can be seen that B shows lower FCG rate in lower tearing energy level and higher rate at higher tearing energy. This observation is in agreement with Young [5] where crossover point between BR and NR curves is reported at 15% strain. N compound shows highest FCG resistance at higher tearing energy level due to strain induced crystallization in NR based compound at moderate to large strain [1]. It is also important to mention here that the local strain at the crack tip is much higher than the global strain level due to strain amplification phenomena as observed by Liu et al. [33]. Higher strain level facilitates the crystallization process near the crack tip. It is also evident from the work of Bruning et al. [34], where they have experimentally measured the degree of crystallinity of NR compound near the crack tip with the help of wide-angle X-ray diffraction (WAXD). From Table 5, it is evident that the compound B exhibited highest crack growth exponent (m) among all the compounds under study.

The effect of blending of BR and SBR with NR on FCG resistance is shown in Fig. 6. It is evident from this figure that introduction of 40 parts of BR (NB) and SBR (NS) in the rubber compounds facilitated to achieve better FCG resistance compared to the single rubber compound (N). Out of these three compounds, NB exhibited superior FCG resistance in the entire tearing energy level and also reflected in terms of lower crack growth parameters, b & m. Whereas, FCG resistance of NS appears in between N and NB compound at lower tearing energy level. However, at higher tearing energy level, NS curve crosses the N curve ~20% strain corresponding to tearing energy of 1,258 J/m^2. Reason for exhibiting superior FCG resistance in NB compound may be due to the selective migration of carbon black to the NR phase as reported by Wunde and Klüppel [35]. Higher carbon black in NR phase facilitates strain induced crystallization and in turn these crystallites hinder crack propagation [36].

3.2.2 Filler

FCG properties of gum and pristine compounds are shown in Fig. 7a–d. It is also evident from FCG properties that pristine compounds of non-crystallizable rubbers

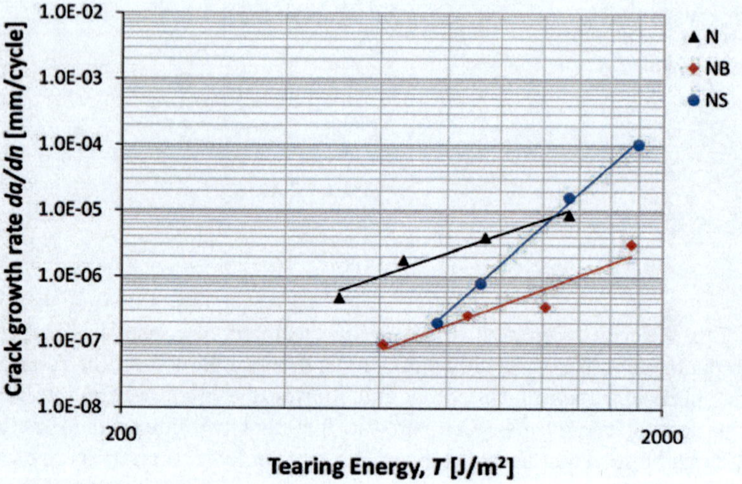

Fig. 6 FCG curves of N, NB and NS compound

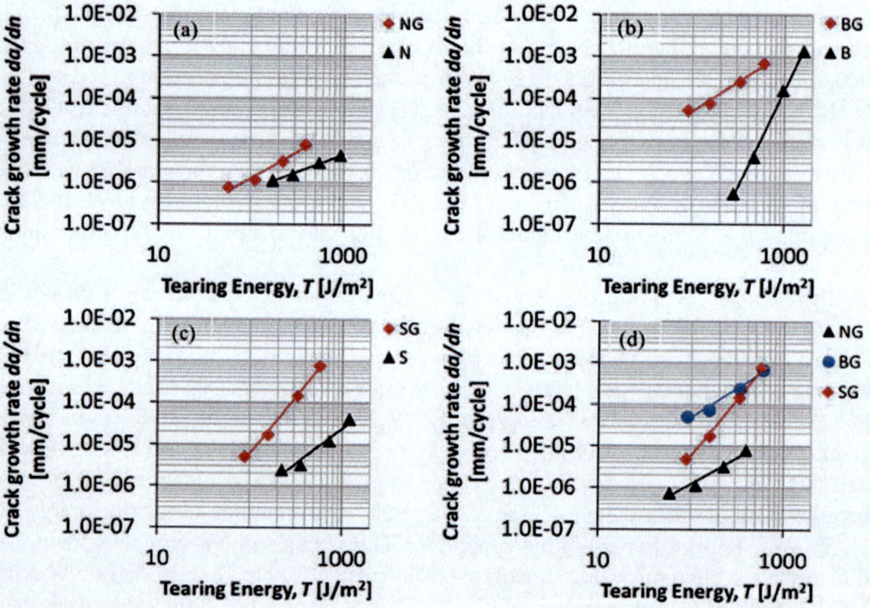

Fig. 7 FCG curves of gum and pristine compounds, **(a)** NG & N **(b)** BG & B **(c)** SG & S **(d)** NG, BG & SG

(BR and SBR) exhibited many fold increase in FCG resistance compared to their gum counterpart. Whereas, in the case of pristine NR compound (Fig. 7a), there is an increase in FCG resistance compared to the gum compound but magnitude is much less. This observation corroborates the findings made in mechanical property Sect. 3.1 for strain crystallized and non-crystallized rubbers.

Topography of fatigue fracture surfaces of both gum and pristine compounds of NR, BR and SBR is depicted in Fig. 8a–f. Topography of fatigue fracture surface has been captured by using a Field Emission Scanning Electron Microscope (ApreoS, FEI, USA) with an accelerating voltage of 5 kV. The samples were coated with Platinum by using a sputter coater Q150T ES (Quorum Technologies, UK) to avoid the charging effect. The fractured surface for compound N is very rough in appearance and distinctly different from corresponding images of B and S compounds. Rough appearance may be linked to the strain induced crystallization characteristics of NR and clearly visible in both filled and gum compound. This supports previous observation on mechanical property enhancement factor for NR gum vs pristine compounds as well as in FCG rate compared to BR or SBR compounds. In the case of BR and SBR, gum compounds exhibited rough appearance in comparison with their pristine counterparts and this is in agreement with the observation made by Sridharan et al. [37]. According to them, this could be attributed to the lower strength of gum compounds and that can lead to removal of large quantity of material in the form of a chunking. Interestingly, fatigue fractured surface of BR pristine compound is very smooth in appearance compared to the rest of the compounds.

Furthermore, roughness quantification is carried out using an Atomic Force Microscope (NX10, Park Systems, Korea) in tapping mode utilizing a silicon nitride tip of 10 nm radius. The root mean square (RMS) roughness of fracture surfaces was calculated from 20 μm X 20 μm images with the help of Smart Scan software. AFM images of both gum and pristine compounds of NR, BR and SBR are presented in Fig. 9a–f. RMS roughness values of these compounds are also computed and given in Table 6. RMS roughness of NR gum and pristine compound shows higher value in comparison with corresponding BR and SBR compounds. This may be attributed to inherent green strength of NR that promotes stick-slip type of fracture mechanism that leads to generation of rough fracture surface [38]. It was also observed that the roughness of NR pristine compound is higher than gum in contradiction to the BR and SBR where pristine compound displayed more smooth appearance than the gum counterpart as seen in SEM images and the reasons are explained thereon.

Influence of various carbon black grades on FCG rate is shown in Fig. 10. It is evident that FCG resistance is seen to be decreased with increase in particle size of the carbon black. This could be associated with the better reinforcement ability of lower particle size carbon black. Moreover, compound with lower particle size will have higher hysteresis, this in turn helps in achieving higher FCG resistance by dissipating more energy at the crack tip.

3.2.3 Temperature

The influence of temperature on FCG rate of different rubber compounds based on single rubber as well as blend compounds is shown in Fig. 11 (a–d). Comparison of FCG rates between 30 and 70°C of N, NB and NS is shown in Fig. 11a–c, respectively. It can be observed that FCG rate is higher at 70°C compared to 30°C

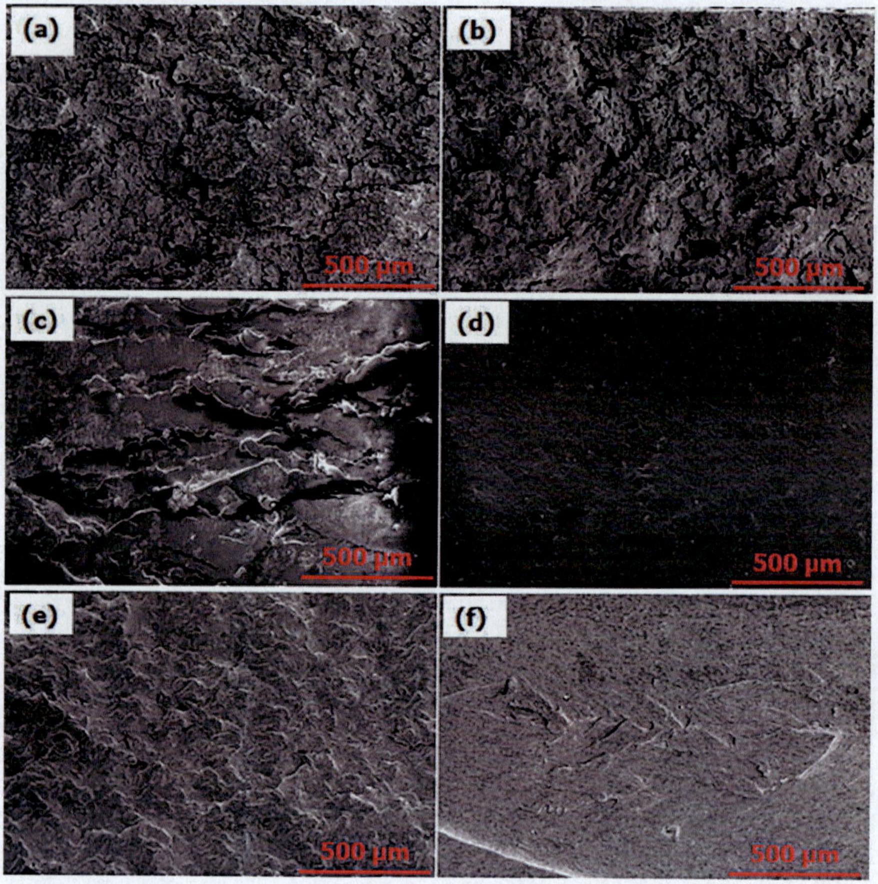

Fig. 8 Topography of fatigue fracture surfaces of gum and filled compounds (a) NG (b) N (c) BG (d) B (e) SG (f) S

in all the compounds. However, effect is much higher in blend compounds (NB and NS) compared to the NR compound which is also evident from Figs. 11d and 12 depicts the FCG of N, S and 60–40 NR/SBR blend NS to elucidate the effect of temperature on FCG in crystalline and amorphous rubber. It has been observed that at 70°C, FCG rate of compound S is higher than compound N and their blend compound exhibited intermediate FCG characteristics. From these results, two major inferences can be drawn here. Firstly, from Fig. 11, sensitivity of FCG rate towards temperature increase is comparatively less in NR (N) than BR (B) and SBR (S). Secondly from Fig. 12, though the blend compound NS showed higher temperature sensitivity than pristine N, it is lesser than pristine S compound. This requires discussion on underlying mechanism responsible for temperature sensitivity of FCG characteristics.

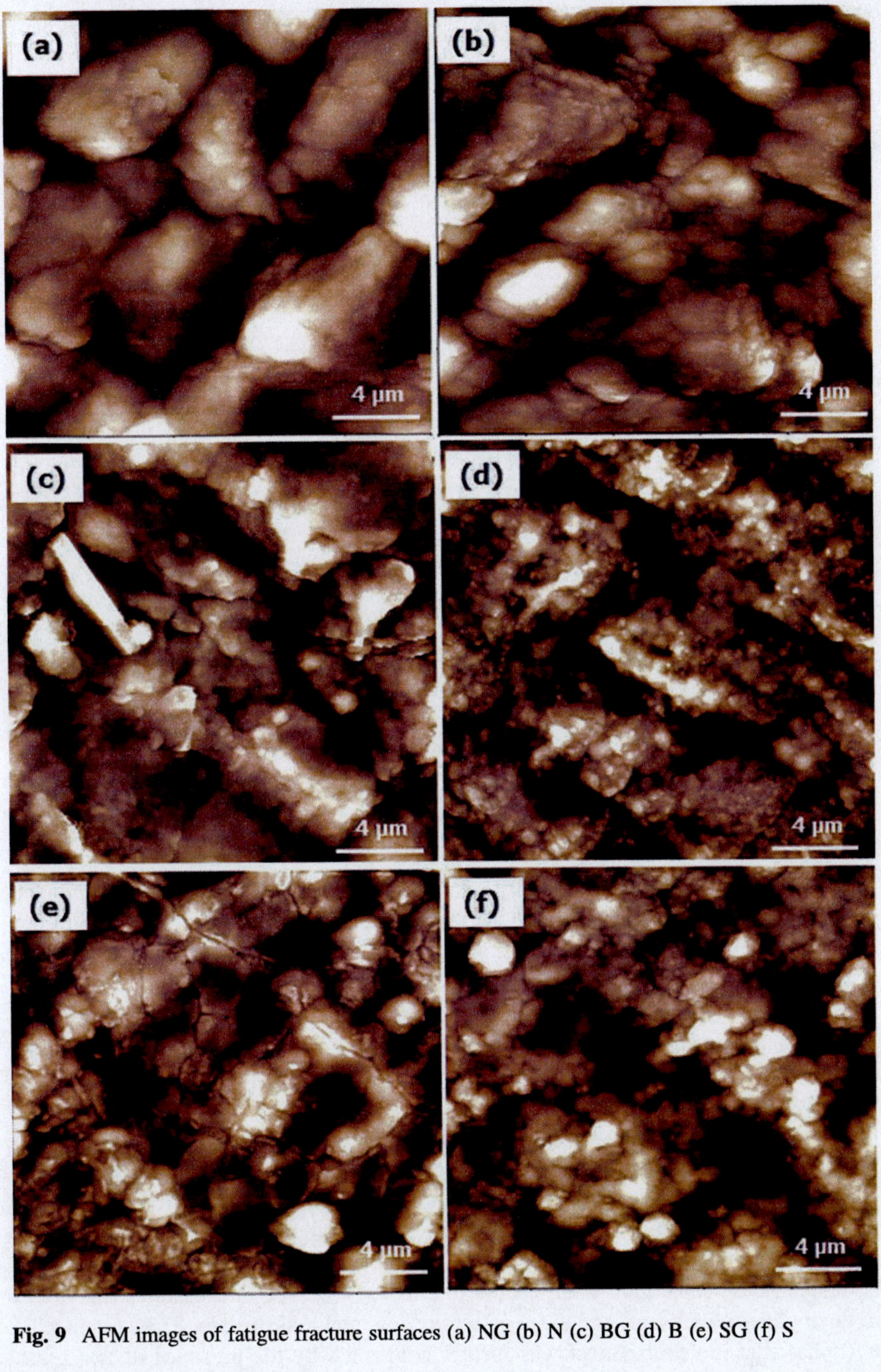

Fig. 9 AFM images of fatigue fracture surfaces (a) NG (b) N (c) BG (d) B (e) SG (f) S

Table 6 RMS roughness of gum and filled compound

Compound	RMS roughness (nm)
NG	352
N	660
BG	248
B	151
SG	192
S	173

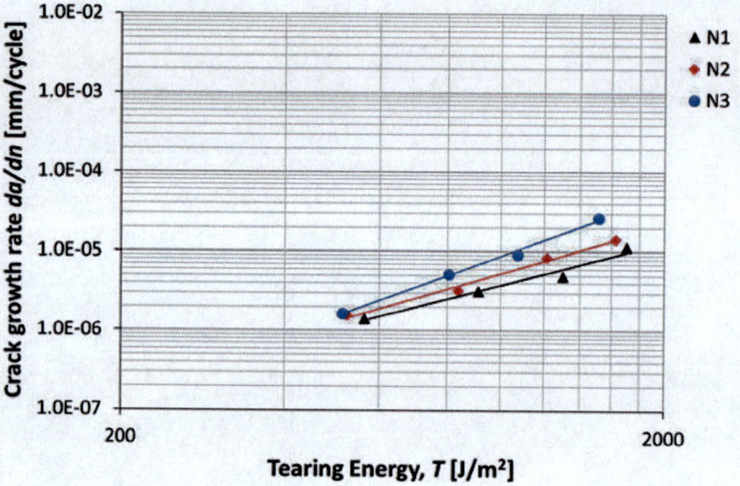

Fig. 10 FCG curves of N1, N2 & N3 compounds

In crack growth process of elastomeric material, energy dissipation takes place through various means that include cavity formation, bond breaking, chain pull out and viscoelastic dissipation [39]. Out of these, viscoelastic dissipation in front of the crack tip contributes significantly towards fracture resistance of the polymer. Temperature increase facilitates molecular movement in the polymer system and that leads to decrease in viscoelastic dissipation and that is also evident from the sharp reduction of viscoelastic loss modulus with temperature [8]. This in turn ensures more energy availability at the crack tip for crack to propagate and reduces fracture resistance. Kluppel [40] also showed the relation between crack propagation rate and viscoelastic property (scaling exponent of the relaxation time spectra) of elastomers. Higher temperature sensitivity of blend compounds may be attributed to the thermodynamic immiscibility of rubbers in micro level as there is a possibility of having existence of heterogeneous phase structure at room temperature [28]. However, complete phase separation is prevented due to the presence of crosslinks (network) in the rubber vulcanisates and these networks bear residual stresses. The existence of two phases at molecular level was further confirmed by the presence of two glass transition temperatures (*Tg*) of the blend compounds [8]. Regarding higher

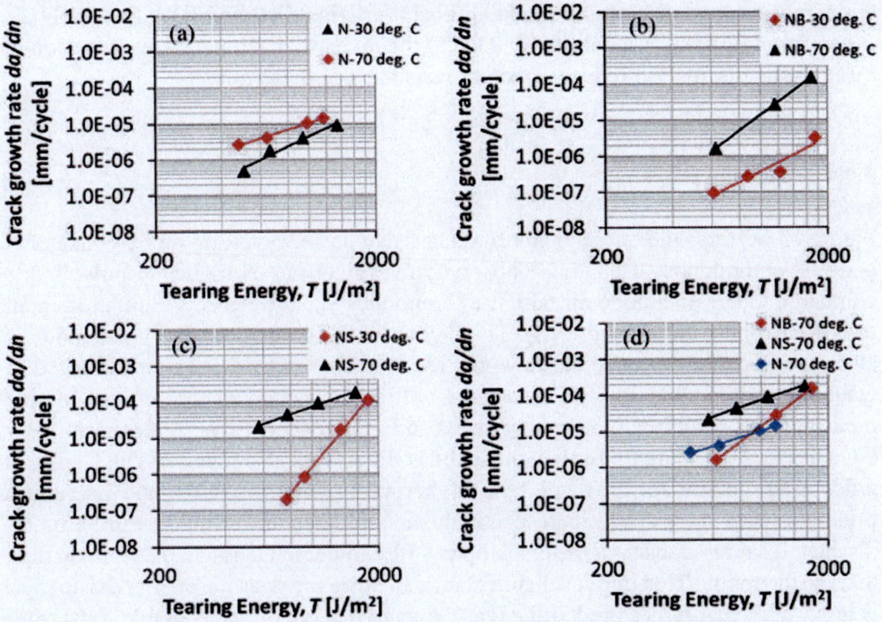

Fig. 11 FCG curves: 30°C vs. 70°C (a) N (b) NB (C) NS (d) N, NB & NS

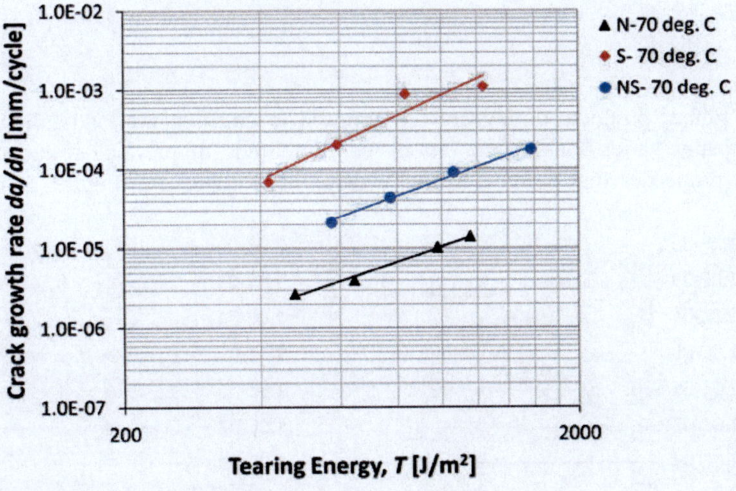

Fig. 12 FCG curves for N, S and NS at 70°C

temperature sensitivity of amorphous rubbers (pristine SBR and BR), these rubbers derive strength from viscoelasticity. Due to the increase in temperature, reduction of viscoelastic dissipation reduces fracture resistance of these rubbers.

3.2.4 Waveform

Figure 13 portrays the effect of sine and Gaussian pulse waveform on FCG rate of N and NB compounds. It has been observed that FCG rate is higher in pulse mode compared to the sine mode for both the compounds and the effect is more prominent in NR/BR blend compound (NB). The higher FCG rate in Gaussian pulse mode is also evident from earlier research works [8, 20–22]. Higher FCG rates observed in Gaussian pulse mode may be associated with higher strain rate compared to sine mode and that resulting into elevated stress and strain energy level. A recent study by Ghosh et al. [41] using finite element analysis (FEA) tool also revealed that Gaussian pulse mode exhibited 22% and 16% higher strain energy density and maximum principal stress at the crack tip, respectively, as compared to sine mode. Furthermore, dissipated energy density at the crack tip is ~50% lower in Gaussian pulse mode than that of sine mode. This implies that in Gaussian pulse mode as the energy dissipation is less, more energy is stored at the crack tip and this energy is available for further crack propagation. Hence energy availability at the crack tip is an important criteria for a crack to propagate further. Similar observation was also made by Wunde et al. [42] and stated that in viscoelastic material, most of the energy dissipates around the crack tip region and leaves small amount of energy for crack propagation.

3.2.5 R Ratio

Many rubber products during service exposed to constant pre-load followed by dynamically varied loading condition. The amplitude of pre-load influences the rubber properties that include stiffness, damping, fatigue, etc. Figure 14 depicts

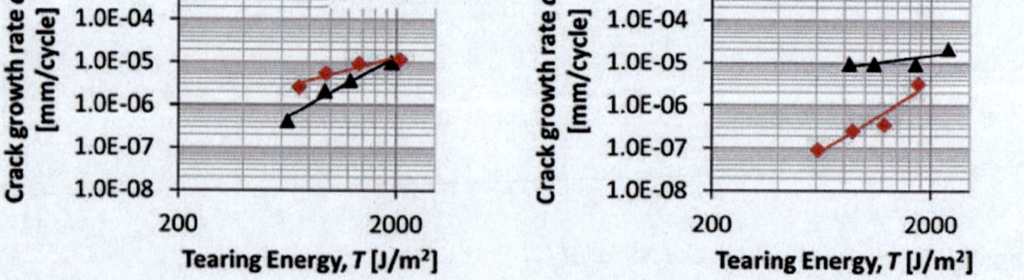

Fig. 13 FCG curves: sine vs. Gaussian pulse (**a**) N- Sine & N-Gaussian Pulse (**b**) NB- Sine & NB-Gaussian Pulse

Fig. 14 Schematic loading sequence

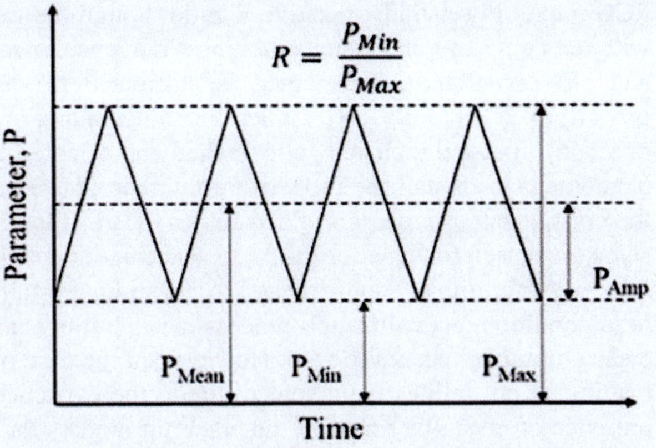

the schematic of loading sequence of rubber product. The loading sequence is termedas *R* ratio (ratio of minimum to maximum amplitude of applied force, strain, stress, etc.). Depending on the application, the pre-load could be negative (constant compression) or positive (constant tension).

Figure 15a–d depicts the influence of positive R ratio on fatigue crack growth rate of different rubber compounds N (100 NR) and blend compounds NB (60–40 NR/BR) and NS (60–40 NR/SBR). It has been observed that in all the compounds,

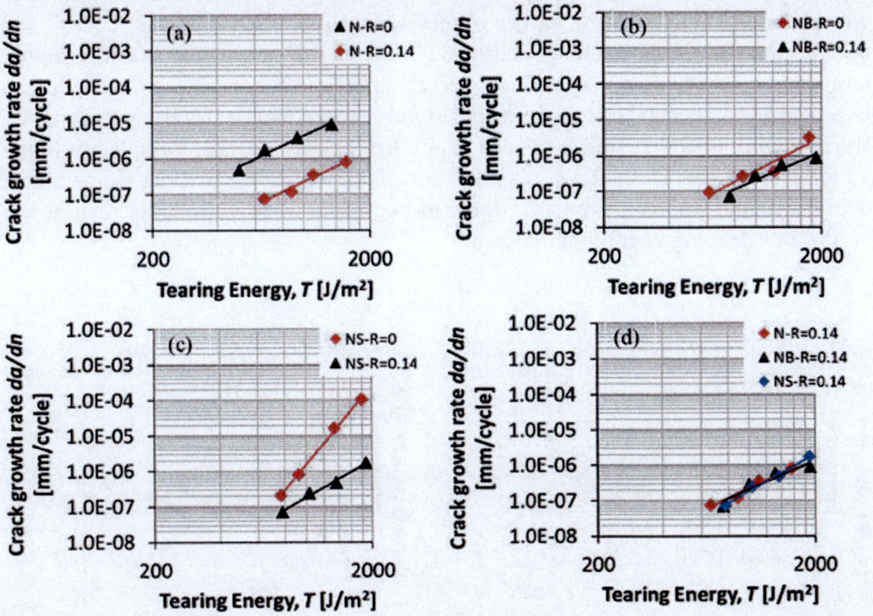

Fig. 15 FCG curves for N, NB and NS at different R ratios: **(a)** N-R = 0 & N-R = 0.14, **(b)** NB-R = 0 & R = 0.14, **(c)** NS-R = 0 & R = 0.14, **(d)** N-R = 0.14, NB-R = 0.14 & NS-R = 0.14

FCG rate is lower under positive R ratio though the magnitude of decrease varies with the composition. N showed highest influence towards R ratio followed by NS and NB compounds. This could be attributed to strain induced crystallization property of NR [23, 26, 34]. Under dynamic loading to a specific amplitude (stress or strain), molecular chains get stretched and oriented that facilitate crystallization phenomena in crystallized rubbers. Furthermore, presence of carbon black augments the crystallization process in filled rubbers [36]. It is also necessary here to understand the reason for superior FCG resistance under positive R ratio compared to zero R ratio. According to Saintier et al. [26], in strain crystallized rubbers, even in zero R ratio condition, crystallization process exists but crystallites melt once the loading cycle amplitude reaches to zero. However, in the case of positive R ratio, complete melting of crystallites is prevented due to the existence of constant pre-load. The presence of crystallites around the crack tip hinders the crack to grow linearly and leads to crack branching and also delay the crack growth process. In the case of blends, NS exhibited higher R ratio sensitivity compared to NB. Figure 15d shows the comparison of FCG rate of N, NB and NS compound at 0.15 R ratio. It can be observed that under this condition (R = 0.15), FCG rate of all these compounds is comparable.

As explained by Bruning et al. [34], strain induced crystallization in a rubber is a function of loading rate, temperature, etc. To capture the influence of temperature on crystallization in positive R ratio, additional experiments were carried out at different R ratio (0, 0.05 & 0.15) at 30°C and at 0.15 R ratio at 70°C and results are plotted in Fig. 16a, b. At 30°C, FCG rate is showing decreasing trend with increase in R ratio which is consistent with the earlier observation. However, FCG rate at 70°C and 0.15 R ratio is same as that of 0 R ratio and 30°C. This clearly explains that at higher temperature, melting of crystallites occurs and hence not able to provide the expected FCG resistance because of positive R ratio effect. It has been reported by Ruellan et al. [43] that strain induced crystallization disappeared completely in the temperature above 75°C. It was also mentioned that crystallite melting temperatures depends on the amount of stretch. Higher test temperature is required for melting the crystallites developed at higher stretch.

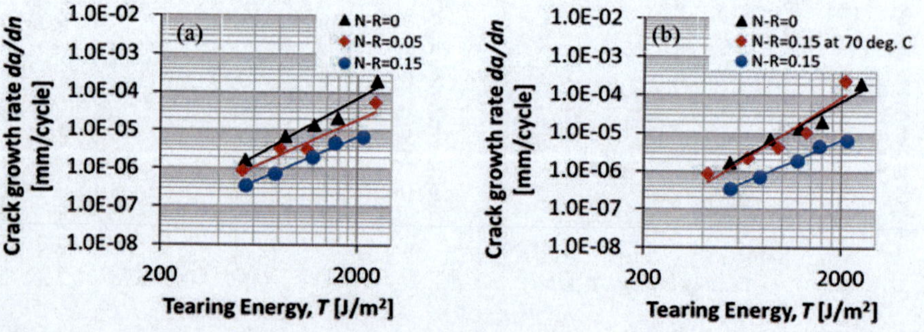

Fig. 16 FCG curves of N at different R ratios and temperature

4 Conclusion

The present work discusses the fatigue crack growth behaviour of tyre rubber compounds. FCG properties of NR, BR, SBR and their blends were demonstrated using a Tear & Fatigue Analyser. Influence of material and major operational conditions on FCG properties were described in detail. While analysing FCG behaviour of single rubber compounds, reversal of ranking has been observed beyond a certain tearing energy level. This can be considered as a very crucial information to the material developer for material selection. Blending of BR and SBR with NR yielded superior FCG resistance up to a certain tearing energy level. Addition of filler (carbon black) improved the FCG resistance in all the compounds. However, the degree of FCG property enhancement is more prominent in BR and SBR compared to NR compound. Temperature has significant influence on FCG property and considerable decrease in FCG resistance at 70°C is observed in all the compounds. Higher temperature sensitivity can be seen in blend compounds. Higher FCG rate was observed in Gaussian pulse mode compared to the sine mode. All the compounds exhibited superior FCG resistance at positive R ratio and highest effect is observed in NR compound followed by NR/SBR blend and NR/BR blend. It has been also observed that at higher temperature (70°C), influence of positive R ratio is diminished due to the melting of crystallites. SEM and AFM micrographs clearly revealed the differences in fractured surfaces of gum and filled compounds of NR, BR & SBR. RMS roughness values for NR compounds are significantly higher than BR and SBR.

Acknowledgements The authors thank the Managing Committee of Hari Shankar Singhania Elastomer & Tyre Research Institute (HASETRI) for giving permission to publish this research work.

References

1. Hamed GR (1994) Molecular aspects of the fatigue and fracture of rubber. Rubber Chem Technol 67:529–536
2. Mangaraj D (2002) Elastomer blends. Rubber Chem Technol 75:366–428
3. Mars WV, Fatemi A (2004) Factors that affect fatigue life of rubber: a literature survey. Rubber Chem Technol 77:391–412
4. Tee YL, Loo MS, Andriyana A (2018) Recent advances on fatigue of rubber after the literature survey by Mars and Fatemi in 2002 and 2004. Int J Fatigue 110:115–129
5. Young DG (1985) Dynamic property and fatigue crack propagation research on tire sidewall and model compounds. Rubber Chem Technol 58:785–805
6. Lee MP (1993) Analysis of fatigue crack propagation in NR/BR rubber blend. Rubber Chem Technol 66:304–316
7. Chung WW, Chang YW (2001) Fatigue crack growth behavior of NR/EPDM blend. Korea Polym J 9:319–326

8. Ghosh P, Stocek R, Gehde M, Mukhopadhyay R, Krishnakumar R (2014) Investigation of fatigue crack growth characteristics of NR/BR blend based Tyre tread compounds. Int J Fract 188:9–21
9. Ghosh P, Mukhopadhyay R, Stocek R (2016) Durability prediction of NR/BR and NR/SBR blend tread compounds using tear fatigue Analyser. KGK-Kautschuk Gummi Kunststoffe 69: 53–55
10. Stoček R, Ghosh P, Machů A, Chanda J, Mukhopadhyay R (2020) Fatigue crack growth vs. chip and cut wear of NR and NR/SBR blend-based rubber compounds. In: Heinrich G, Kipscholl R, Stoček R (eds) Fatigue crack growth in rubber materials. Advances in polymer science, vol 286. Springer, Cham. https://doi.org/10.1007/12_2020_67
11. Stoček R, Stěnička M, Maloch J (2020) Determining parametrical functions defining the deformations of a plane strain tensile rubber sample. In: Heinrich G, Kipscholl R, Stoček R (eds) Fatigue crack growth in rubber materials. Advances in polymer science, vol 286. Springer, Cham. https://doi.org/10.1007/12_2020_78
12. Kim JH, Jeong HY (2005) A study on the material properties and fatigue life of natural rubber with different carbon blacks. Int J Fatigue 27:263–272
13. Nie Y, Wang B, Huang H, Qu L, Zhang P, Weng G, Wu J (2010) Relationship between the material properties and fatigue crack-growth characteristics of natural rubber filled with different carbon blacks. J Appl Polym Sci 117:3441–3447
14. Reincke K, Grellmann W, Kluppel M (2009) Investigation of fracture mechanical properties of filler-reinforced styrene-butadiene elastomers. Kautschuk Gumi Kunstoffe 5:246–251
15. Young DG, Danik JA (1994) Effect of temperature on fatigue and fracture. Rubber Chem Technol 67:137–147
16. Legorju-jago K, Bathias C (2002) Fatigue initiation and propagation in natural and synthetic rubbers. Int J Fatigue 24:85–92
17. Lake GJ, Lindley PB (1965) Cut growth and fatigue of rubbers. II. Experiments on a noncrystallizing rubber. Rubber Chem Technol 38:301–313
18. Stoček R, Kratina O, Ghosh P, Maláč J, Mukhopadhyay R (2017) Influence of thermal ageing process on the crack propagation of rubber used for tire application. In: Grellmann W, Langer B (eds) Deformation and fracture behaviour of polymer materials. Springer, Berlin, pp 305–316. ISBN 978-3-319-41879-7. https://doi.org/10.1007/978-3-319-41879-7_24
19. Harbour RJ, Fatemi A, Mars WV (2008) Fatigue life analysis and predictions for NR and SBR under variable amplitude and multiaxial loading conditions. Int J Fatigue 30:1231–1247
20. Harbour RJ, Fatemi A, Mars WV (2007) The effect of dwell period on fatigue crack growth rates in filled SBR and NR. Rubber Chem Technol 80:838–853
21. Andreini G, Straffi P, Cotugno S, Gallone G, Polacco G (2010) Comparison of sine versus pulse waveform effects on fatigue crack growth behaviour of NR, SBR and BR compounds. Rubber Chem Technol 83:391–403
22. Andreini G, Straffi P, Cotugno S, Gallone G, Polacco G (2013) Crack growth behaviour of styrene-butadiene rubber, natural rubber and polybutadiene rubber compounds: comparison of pure-shear versus strip tensile test. Rubber Chem Technol 86:132–145
23. Lindley PB (1973) Relation between hysteresis and dynamic crack growth resistance of natural rubber. Int J Fract 9:449–462
24. Mars WV, Fatemi A (2003) A phenomenological model for the effect of R-ratio on fatigue of strain crystallizing rubbers. Rubber Chem Technol 76:1241–1258
25. Abraham F, Alshuth T, Jerrams S (2004) The effect of minimum stress and stress amplitude on the fatigue life of non strain crystallising elastomers. Mater Des 26:239–245
26. Saintier N, Cailletaud G, Piques R (2011) Cyclic loadings and crystallization of natural rubber: an explanation of fatigue crack propagation reinforcement under a positive loading ratio. Mater Sci Eng A 528:1078–1086
27. Stadlbauer F, Koch T, Archodoulaki VM, Planitzer F, Fidi W, Holzner A (2013) Influence of experimental parameters on fatigue crack growth and heat build-up in rubber. Materials 6:5502–5516

28. Qazvini N, Mohammadi N, Jalali A, Varasteh A (2002) The fracture behaviour of rubbery vulcanizates: I. Single component versus blend systems. Rubber Chem Technol 75:77–82
29. Eisele U, Kelbch SA, Engels HW (1992) The tear analyzer – a new tool for quantitative measurements of the dynamic crack growth of elastomers. Kautschuk Gummi Kunststoffe 45: 1064–1069
30. Stoček, R.; Heinrich, G. Gehde, M., Kipscholl, R (2013) Analysis of dynamic crack propagation in elastomers by simultaneous tensile- and pure-shear-mode testing. In: W. Grellmann et al. (eds) Fracture mechanics & statistical mechanics, LNACM 70, pp. 269–301, 978-3-642-37909-3
31. Bertenev GM, Zuyev YS (1968) Strength and failure of visco-elastic material. Pergamon Press
32. Gent AN, Lindley PB, Thomas AG (1965) Cut growth and fatigue of rubbers. I. The relationship between cut growth and fatigue. Rubber Chem Technol 38:292–300
33. Liu C, Dong B, Li Z, Zheng Q, Wu Y (2015) Influence of strain amiplification near crac tip on the fracture resistance of carbon black filled SBR. Rubber Chem Technol 88:276–288
34. Bruning K, Schneider K, Roth SV, Heinrich G (2013) Strain-induced cristallization around a crack tip in natural rubber under dynamic load. Polymer 54:6200–6205
35. Wunde M, Klüppel M (2016) Influence of phase morphology and filler distribution in NR/BR and NR/SBR blends on fracture mechanical properties. Rubber Chem Technol 89:588–607
36. Lee DJ, Donovan JA (1987) Microstructural changes in the crac tip region of carbon black filled natural rubber. Rubber Chem Technol 60:910–923
37. Sridharan H, Chanda J, Ghosh P, Mukhopadhyay R (2019) Rubber blend and filler effects on damage mechanisms under monotonic and fatigue loading. Rubber Chem Technol 92:415–430
38. Lake GJ (1995) Fatigue and fracture of elastomers. Rubber Chem Technol 67:137–147
39. Persson BNJ, Albohr O, Heinrich G, Ueba H (2005) Crack propagation in rubber like materials. J Phys Condens Matter 17:R1071–R1142
40. Kluppel M (2009) Evaluation of viscoelastic master curves of filled elastomers and applications to fracture mechanics. J Phys Condens Matter 21:1–10
41. Ghosh P, Mukhopadhyay R, Krishna Kumar R (2020) Influence of waveforms on fatigue crack growth characteristics of Tire tread rubber using finite element analysis. Tire Sci Technol. https://doi.org/10.2346/tire.20.190215
42. Wunde M, Plagge J, Klüppel M (2019) The role of stress softening in crack propagation of filler reinforced elastomers as evaluated by the J-integral. Eng Fract Mech 214:520–533
43. Ruellan B, Le Cam J-B, Jeanneau I, Canevet F, Mortier F, Robin E (2019) Fatigue of natural rubber under different temperatures. Int J Fatigue 124:544–557

Adv Polym Sci (2023) 289: 317–344
https://doi.org/10.1007/12_2022_127

Published online: 3 September 2022

Mechanistic and Kinetic Studies on Degradation Processes of Rubber Types

Ulrich Giese, Stephanie Kautz, Corinna Schwarzendahl, and Sabine Thust

Contents

Abstract Modern rubber materials face high demands in terms of oxidation stability as well as thermal resistance and, in some cases, enhanced oil resistance. At the same time, their mechanical properties have to maintain a high level. High-performance synthetic rubbers with highly sophisticated properties – like nitrile rubber (NBR), hydrogenated nitrile rubber (HNBR) and peroxide-crosslinked ethylene – propylene – diene rubber (EPDM) – are thus becoming more and more important. Nevertheless, irreversible aging processes limit lifetime and result in the failure of these elastomer materials. Knowledge about chemical aging mechanisms, kinetics and the weak elements in the polymer chain and in the chemical network are fundamental to understanding and predicting lifetime as well as in arriving at appropriate measures for stabilizing the polymer matrix. Alongside the polymer's microstructure, other determining parameters are temperature, exposure to oxygen, and its diffusion into the bulk in competition to reaction speed, which is responsible for the spatial changing of the material from the surface inward. Investigations by

U. Giese (✉), S. Kautz, C. Schwarzendahl, and S. Thust
Deutsches Institut für Kautschuktechnologie e. V., Hannover, Germany
e-mail: ulrich.giese@dikautschuk.de

means of selected methods like nanoindentation, FT-IR spectroscopy, chemiluminescence, and physical testing in combination with artificial aging are used to describe quantitatively the mechanistic role of the polymer's microstructure and of crosslinking during thermal-oxidative aging.

Keywords Characterization · EPDM · HNBR · NBR · Reaction mechanism · Residual peroxide · Thermal-oxidative degradation

1 Introduction

Materials on the basis of crosslinked rubbers (elastomers) are widely used for dynamically stressed, temperature- and media-resistant components – e.g., anti-corrosion coatings on metals, everyday products, tires, bushings, engine mounts, vibration dampers, hoses, radial shaft seals, O-rings, etc. Elastomer components enhance safety and comfort, specifically in the automotive sector. Important requirements for all such components are long-term stability of physical and chemical properties like strength, elasticity, compression set or tear resistance as well as functional reliability over the longest possible service life. The latter is also important from an economical perspective. The lifetime of elastomer components is determined largely by aging processes. Over the course of time, mechanical-dynamic load (fatigue), oxidation processes, temperature, exposure to UV light (photo-oxidation), ozone, and aggressive media lead to irreversible changes in the physical and chemical properties of an elastomer material, often resulting in premature functional failure of the component concerned [1, 2]. Already in the course of mixing and vulcanization during manufacture as well as in a subsequent post-curing operation, as is employed in some production processes, the material undergoes thermo-oxidative stress, resulting in a certain degree of prior deterioration. In DIN 50035, or according to the International Union of Pure and Applied Chemistry (IUPAC), aging is defined as a strong irreversible change in the material's chemical and physical properties over time [1, 2]. The aging process leads to the failure of elastomer parts in terms of their application and effects damage of various kinds. This can be attributed more or less to this cause by means of visual assessment, high experience, and chemical and physical analyses [3]. As regards the aging mechanisms, a distinction is made between (1) pure thermal aging (elevated temperature, absence of oxygen), (2) oxidation or thermal-oxidation, (3) photo-oxidation, (4) ozone attack, (5) mechanical load (static or dynamic), and (6) chemical attack by media [4]. In actual practice, it is most often a combination of different mechanisms. In view of the ubiquitous presence of oxygen and temperature in the frequent application of elastomer products, thermal-oxidative aging is one of the most relevant topics with respect to their lifetime. A particular difficulty in understanding and investigating thermal-oxidative aging mechanisms is that of distinguishing the effect of temperature, oxygen, and energy loading during dynamic load insofar as energy dissipation leads to an increase in temperature, usually abetted by the presence of oxygen. The objective of the experimental work and research described in this chapter is to understand the mechanisms and kinetics of several chemical reactions during the thermal-oxidative aging process and to gather knowledge on their foremost contribution in altering physical properties. Key parameters of interest are the influence of polymer structures, the type and density of crosslinking and selected fillers as well as aging parameters like temperature, time, and oxygen concentration. The investigations focused on nitrile rubber (NBR), hydrogenated nitrile rubber (HNBR) – and ethylene – propylene – diene rubber (EPDM)-based systems due to their high technical relevance as oil-resistant material (NBR, HNBR) and as sealings (EPDM, peroxidic or sulfur crosslinked). In the case of NBR- or HNBR-systems, the work focuses in detail on an investigation of the following aspects influencing aging behavior:

- Impact of polymer constitution on the thermal-oxidative stability of NBR/HNBR
- Role of the acrylonitrile group
- Effect of pre-aging
- Impact of crosslinking density in the case of NBR
- Crosslink structure and type in the case of NBR (peroxide, sulfur-- semi-efficient vulcanizationsystem (SEV), sulfur-efficient vulcanization system (EV)
- Determination of quantitative data for potential use in simulation models

The investigations into the aging behavior of peroxide-crosslinked EPDM-elastomers, which, unlike NBR, have a saturated polymer backbone, focused on the following aspects:

- EPDM constitution: diene type and concentration, ethylene content
- Peroxide types and concentration for crosslinking
- Crosslinking time and residual peroxide content

Analytical methods for characterization were, moreover, optimized and selected so as to accelerate the aging process so as to aid in the prediction of material changes.

2 Fundamentals and Overview of Aging Processes

A clear distinction is to be made between material damage caused by ozone as well as by chemical attack from media like lubricants or their additives, on the one hand, and oxidative aging, on the other. Ozone poses a major problem for diene rubber types like NR, SBR, or NBR in particular due to their microstructure. Ozonization is an electrophilic reaction and starts at the point with the highest electron density in the polymer chain, generating a primary ozonide by a dipolar cycloaddition of ozone to the double bond. In further reactions, the ozonide cleaves and forms aldehydes, ketones, or unstable polymeric peroxides as intermediates. These unstable molecules cleave into various products like acids, esters, ketones, and aldehydes. What is more, oxygenated products are formed via scission of intermediate ozonides [5, 6]. In contrast to this mechanism, thermal-oxidative aging, or the photo-oxidation process, consists of a three-phase radical mechanism, described early on and well known as the Bolland mechanism [7–15]. Initially, radicals form on organic components or on polymers, triggered by heat in combination with atmospheric oxygen or by light (photo-oxidation) [8–10]. In the second step (propagation), the radicals formed in the first step react with oxygen, resulting in peroxo radicals capable of hydrogen abstraction on a polymer molecule. This results, on the one hand, in the creation of macroradicals, and, on the other hand, in unstable hydroperoxides due to reactive decomposition into radicals. The chain reaction is terminated by recombination reactions, with stable compounds being formed, such as, for instance, the constitution of C–C or C–O–C bonds from two macroradicals, which is tantamount to an increase in crosslinking density. The chain reaction mechanism is summarized as follows [15]:

Initiation:	R-H	$\xrightarrow[\text{Shear}]{\Delta T,\ h\bullet\nu}$	R• + H•	
	R-R	$\xrightarrow[\text{Shear}]{\Delta T,\ h\bullet\nu}$	2R•	
	R• + O_2	$\longrightarrow$	ROO•	
Propagation:	ROO• + RH	$\longrightarrow$	R• + ROOH	
	ROOH	$\longrightarrow$	RO• + OH•	(acceleration)
	2 ROOH	$\longrightarrow$	RO• + R• + H_2O	
	RO• + RH	$\longrightarrow$	ROH + R•	
	OH• + RH	$\longrightarrow$	HOH + R•	
Termination/ Recombination:	ROO• + R•	$\longrightarrow$	ROOR	(unstable)
	RO• + R•	$\longrightarrow$	ROR	
	R• + R•	$\longrightarrow$	R - R	
Termination/ chain scission	$R_{p+1}O\bullet$	$\longrightarrow$	$R_p=O + \bullet R_1$	

The terminal recombination reaction results in an embrittlement of the material due to crosslinking. What is more, polar oxygenic side groups can be formed on the polymer. These likewise have a stiffening effect due to inter- and intra-molecular interactions. As a function of aging conditions and the microstructure of the polymers, terminal chain scission can also be observed, accompanied by stickiness and plastication of the rubber or elastomer [8, 11]. The two reaction channels compete

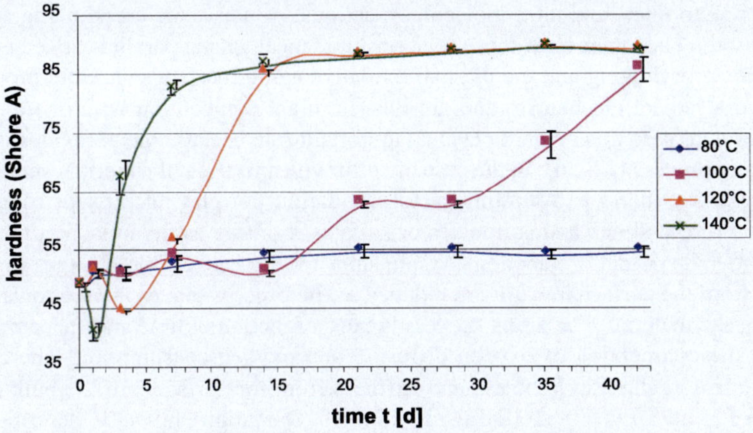

Fig. 1 Changes in hardness due to the different reaction mechanisms in the case of an SBR-compound as a function of time and temperature [20], chain scission with softening effect below 3 min. For aging at 40°C and below 7 min. For aging at 120°C, post-crosslinking with increase in hardness at higher aging times for 140, 120, and 100°C

with one another, with the polymer configuration (double bonds of the main chain, side groups) playing a more significant role [16–19]. The Fig. 1 [20] below clearly shows competition of the different reaction paths in the case of a sulfur crosslinked SBR-vulcanizate (unfilled) as a function of time and temperature.

As for the molecular structure of the polymer backbone, the bonding energies between carbon and hydrogen or the situation of primary, secondary, or tertiary C-atoms in combination with substituents or chain branching have an important role to play in thermal-oxidative stability, because the hydrogen abstraction tends to take place at low bonding stabilities. The bonding energy of C–H-bonds decreases from the primary, on through the secondary to the tertiary C-atom, going from 421 kJ/mol, then 411 kJ/mol, and finally 381 kJ/mol. At 368 kJ/mol, the allylic position in unsaturated structures such as those found in polydienes with a 1.4-butadiene-linked monomer or comonomer, like BR, NBR, or SBR, is one of the weakest [21]. In the case of NBR or HNBR, the inductive effect of the nitrile group leads to an extremely low bonding energy of 362 kJ/mol for the hydrogen in the case of the nitrile-substituted tertiary C-atom [21]. In keeping with the Bolland mechanism and in view of the hydrogen abstraction for NBR and HNBR, this tertiary C-atom is favored for chemical attack during oxidative aging. Overall, this mechanism accounts for the variations in oxidative stability for different polymers. Additional inductive effects I (+I or –I – effects) and mesomeric effects of substituents like phenyl rings in the case of SBR can stabilize the radical, rendering the abstraction of a hydrogen very easy in such cases. The radicalic attack of peroxy radicals on the polymer backbone coincides with decreasing bonding energies on the part of the different structures. An embrittlement of the material is a consequence of the macroscopic process. Polar oxygenic side groups can be formed on the polymer. These likewise have a stiffening effect due to inter- and intra-molecular reactions. As far as the crosslinking system used by the elastomer from these relationships, mechanistic parallels arise between peroxide crosslinking and the thermal-oxidative aging mechanisms. Both processes involve a radical mechanism and are thus in direct competition with one another. The residual peroxide content plays an important role because the peroxides initiate the aging process [22–26]. In the case of sulfur vulcanization, the thermal stability of the weak S–S bond is a limiting factor, oxidation not playing a major role, with scission of crosslinks and sulfonates or sulfates resulting as final reaction products [27, 28]. The oxidation reactions, responsible for the irreversible changes in properties from the surface inward, are reduced as the process moves inward toward the bulk of the material. The aging process is thus a function of time and space brought on by the competition of oxygen diffusion and oxygen consumption. The consequence is a gradient in hardness and stiffness from the surface into the bulk of the material. The DLO effect (Diffusion Limited Oxidation) [29, 30] describes the phenomenon.

Alongside the use of stable polymers, which cannot be used in every application due to the mechanical requirements the elastomer material has to meet, a measure frequently used is that of compounding with anti-aging agents – e.g., p-phenylenediamines or substituted phenols as well as hindered amines (HALS) [28, 31–35].

3 Experimental

3.1 Materials and Methods

3.1.1 Materials: NBR and HNBR-Systems

With regard to technical applications, five nitrile butadiene rubber (NBR) types, which are copolymers of acrylonitrile and butadiene, were selected with differing ACN contents of 18, 22, 28, 34, and 39%, which additionally means that the 1.4-unit content decreases proportional to the acrylonitrile (ACN) content. The NBRs are identified as NBR 1 – NBR 5. The crosslinking systems consist of a sulfur semi-efficient (S-SEV) system (N-cyclohexyl-2-benzothiazole sulphenamide/sulfur (CBS/S) (1.5:1 phr), sulfur-efficient (S-EV) system (CBS/S (5.0:1 phr)) and a peroxide system using dicumyl peroxide (DCP, 2.0 phr). Alongside two levels of crosslink density, there is also variation in crosslink structure. In the case of DCP, only C–C or C–O–C-bonds are expected, and the sulfur systems will show differences in the relationship between mono-, di-, and polysulfidic crosslinks. The crosslink densities were normalized with regard to the different structures. For investigations into the effect of the ACN group on the stability of HNBR, types with an ACN content of 34% and 39% were used. As model polymer, a special methylated M-HNBR (31% ACN) (actually ARLANXEO from Lanxess Deutschland, prepared by K. Kubaba) with the following structure was used to study the influence of inductive effects and reaction mechanisms on the radical abstraction of hydrogen in the polymer chain:

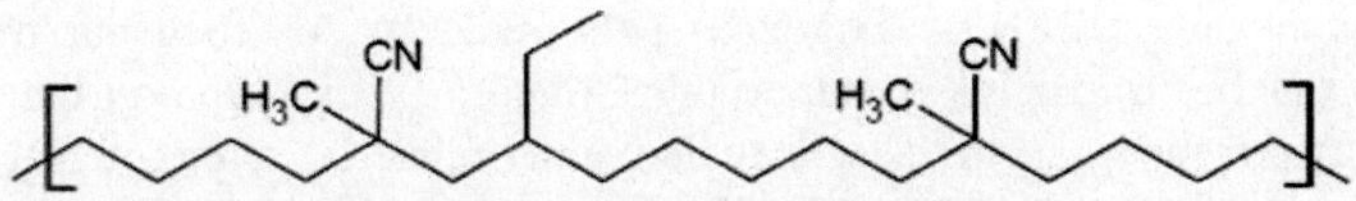

The tertiary C-atom of the polymer backbone near the CN group is substituted with the methyl group, so that there is no H-atom that can be abstracted in the alpha position, and the inductive effect of the nitrile group on the β-hydrogen is changed. All the compounds were prepared by means of a lab mixing system (Thermo Fisher lab mixer, 300 ml vol., tangential rotor geometry). The vulcanization behavior of the compounds was characterized by means of rheometry according to DIN 53529 (refer to Sect. 4.1.3 for details). Test plates (2 mm) were vulcanized in an electrically heated press at 170°C up to a vulcanization time of t_{90}. (= time until 90% of the torque maximum is attained).

3.1.2 Materials: EPDM/EPM Systems

The studies were performed using EPDMs of different types and one type of EPM, as listed in Table 1. The selection was made with a view to the varying reactivity of the diene component to radical mechanisms and the ethylene/propylene ratio, with the propylene content of particular interest in view of the tertiary C-atom and the weaker C–C- or C–H-bonds.

Table 1 Used EPDM rubbers and characteristic data

Rubbers	Ethylene content, %	Diene component	Diene content, %	Classification
EPDM 1	48	Ethylene-norbonene	4.5	Amorphous
EPDM 2	70	Ethylene-norbonene	4.5	Partially crystalline
EPDM 3	50	Vinylidene-norbonene	3.0	Amorphous
EPDM 4	58	Dicyclopentadiene	4.5	Amorphous
EPDM 5	55	Ethylene-norbonene	4.3	Amorphous
EPM	52	n.a.	n.a.	Amorphous
EPDM	55	Ethylene-norobonene		Amorphous
EPDM	66	Ethylene-norobonene		Amorphous
EPDM	56	Ethylene-norobonene		Amorphous

For crosslinking, three types of peroxides were used overall: (1) dicumyl peroxide (DCP – Perkadox BC FF, AKZO), (2) 1,4-bis (tert.-butylperoxi-isopropyl)-benzene (DIPP– Perkadox -14 FF, AKZO) and (3) 2.5-dimethyl-2.5-di-tert.-butylperoxohexan (DBPH – Trigonox 101, AKZO). As coagents triallyl-iso-cyanurate (TAIC) or triallyl cyanurate (TAC) and a special type on the base of a polyorganosiloxane – methylacrylate- derivate (industrial experimental product, Schill + Seilacher "Struktol" GmbH, Hamburg, see patent: EP 2531512 B1) were selected. The model compounds were mixed by means of internal lab mixer (Thermo Fisher lab mixer, 300 ml vol., tangential rotor geometry).

The peroxide concentrations were varied in the steps 2.5, 5.0, and 10.0 phr (relative to the pure substance). The 2 mm plates were vulcanized at a standard temperature of 170°C in an electrically heated press, with the vulcanization times varied between t_{50}. and t_{90} + 120 min, normalized to rheometric measurements.

3.2 Methods

Artifical Aging and Mechanical Tests The artificial aging of the materials (sample geometry S 2 stripes, 2 mm) was performed in circulating air cabinets with variation of temperature and time. Afterwards, the effects of aging on the mechanical and physical properties were tested on the standardized test specimen. Elongation at break and tensile strength were thus determined to DIN 53504 and hardness to DIN 53505.

ATR_FT-IR-Spectroscopy The status of oxidation was controlled by means of ATR-FT-IR spectroscopy (Nicolet Nexus FT-IR 6700 with Nicolet Omnisampler^{Tm}) using a diamond crystal.

Rheometry and Crosslink Density The degree of crosslinking and the crosslink efficiencies were characterized by means of rheometry (MDR 2000 E, Alpha Technology), with torque registered as a function of time at 170°C at 1 Hz oscillating deformation of 1.5° (phase angle) to DIN 53529. Furthermore, relative crosslink densities were characterized on selected systems using equilibrium swelling in accordance with Flory-Rehner and by means of low-field NMR relaxation measurements using the transversal relaxation time (T_2) (equipment: IIC-crosslink density spectrometer), but these did not provide additional information.

Chemiluminescence Aging stability and kinetics were determined by means of chemiluminescence (CL)-apparatus (Atlas Analytical Instruments, CL400 ChemilumeTM Analyzer), which detects directly the light emission of the stabilization reaction of an excited triplet oxygen generated by the reaction of peroxo radicals (intermediate of oxidation reaction) at 450 nm wavelength corresponding to 315 kJ/mol. The number of emitted light quants is proportional to the number (mol) of excited carbonyls, alcohols, or oxygen from the peroxo radical stabilization reaction. Aging in the CL unit was performed at elevated temperature in pure oxygen. The oxygen induction time (OIT), the position of the signal maximum as a function of time, and the slope of the signal increase were the data used to calculate the reaction speed constant as per first order reaction kinetics [36]. Activation energies could be obtained by evaluation as per the Arrhenius principle.

Residual Peroxide Determination by Means of HPLC In the case of the EPDM systems, the residual peroxide content of the compounds as a function of vulcanization time and crosslinking system was characterized by RP-18-HPLC-UV (Finnigan Surveyor pump system, PDA Plus /UV-detector, 220 nm wavelength, column: Macherey und Nagel – 4.6 Nucleosil 100–3 C 18) using a gradient elution (acetonitrile/methanol, flow 0.7 ml/min). Extracts were prepared at room temperature with extraction times of between 3 and 4 days in order to obtain complete extraction without decomposition of residual peroxide. An internal standard di-tert.-butylbenzene was used for quantification. The calibration was carried out using standard solutions with concentrations of the peroxide to be analyzed in the range of 0–100 ppm in methanol. The limit of detection (LOD) attained and the limit of quantification (LOQ) were determined to DIN 32645 with LOD_{DCP} 320 mg/kg rubber, LOD_{DIPP} 360 mg/kg rubber; LOQ_{DCP} 960 mg/kg rubber, and LOQ_{DIPP} 1,080 mg/kg. The relative standard deviation was below 1%.

Micro-Indentation The DLO effect and the spatial development of the oxidative aging effect were characterized by means of micro-indentation using “LNP nano touch” (Ludwig Nano Präzision GmbH) in the constant force modus (20 mN) and with a 25 μm needle as standard parameter. The samples were pre-aged up to 1,000 h at elevated temperature as 2 mm plates. The specimens were prepared as cross-section material using a microtome and cryotechnique (liquid N_2).

4 Results

4.1 Investigations on NBR and HNBR Materials

4.1.1 Influence of Polymer Constitution: NBR Rubbers

Chemiluminescence measurements for the thermal-oxidative stability were carried out as a function of the ACN content on the uncured rubbers. The materials were extracted to avoid the influence of stabilizers added by the polymer supplier. Figure 2 shows the CL curves, which can be evaluated using two methods: (1) the OIT value is defined by the tangent at the increasing part of the curves infront and its intercept with the *x*-axis (time); (2) aging stability is defined by the position of the maximum of the curves related to the *x*-axis (time). If the shape of the curves is not changing significantly, the OIT is used for comparisons, otherwise the position of the curve maximum is more appropriate.

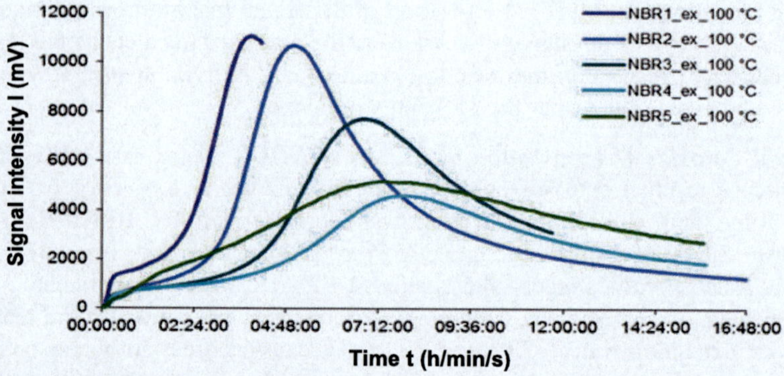

Fig. 2 CL curves of uncured extracted NBR rubbers specimens with variation in the ACN content between 18 and 39%, measured in pure oxygen at 100°C

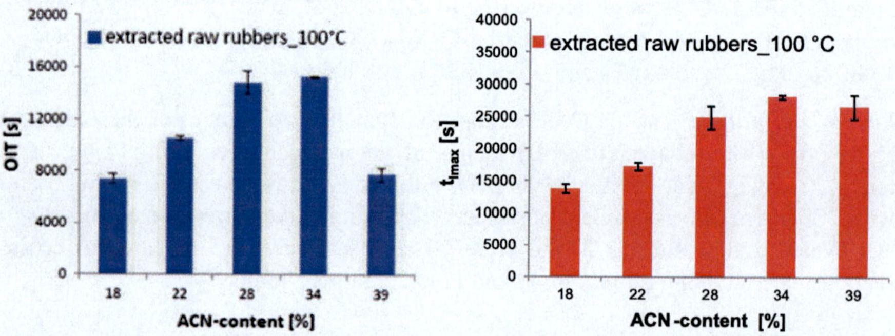

Fig. 3 OIT values (left) and t_{max}-values (right) obtained from CL curves of uncured extracted NBR rubbers with variation of the ACN content between 18 and 39%, measured in pure oxygen at 100°C

The results show that the OIT value clearly depends on the C=C-double-bond concentration, which is reciprocal to the ACN content. Figure 3 summarizes the evaluation of OIT values as well as the position of the signal maximum as a function of time.

The OIT value from the tangential extrapolation of the peaks is not useful for the comparison of the systems due to the strong change of the signal shape. Higher OIT values or t_{max}-time mean greater stability.

CL measurement was carried out on all the NBR types in order to determine the temperature dependency of the oxidation reaction and the activation energies with variation of the ACN content. The temperature dependency and the evaluation of the CL measurements result in a linear dependency of the OIT on temperature measured between 105 and 120°C on extracted and original NBRs. Prediction of the OIT value is possible (see Figs. 4 and 5).

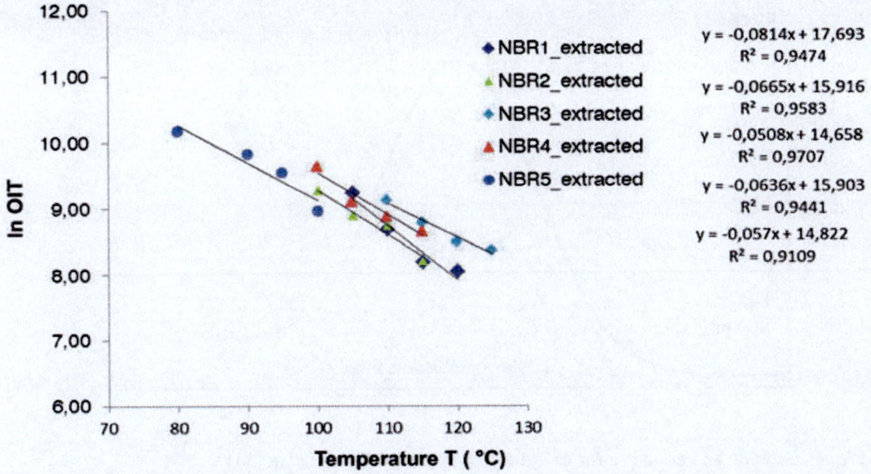

Fig. 4 OIT values obtained from CL curves of uncured extracted NBR rubber specimens with variation of temperature and of the ACN content between 18 and 39%, measured in pure oxygen at 105–120°C

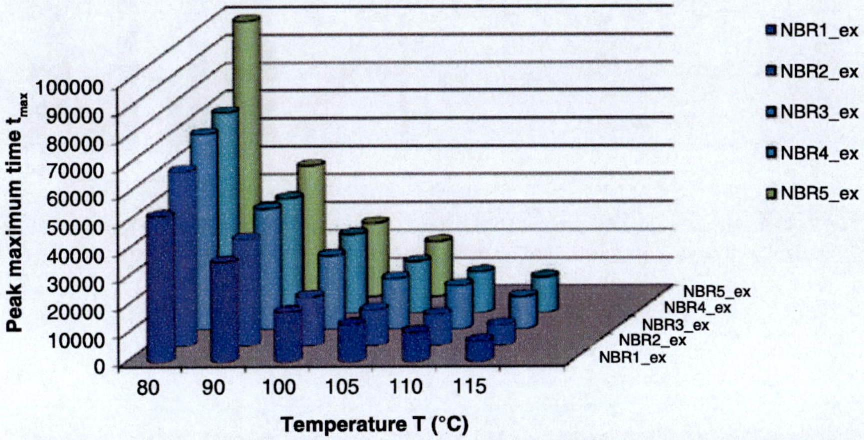

Fig. 5 Peak maximum (t_{max}) values (below) obtained from CL curves of uncured extracted NBR rubber specimens with variation of temperature and of the ACN content between 18 and 39%, measured in pure oxygen at 105–120°C

The temperature dependency of the aging process is quantified by calculating the reaction speed constants from the OIT measurements, which are used in the Arrhenius equation for the activation energy. The calculations were carried out as per the following scheme and Eq. (1) using the CL curves (Fig. 6):

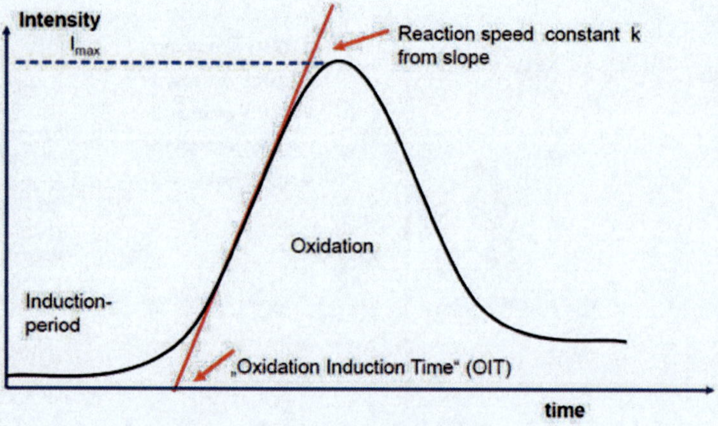

Fig. 6 Scheme and equation for calculation of reaction kinetic data [34]

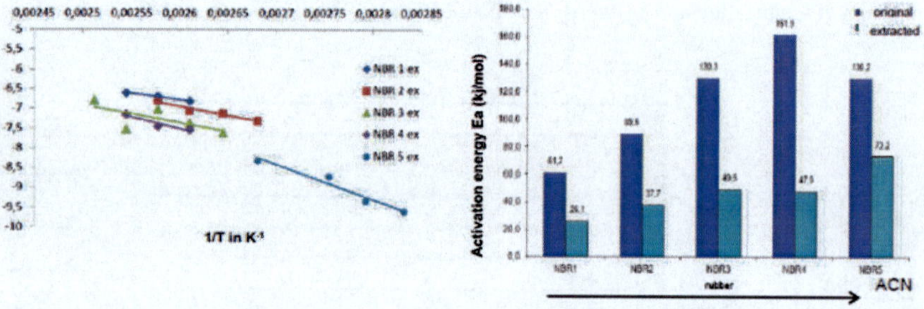

Fig. 7 Left: Arrhenius plot for extracted NBR types with variation of ACN content between 18 and 39%, right: activation energies (Ea) for extracted and original NBR types

$$\ln \frac{I_t}{I_{\max} - I_t} = \ln \frac{[\mathrm{B}]_0}{[\mathrm{A}]_0} + k[\mathrm{A}]_0 t \qquad (1)$$

$$\frac{d[\mathrm{B}]}{dt} = k[\mathrm{A}] \times [\mathrm{B}]$$

where I_t: CL intensity at time t, I_{max}: CL intensity at maximum, [A]: polymer concentration, [B]: hydroperoxide concentration, [A] >> [B], kinetic reaction used: 1st order reaction, k:reaction speed constant from slope of extrapolated tangent at CL-signal.

With the exception of the NBR with ACN content of 39%, the activation energies are all within a range of 26–73 kJ/mol for extracted NBR rubbers and 61 to 161 kJ/mol for the original NBR rubbers, as is shown in Fig. 7.

The original NBR types contain stabilizers like steric hindered phenols, added by the polymer manufacturer, which result in much higher values of OIT. As far as the activation energy (Ea) is concerned, it should be noted that while high values are indicative of the reaction's high temperature dependency, a small change in temperature having a major effect, they likewise have the obvious effect of enhancing the system's stability. To value these results, it is interesting to compare them with, e. g., analog measurements on solution styrene butadiene rubbers (s-SBRs) with Variation of the vinyl- and styrene content. The lowest activation energy (Ea) could be determined for SBR with low vinyl and low styrene content, which means a higher number of butadiene units in the chain. So for an s-SBR with 15% styrene and 30% vinyl, the activation energy was determined to 65 kJ/mol, whereas s-SBRs with 21% or 20% styrene and 63% or 55% vinyl showed 157 kJ/mol or 135 kJ/mol [28].

4.1.2 Influence of Polymer Constitution: HNBR Rubbers

The role of the ACN group and the position of oxidation reaction in the polymer chain were investigated in detail for different HNBRs. The oxidation reaction was analyzed by means of ATR-FT-IR spectroscopy, using a semi-quantitative evaluation with calculations normalized to signal intensities (ΔI) as per Eq. (2):

$$\Delta I = 100 \times I_{\text{analyte}}/I_{\text{reference signal}} \tag{2}$$

Figure 8 provides an overview of the ATR-FT-IR spectrum of an HNBR sample after aging for 336 h in a ventilated cabinet at 160°C:

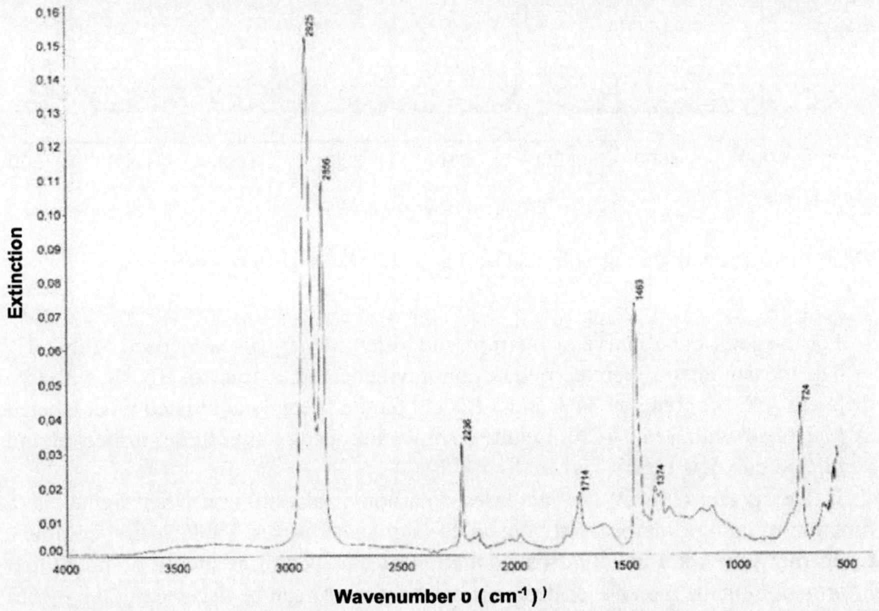

Fig. 8 Overview of the ATR-FT-IR signals of an HNBR after aging at 160° for 336 h in ventilated air (C=O at 1714 cm^{-1}, C–OH at 1,200 cm^{-1} and C–O–C at 1,100 cm^{-1} signals generated by oxidation)

Table 2 Overview of signals used for semi-quantitative evaluation of ATR-FT-IR spectra

Signal of analyte at wavenumber in cm^{-1}	Reference signal at wavenumber in cm^{-1}	Reason and relationship to aging reaction
2,235 nitrile group	2,925 CH_2 -group	Rubber (educt)
1,713 carbonyl group	1,464 CH_2, CH_3 group	Oxidation, oxygen bonding
1,375 CH_3 group	1,464 CH_2, CH_3 group	Chain scission
1,346 CH group	1,464 CH_2, CH_3 group	Rubber (educt)
1,178 C–OH (alcohol)-group	1,464 CH_2, CH_3 group	Oxidation, oxygen bonding
1,120 C–O–C (ether)-group	1,464 CH_2, CH_3 group	Oxidation, oxygen bonding, crosslinking

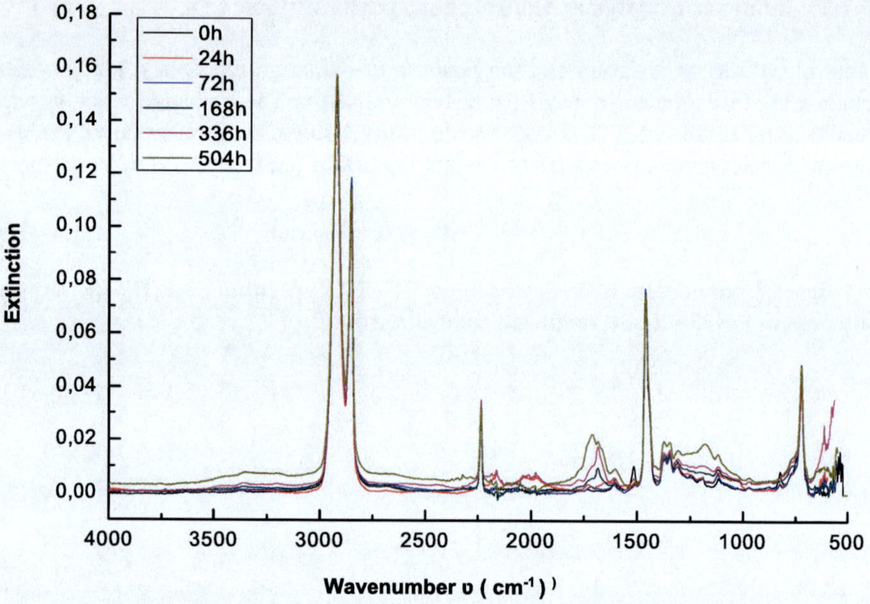

Fig. 9 HNBR sample after aging in ventilated air at 160°C for 0 to 504 h in 6 steps

For the semi-quantitative evaluation, the following signals were used (Table 2):

The measurements were carried out on unvulcanized, extracted. HNBR with two different ACN contents of 34% and 39% and on the special methylated type, labeled as M-HNBR with 31% ACN. Figure 9 shows the results after aging in heated and ventilated cabinets (160°C for up to 504 h).

In the spectra (Fig. 9) the increase of carbonyl, alcohol and ether signals as a function of aging time is clearly observed (signals, compare Table 2, first column). Concerning the semi-quantitative evaluation of the spectra, it should be noted that information on the increase or decrease of components can be derived not only from the increase of a related and normalized signal but also from the difference in the change of signal intensities relative to one another at varying speed as a function of time. Thus the decrease in the signal of methine groups (CH) during aging is minor

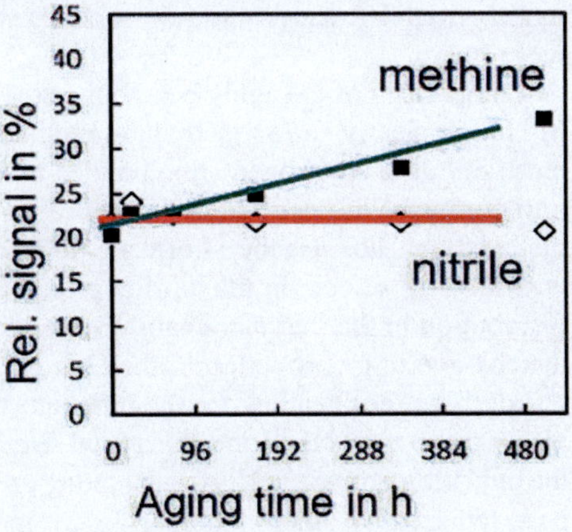

Fig. 10 Change in relative signal intensities of CH groups relative to CH_3 groups and of the nitrile group during aging of HNBR in ventilated air up to 504 h at 160°C

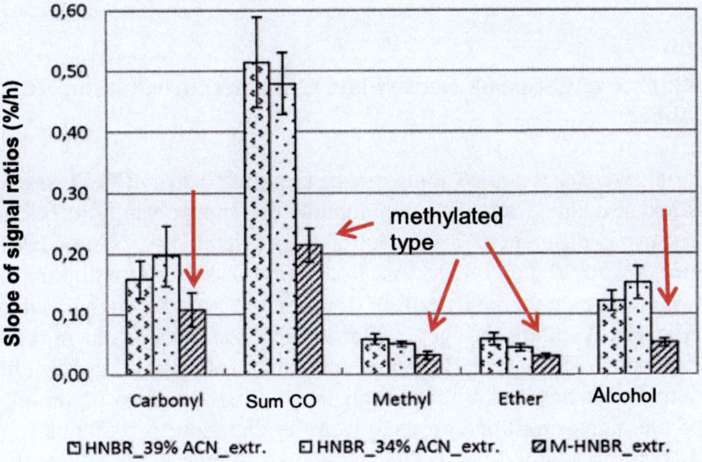

Fig. 11 ATR-FT-IR measurements on extracted HNBR rubber samples and semi-quantitative evaluation of normalized signals of C=O, C–O–C, C–OH and the sum of all C–O-containing groups, and methyl groups in %/h. The signals represent the slope of relative intensity change over time (0–504 h) at 160°C in ventilated air

in comparison to the CH_2 signal intensities used as reference. During the aging process, the relative signal for the nitrile group is thus stable, whereas the methine group signal increases relative to the CH_3 signal. Overall, both signals decrease during oxidation, but at different rates (see Fig. 10).

There is, moreover, no observable shift in the nitrile signal at 2,235 cm^{-1}. The bonding of oxygen happens predominantly at the secondary C-atoms in the chain. Figure 11 summarizes the results of the same measurements on the two types of HNBR and on the model (methylated) M-HNBR using the slope of relative signal intensity over time for different groups from oxidation. In all cases, the methylated model M-HNBR shows the fewest changes in signal intensities, indicating that it is more stable.

Comparison of the aging behavior of the HNBR types and the special methylated M-HNBR clearly shows methylation to result in a stabilizing effect indicative of the reactivity of the hydrogen atom bonded to the quaternary C-atom substituted for the nitrile group, with negligible differences in ACN content. In M-HNBR, moreover, methylation leads, via the +I-effect of the methyl group, to a compensation of the e^--withdrawing effect of the nitrile group by means of a more homogenized e^- distribution in the polymer chain. This additionally results in higher aging stability, thereby averting abstraction of the H-atom in ß-position to the nitrile group.

All these results allow for the inference that there is no detectable reaction of the nitrile group under the conditions used. Here the ACN group is not itself involved in the oxidation process, e.g., by generating an imine via intermolecular reaction, as has been reported by other authors [37, 38]. In such a case, a signal shift for the nitrile group [39] and a new signal at appr.1,640 cm – typical for the imine-group – should be observed. This was not the case in the investigations carried out.

4.1.3 Influence of Crosslink Density and Crosslink Structure on NBR Rubber

NBR materials were crosslinked adjusting the crosslink density (XLD) at two levels. The crosslink structure (ratio between mono-, di- and polysulfidic bridges) was changed by using different ratios of sulfur to accelerator (EV and SEV system) and by using a peroxide in comparison. In the case of sulfur crosslinking, a higher ratio between accelerator and sulfur (EV-system) affects higher amounts of monosulfidic and disulfidic bridges, whereas a high sulfur content in relationship to the accelerator (SEV) leads to more polysulfidic structures in the crosslinks. Different structures have an influence on the thermal stability of the elastomer, because of the increasing bonding energy in the direction C–S–C bonds > C–S–S–C bonds > C–S_x–C-bonds [27]. The C–C-bonds in crosslinking using the peroxide has the highest thermal stability. The CBS/S ratios were for the SEV system CBS/S (1.5:1) and for the EV system CBS/S (5.0:1) (Fig. 12).

On the basis of the results of the chemiluminescence measurements, it is obvious that the uncrosslinked material, containing (unextracted) stabilizer from polymer production, exhibits the highest stability (see Figs. 13 and 14). The vulcanization process affects a shorter OIT value due to an aging effect by consuming the polymer stabilizer. The crosslink densities, which were normalized by adjusting the concentration of the crosslinking systems to two levels due to the rheometer measurements, have less of an effect than the crosslink structure, as it can be seen from the CL curves and the evaluation on the OIT values in Fig. 13. The crosslink structure was varied by changing the accelerator/sulfur ratios and by using the peroxide as a third alternative. The peroxide system in particular is extremely sensitive to aging, which is caused by the peroxide itself. Small amounts of residual peroxide are an effective initiator of the radical aging mechanism. The EV system in sulfur crosslinking shows

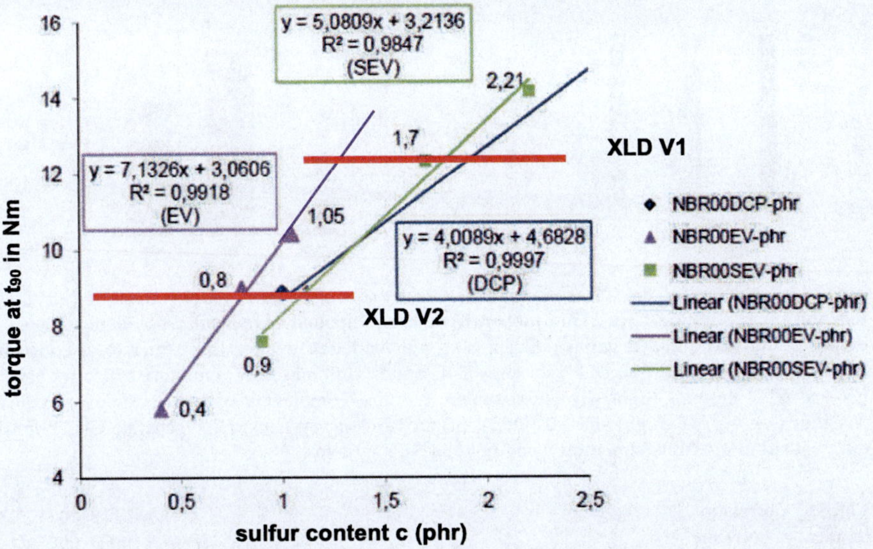

Fig. 12 Normalization of crosslink density using different CBS/S and dicumyl peroxide concentrations at two levels as results from rheometry [40]. The used concentrations are corresponding to the horizontal lines in the diagram. The variation of accelerator to sulfur ratio (EV, SEV) characterizes the crosslink structure

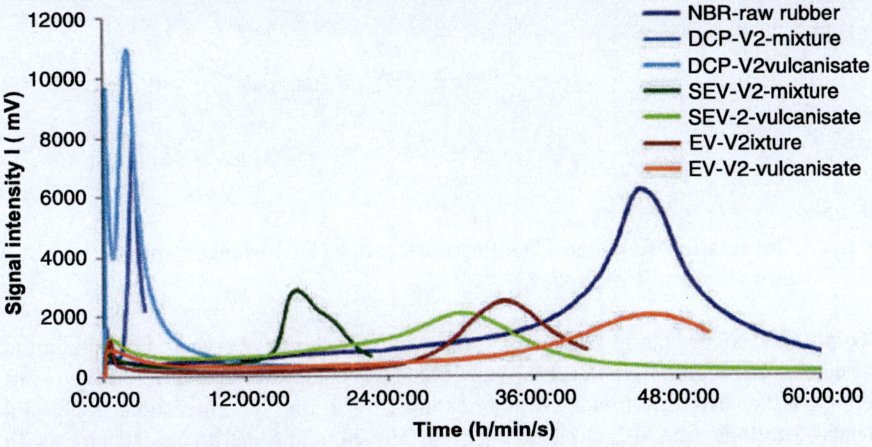

Fig. 13 Signal intensity in mV over storage time from CL measurements (original Cl curves) of NBR 28% ACN at 130°C, original rubber, unvulcanized compound, vulcanizates with SEV, EV, and peroxide crosslinking systems [40]

slightly higher values in the OIT thanks to the enhanced thermal stability of the monosulfidic bridges.

Figure 14 shows the complete evaluation for different crosslink densities or different crosslink structures.

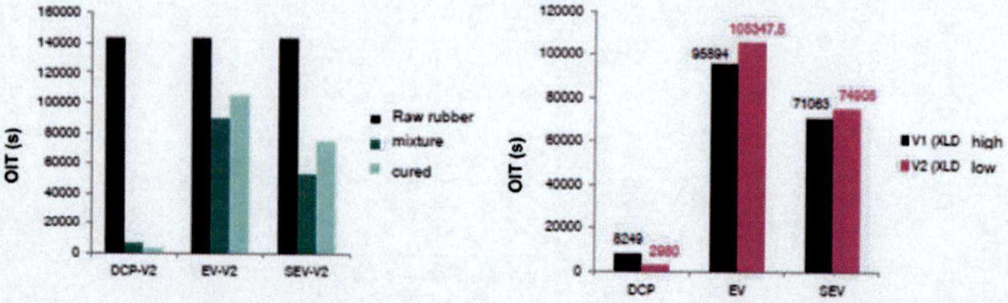

Fig. 14 Evaluation of the CL measurements OIT in sfor NBR 28% ACN content at 130°C as function of crosslinking, Left: (i) raw rubber,(ii) rubber compound mixed with crosslinking system – uncured, and (iii) cured material (vulcanizate) with variation of crosslink structure at constant crosslink density (level V2). DCP generates C–C-bonds; different sulfur crosslink structures using EV and SEV system. Right: (i) Cured rubber systems, comparison of two levels of crosslink densities (XLD – V1 (high) and V2 (low)); (ii) crosslinking systems: DCP generate C–C-bonds; different sulfur crosslink structures using EV and SEV system

Fig. 15 Correlation (tendency) between elongation at break and t_{max}-values from CL measurements at different aging states of NBR with 28% ACN at three different crosslink levels (SEV systems). Aging temperature was 130°C

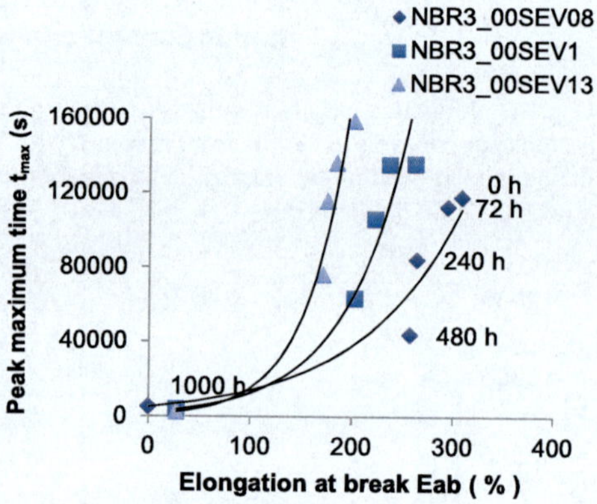

4.1.4 Correlation Between Chemiluminescence (CL) Measurements and Physical Properties

To arrive at a correlation between the maximum of the CL curve and elongation at break for crosslinked NBR (28% ACN content), materials with different levels of crosslink density were first aged in a heating cabinet and then measured in CL and tested for elongation at break (see Fig. 15). The elongation at break decreases with oxidative aging because of an increase in crosslink density due to the recombination of macroradicals, which can be shown by means of parallel NMR relaxation measurements.

As expected, the CL values decrease differently with the aging status depending on the crosslink density. The aging status can thus be quantified by means of CL measurements, which can be carried out in a few hours as opposed to days or weeks of storage in a cabinet.

4.1.5 Characterization of the Diffusion Limited Oxidation (DLO) Effect on NBR

Because of the ubiquitous high presence of oxygen near the surface and the aforementioned balance between diffusion and oxygen consumption, the aging process occurs heterogeneously, as can be seen in Fig. 16:

The spatial development of crosslink density, increasing inward from the surface to the bulk, was characterized by means of micro-indentation after aging at 130°C for 1,000 h. The measurements were carried out on SEV-crosslinked (0.8 phr S/1.2 phr CBS) NBR (28% ACN).

This simple measurement (Fig. 17) shows that it is possible to quantify the spatial distribution of the aging effect. If the aging temperature is increased, the difference in the indentation depth or local hardness between bulk and surface can be expected to be higher, the "slope" is more extreme because of the high increase in the oxidation reaction speed relative to diffusion. The consequence is a higher sensitivity to cracks at the surface during dynamic deformation. For prognosis purposes, crack initiation and propagation should be investigated under dynamic load as a function of the shape of the DLO effect from the surface to the bulk.

Fig. 16 Cross-section of HNBR specimen, aged at 160°C for 72 h

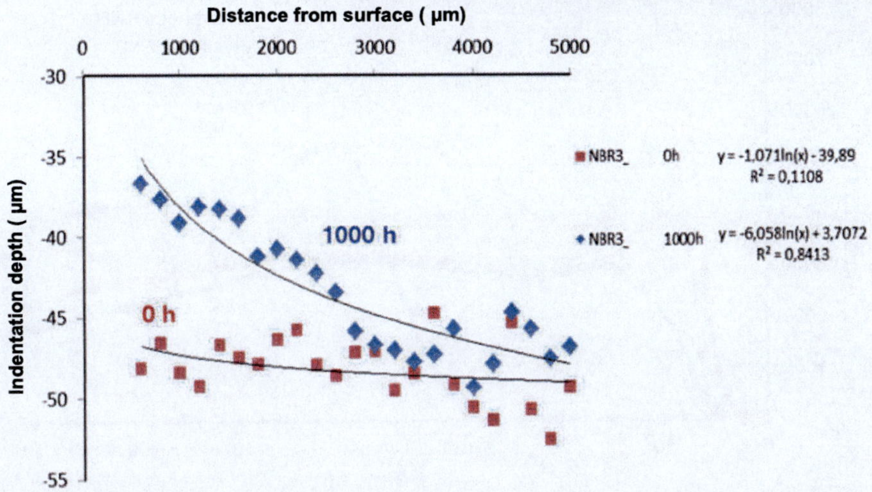

Fig. 17 DLO effect measured by means of micro-indentation on reference and pre-aged SEV-NBR material (1,000 h at 130°C)

4.2 Investigations on EPDM Materials

4.2.1 Investigations of Aging Behavior on Uncured EPDM/EPM

With respect to the different radical stabilities and the combined potential for a radical attack during the aging reaction, EPDM with different ethylene content (amorphous (EPDM 1), partial crystalline structure (EPDM 2) and ethylene-diene (ENB) as a constant termonomer, the vinylidene-norbonene (VNB)- type (EPDM 3) which has a different termonomer, and the amorphous ethylene-propylene rubber (EPM) were selected for investigations of the influence of the microstructure of the uncured rubbers on aging behavior. More detailed information on the structure of the used rubbers was given in Table 1. The measurements were carried out using the CL method at 130°C. Because of the stabilizing effect of the stabilizer added by the manufacturer, the materials were extracted by means of pretreatment with methanol in a soxhlet for 24 h. The CL curves are shown in Fig. 18.

The shape of the CL curves shows that the oxidation reaction – in particular as concerns the CL principle on the detection of the scission of peroxy radicals via triplet oxygen components as intermediates – is the result of several different reactions. The reason could well be the microstructure of EPDM, where the radical attack can start with the methylene or propylene groups or with the diene component at different positions, as described by v. Duin et al. for peroxide crosslinking of EPDM [41]. Evaluation of the CL curves results in the OIT values in Table 3.

The high stability of the EPDM 1 in comparison to EPDM 2 is remarkable. Normally it is to be expected that EPDM 2 should have the higher stability thanks to the reduced number of tertiary C-atoms from the propylene units. It might be that the

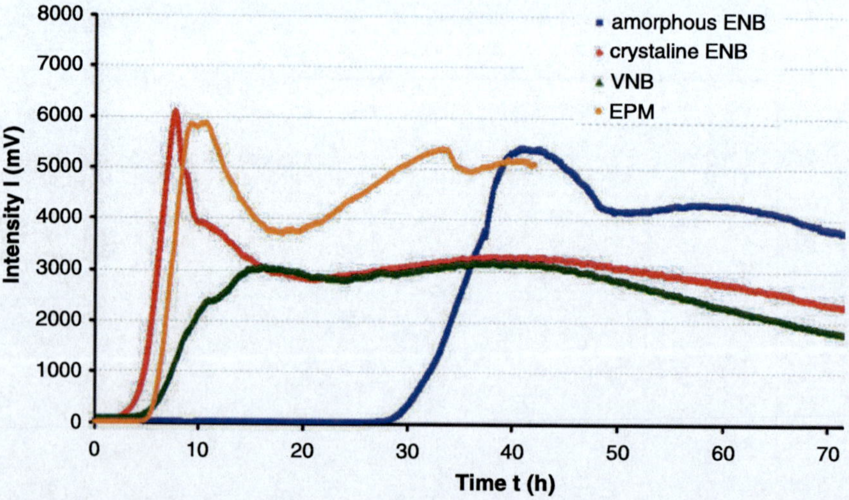

Fig. 18 Original CL curves of uncured and extracted EPDM 1 (amorphous), EPDM 2 (crystalline), EPDM 3 (VNB-type) and EPM

Table 3 OIT values for the EPDM/EPM rubber specimens with variation of the microstructure

EPDM-type	OIT in h. min
EPDM 1 (ENB, amorphous)	30:53 ± 0:32
EPDM 2 (ENB, partial crystalline)	4:25 ± 0:12
EPDM 3 (VNB-type)	4:55 ± 0:04
EPM	5:41 ± 0:07

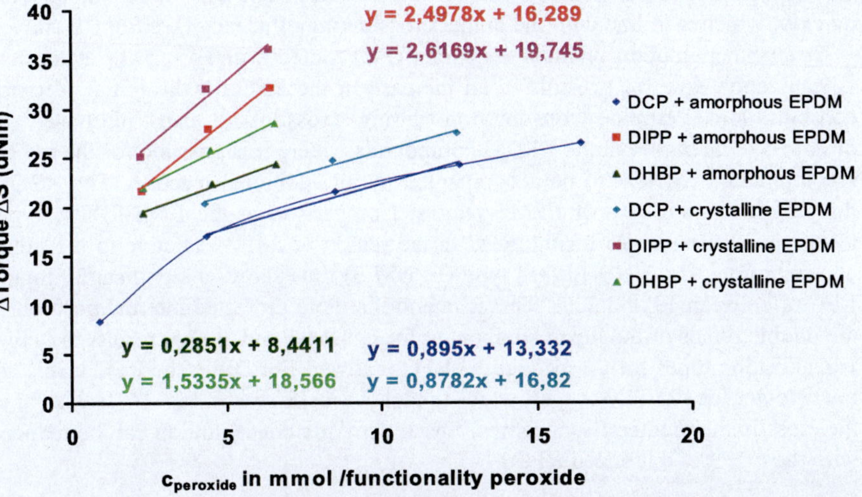

Fig. 19 Crosslinking efficiency of different peroxides in amorphous and crystalline EPDM (EPDM 1 and 2) measured using the rheometer, (vulcanization time t_{90}, ΔS is defined as torque difference between the beginning of crosslink reaction and maximum of the curve, compare DIN 53529), concentration of peroxide is normalized to its functionality(number of perox groups per molecule)

EPDM 1 has more long-chain branching or a different catalyst system. No detailed information about this was available.

4.2.2 Crosslinking Efficiency of Different Peroxides in Amorphous and Crystalline EPDM Rubbers

To estimate the residual peroxide content in conjunction with the crosslinking density attainable, the crosslinking efficiency of various peroxides was studied in two different EPDMs (Fig. 19). For the peroxides DCP, DIPP and DHBP and for the ENB-EPDM types EPDM 1 (48% ethylene content, 4.5% ENB, amorphous) and EPDM 2 (70% ethylene content, 4.5% ENB, crystalline), use of the rheometer results in the following sequence for crosslinking efficiency on the basis of the slopes in diagram (Fig. 19) measured up to t_{90}: DIPP + EPDM 2 > DIPP + EPDM 1 > DHBP + EPDM 2 > DHBP + EPDM 1 > DCP + EPDM 1 > DCP + EPDM 2.

4.2.3 Influence of the Type of Peroxide on Its Residual Content

The effect of the residual peroxide content on aging behavior was investigated under the aspect of the influence of polymer structure, peroxide functionality and concentration and crosslinking time. Figure 20 below shows some representative results for the relative residual peroxide content (RPC) as a function of the vulcanization time for EPDM 1 crosslinked with 5 phr DCP or 5 phr DIPP. The DIPP is consumed more quickly, which is in line with the higher crosslinking efficiency (see Fig. 19 above).

Determination of the residual peroxide (RPC) content with HPLC-UV at constant vulcanization time (t_{90}) results in an increase in the RPC for the initial peroxide concentration. Peroxide consumption during crosslinking can, moreover, be observed to be higher in the EPDM compounds, where it is possible for the radical recombination reaction to be accompanied by an additional reaction. The RPC is thus higher in the case of EPM systems. Comparison of the ENB-EPDM types results in similar values for the RPC in the region of 8–11% relative to the initial concentration. The VNB-EPDM type (EPDM 3) thus shows a significantly higher RPC of between 19 and 22%. This is due to the more protected internal position of the double bonds of the diene component for EPDM 1 and 2. This results in longer vulcanization times for t_{90}, so that the RPC is lower. For DIPP, the RPC levels are much lower for all EPDM types, which tendency can be seen in Fig. 16. But as far as the crosslinking degree is concerned, initial peroxide concentration can be reduced with the effect of a reduced RPC.

4.2.4 Influence of Peroxide Type and Residual Peroxide Content on Aging Behavior

The influence exerted by the peroxide type was analyzed on unfilled EPDM 1 vulcanizates cured to t_{90} S2 specimens were prepared from test plates and characterized for their stress-strain behavior to DIN 53504 after different aging times in ventilated air at 130°C. The stress-strain measurements were carried out at room temperature. Figure 21 (left hand) shows the results for elongation at break.

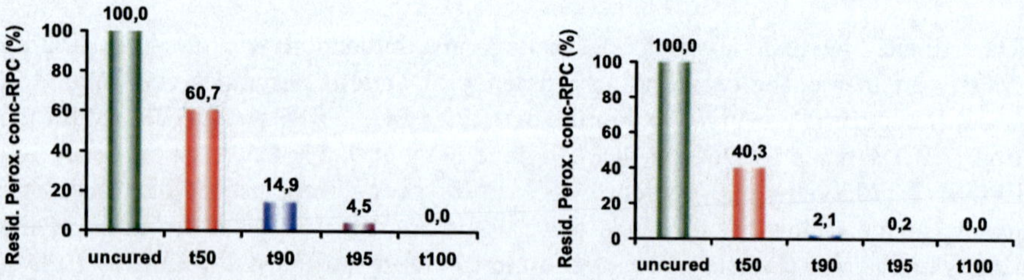

Fig. 20 Residual content of peroxide DCP (left) or DIPP (right) relative to the initial concentration of 5 phr as a function of vulcanization time for amorphous ENB-EPDM (Type 1) determined by means of RP-HPLC-UV (rel. standard dev. of the method <1%)

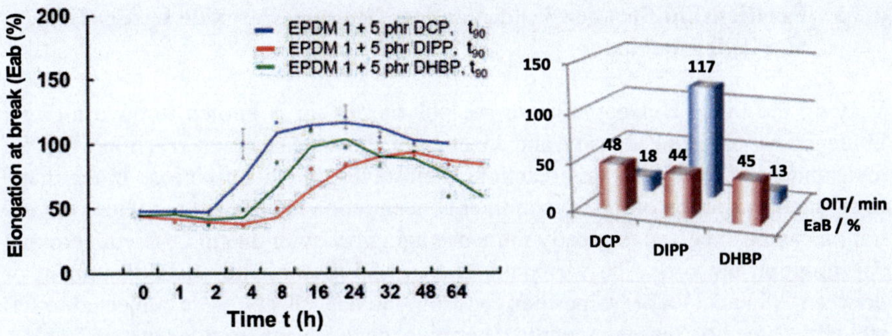

Fig. 21 Left: Elongation at break (Eab) over aging time in ventilated air at 130°C. Right: initial elongation at break in comparison to OIT values from CL measurements on original materials at 130°C

It is obvious that elongation at break increases with aging time during the first few hours, due to the chain scission effect. At longer times higher than appr. 4 h, however, post-crosslinking via recombination reactions of macroradicals takes over, with elongation at break decreasing and materials becoming more brittle. The interesting aspect is that the rise and the maxima of the curves shift to higher times due to the peroxides used – DCP < DHBP < DIPP. The OIT values simultaneously measured by means of CL are presented vis-à-vis the initial elongation at break, which is at the same level for all materials, showing different values, with the DIPP system in particular exhibiting the greatest stability. Thus, as far as the trend in aging behavior is concerned, there is a good correlation between the OIT measurements, which it was possible to conduct in pure oxygen in a short time period of less than 200 min. By contrast, much more time was required for aging in the ventilated air cabinet in conjunction with stress-strain measurements – up to 64 h, in fact. The enhanced stability of the DIPP system cannot be explained on the basis of functionality, which is "two" for the DHBP as well. Both peroxides show the highest crosslinking efficiency in a similar way (see Fig. 18). The lesser stability of the DCP and DHBP crosslinked systems overall cannot be explained by their short half-life times of 124 s (DCP) or 256 s (DHBP) at 170°C. The DIPP has a half-life time of 256 s. The half-life times were derived from data sheets provided by the manufacturer.

Nevertheless, the peroxide type does have an influence overall. This is not a question of the crosslink structure, however, as it is clear from the mechanism that the crosslinks consist of C–C-bonds in all cases. What is more, crosslink density should be very similar, what with the initial elongation at break value being the same. This means that different residual peroxide concentrations after t_{90} – vulcanization time – have to be taken into account. With regard to the aforementioned rheometric results, it should be noted that the DIPP has the highest crosslink efficiency and, with respect to peroxide structure, has a higher mol mass than the others. All are used with 5 phr (parts per hundred rubber in mass) and not in equimolar concentrations.

4.2.5 Relationship Between Vulcanization, Residual Peroxide Content, and Aging Behavior

Varying the diene concentration in the polymer chain is known to have a direct influence on crosslink density and kinetics of the vulcanization reaction [41]. So independent of polymer chain reactions themselves due to an increase in the diene component, variation of the termonomer concentration at constant peroxide concentration can be expected to greatly influence aging behavior due to different peroxide consumption up to t_{90}. To verify this theory, 4 EPDMs with an ENB content of between 2.3 and 11.4% (values derived from the data sheets) were subjected to CL and HPLC-method measurement to determine the residual peroxide content. Table 4 shows the correlation between residual peroxide content, OIT and torque from rheometer measurements for t_9.

ENB content greatly influences aging stability overall. At low ENB concentration, 2.3% or 4.4%, the positive effect on the OIT can be traced back to the reduced residual content of peroxide at t_{90}, which was similar. At higher ENB concentrations, the residual peroxide content is reduced even more. But the OIT increases considerably. This huge effect cannot, therefore, be caused solely by the decrease in residual peroxide concentration. The high ENB concentrations result in an altered network, because the peroxide attack during crosslinking is presumably directed more at the diene component, while being reduced on the polymer chain directly as a consequence of the different reactivity [41]. This means that fewer diene groups remain when the radicals attack during aging, and the aging process is thus focused on the more stable EPDM backbone. What is more, the torque values from rheometry show a higher network density, which means more stable C–C-bonds. In sum, the high diene content in combination with the crosslinking system leads to greater stability.

4.2.6 Influence of the Peroxide Type and Residual Peroxide Content on Aging Behavior

The results presented in Sects. 4.2.4 and 4.2.5 clearly show the residual peroxide content to have a key influence on the elastomers' thermal-oxidative stability in the case of the EPDM types used. Figure 22 below shows the relationships between the OIT value from chemiluminescence, residual peroxide content and vulcanization time for the system DCP 5 phr in EPDM 1. Quantifying this effect is thus of the utmost interest. Determination of OIT values as a function of crosslinking time

Table 4 Correlation between ENB content of EPDM, residual peroxide content (RPG) at t_{90}, and OIT from CL measurements at 130°C

ENB content	ΔS in dNm	RPG in wt.%	OIT in min
2.3	17.5 ± 0.04	14.4 ± 0.0005	52 ± 6
4.5	17.7 ± 0.10	8.7 ± 0.0006	67 ± 10
9.0	18.8 ± 0.07	7.6 ± 0.0003	143 ± 12
11.4	19.7 ± 0.24	7.4 ± 0.0003	1,454 ± 173

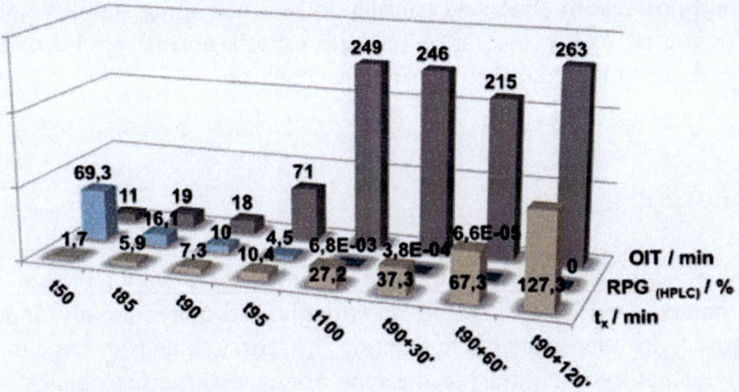

Fig. 22 Relationship between crosslinking time (t_{90}), residual peroxide content (RPG) and OIT value from CL measurements at 130°C for the system 5 phr DCP in EPDM 1

Fig. 23 Comparison of theoretically calculated OIT and the experimental value measured using EPDM 3

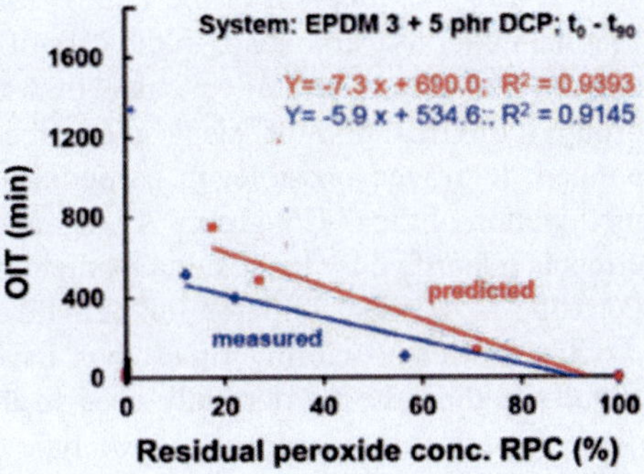

shows that at t_{90} times, the OIT values are so low that there is practically no aging resistance. The crosslinking time of t_{95} shows an improvement in resistance and from t_{100} on up, a significant rise in OIT values – and thus in aging resistance – is achieved, which does not then exhibit any further change – within the range of error – between t_{100} and $t_{90\,+\,120'}$.

VNB-EPDM's (EPDM 3) terminal double bond (EPDM 3) is markedly more reactive, for which reason there are shorter vulcanization times with reduced peroxide consumption, resulting in a higher residual peroxide content and lesser aging resistance. Comparison of different peroxides in the crosslinked systems reveals a slight trend in the direction of improved aging resistance when DIPP is used instead of monofunctional or aliphatic peroxide (DCP and DHBP).

Using reaction kinetic data from the different peroxides from crosslinking reactions and residual peroxide concentrations, it proved possible to arrive at an estimation and prediction of OIT values as a function of vulcanization time. Figure 23 shows the correlation between measured OIT value and the residual peroxide content up to t_{95} for the EPDM 3 compound and the expected behavior calculated from kinetic data.

It was not possible to observe a correlation between aging stability and the half-life time of the peroxides insofar as half-life time is normally related to the pure peroxide and not to the complex crosslinking reaction.

5 Conclusions

The study of aging mechanism of NBR in correlation with CL shows that the stability characterized by OIT values is strongly and quantitatively dependent on C=C-bonds, ACN content and temperature. The crosslink density has a minor effect on oxidative stability in contrast to the type of crosslinking (especially peroxides). The spatial development of the DLO effect and the heterogeneity of the post-crosslinking reaction from oxidation beginning at the surface is characterized quantitatively by micro-indentation. The gradient in stiffness or brittleness from the surface into the direction of the bulk is likely important for crack propagation and lifetime under dynamic load. In the case of HNBR, the aging mechanism was found to consist mainly in the radical attack on β-H in the backbone. No reaction of the CN group (compensation of electronic effects) was detected under the conditions applied. It proved possible, in particular, to describe a high stabilizing effect by methylation of the C-CN group. As regards crosslinking systems, an extremely high effect is quantified by the residual content of peroxide in the crosslinking of EPDM. Peroxide is an ideal initiator for oxidative aging, with even small traces already proving sufficient. Curing times thus have to be clearly in the region of t_{100} or slightly higher; the t_{90} normally used in the field is not sufficient. In addition the diene content (ter-monomer) and the type of peroxide indirectly influence thermal-oxidative aging of EPDM through different crosslinking reactivity.

This behavior is not observed in sulfur crosslinked polydienes like NR, BR, or SBR, because the peroxide initiator is not part of the system. Different numbers of C=C-double bonds or allylic hydrogen at the backbone of the polymer cause different crosslink densities, but the effect on the thermal-oxidative stability is small shown in this work for NBR -materials, for example.

References

1. DIN 50035 (1972)
2. IUPAC (1996) Recommendations – definitions of terms relating to degradation ageing and related chemical transformations of polymers
3. Verein Deutscher Ingenieure, VDI 3822 (2011) Failure analysis – fundamentals and performance of failure analyses, Beuth Verlag, Düsseldorf
4. Chakraborty S, Dasgupta KS, Mukhopadhyay R (2010) Effect of ozone, thermo, and thermo-oxidative aging on the physical property of styrene butadiene rubber – organoclay nanocomposites. J Elastomers Plast:443–452
5. Fisher TJ, Dussault PH (2017) Alkene ozonolysis. Tetrahedron:1–26

6. Huntink NM, Datta RN, Noordermeer JWM (2004) Addressing durability of rubber compounds. Rubber Chem Technol:476–511
7. Scott G (1990) Mechanism of polymer degradation and stabilisation. Elsevier Applied Science, London, p 170
8. Hoff S, Jacobson S (1982) J Appl Sci 27:2539
9. Scott G (1981) Developments in polymer stabilisation. Appl. Science Publ. Ltd, London. 145
10. Snijders EA, Boersma A, Baarele BV, Noordermeer JWM (2005) Polym Degrad Stab 89:200
11. Billingham N, Dudler V (2001) Advanced analytical techniques for studying polymer oxidation. In: Mallinson LG (ed) Aging studies life time extension. Kluwer/Plenum Publishers, New York, p 253
12. Röthemeier F, Sommer F (2001) Kautschuktechnologie. Hanser Verlag
13. Naskar K, Noordermeer JWM (2004) Multifunctional peroxides as a means to improve properties of dynamically vulcanized PP/EPDM blends. Kautschuk Gummi Kunststoffe 57:235
14. Coran A, Vulcanization Y (1994) Mark JE, Erman BF, Eirich R (eds) Science and technology of rubber. Academic Press, p 372
15. Bolland JL (1949) Kinetics of olefin oxidation. Q Rev Chem Soc:1–21
16. Keller RW (1985) Oxidation and ozonisation of rubber. Rubber Chem Technol 58:637
17. Norling PM, Lee TCP, Tobolsky P (1965) Rubber Chem Technol 38(5):1198–1213. https://doi.org/10.5254/1.3547109
18. Santoso M, Giese U, Schuster RH (2007) Investigations on initial stage of aging of tire rubbers by chemiluminescence spectroscopy. Rubber Chem Technol 81:762
19. Santoso M, Giese U, Schuster RH (2007) Thermal oxidation aging of rubbers – characterization by chemiluminescence. Kautschuk Gummi Kunststoffe 60:192
20. Giese U (2014) Aging behavior of elastomers. In: Encyclopedia of polymeric nanomaterials. Springer, Berlin, pp 1–7. https://doi.org/10.1007/978-3-642-36199-9_284-1
21. Luo Y (2003) Handbook of dissociation energies in organic compounds. CRC Press LLC, Boca Raton
22. Nakayama K, Watanabe Ohtake Y, Furukawa M (2008) Influence of residual peroxide on the degradation of peroxide-crosslinked ethylene–propylene–diene rubber. J Appl Polym Sci 108:2578–2586
23. Giese U, Thust S (2017) Aging of rubbers with saturated backbone – influence of crosslinking using peroxides and measures, paper E2 – proceedings, ACS Rubber Division, 192th technical meeting, Cleveland
24. Klotz S (2015) Characterization of aging stability of peroxide cured elastomers, PhD thesis, DIK/Leibniz University of Hannover
25. Giese U, Santoso M, Schuster RH (2009) Special aspects on aging and characterization of elastomers. Proceedings "International rubber conference", Nürnberg
26. Giese U, Santoso M, Schuster RH, Navarro Torrejón Y, Welzel C (2009) Selected aspects on aging of elastomers, proceedings 176th ACS-meeting, Pittsburgh
27. Modrow H, Zimmer R, Visel F, Hormes J (2000) Monitoring thermal oxidation of sulfur crosslinks in SBR-elastomers using sulfur K-edge XANES: a feasibility study. Kautschuk Gummi Kunststoffe 53:328
28. Homeier I, Navarro Torrejón Y, Giese U (2017) Aging processes of crosslinked polydienes and measures – mechanisms and quantitative characterization. Kautschuk Gummi Kunststoffe 1–2: 44–51
29. Ellwood KRJ, Baldwin J, Bauer DR et al (2006) Numerical simulation of thermal oxidation in automotive tires. Rubber Chem Technol 79:249–266
30. Wise J, Gillen KT, Clough RL (1997) Quantitative model for the time development of diffusion-limited oxidation profiles. Polymer:1929
31. Engels HW, Hammer H, Brück D, Redetzky W (1989) Effectiveness of new alkyl-aryl-*p*-phenylenediamines which can be chemically bound to polymers – model study. Rubber Chem Technol 62:609

32. Sampers J (2002) Importance of weathering factors other than UV radiation and temperature in outdoor exposure. Polym Deg 76:455
33. Parra DF, De MT, Freire A, de Paoli MA (2000) Diffusion of amine stabilizers in vulcanized natural rubber compositions used in tires. Appl Polym Sci 75:670
34. Boxhammer J (2000) Determination of oxidative stability of materials by chemiluminescence a new industrial analytical tool. Mater Test Prod Tech News (Atlas Sun SP) 30:1
35. Krüger RH, Boissiere C, Klein-Hartwig K, Kretzschmer HJ (2005) New phenylenediamine antiozonants for commodities based on natural and synthetic rubber. Food Addit Contam 22: 968
36. Zlatkevich L (1989) Chemiluminescence in evaluating thermal oxidative stability in luminescence techniques in solid state polymer research. Marcel Dekker, Inc., New York
37. Bhatterjee S, Bhowmik AK, Avasthi BN (1991) Degradation of hydrogenated nitrile rubber. Polym Degrad Stab:71–87
38. Campomizzi E, Bender H, von Hellens W (2001) Improving the heat resistance of hydrogenated nitrile rubber compounds-part 2. Kautschuk Gummi Kunststoffe:114–121
39. Ivan G, Giurginca M, Basuc S (1998) Structural modification by thermo-oxidation of NBR and HNBR elastomers. Revue Roumaine de Chimie:231–234
40. Krause K (2013) Unpublished report, Leibniz Universität Hannover/DIK
41. van Duin M, Dikland HG (2004) Drittmonomere und ihre Wirkung auf die Vernetzung von EPDM mit Peroxiden. Gummi Fasern Kunststoffe 10:663